D0162382

INTRODUCTORY COMPLEX ANALYSIS

RICHARD A. SILVERMAN

Based, in part, on material by A. I. Markushevich

DOVER PUBLICATIONS, INC.
NEW YORK

This Dover edition, first published in 1972, is an unabridged and corrected republication of the work originally published by Prentice-Hall, Inc., in 1967.

International Standard Book Number
ISBN-13: 978-0-486-64686-2
ISBN-10: 0-486-64686-6

Manufactured in the United States by LSC Communications
64686613 2017
www.doverpublications.com

PREFACE

During the years 1963–1965, a substantial sector of my activities as editor of Prentice-Hall's series "Selected Russian Publications in the Mathematical Sciences" was devoted to the translation and revision of Professor A. I. Markushevich's magnum opus "The Theory of Analytic Functions," Moscow (1950). The English edition takes the form of three volumes, collectively entitled "Theory of Functions of a Complex Variable," the first two published in 1965 and the third in 1967. Out of my strong interaction with the Russian original, there emerged an English edition markedly different from its forebear, a fact duly recognized by various reviewers.

Because of the grand scale of the master three-volume course, which runs to well over a thousand pages, I decided to distill it down into a shorter one-volume course, matching the usual one-year graduate course on complex analysis given in the United States. The result is the present volume, called "Introductory Complex Analysis" (hereinafter abbreviated as ICA), whose debt to the master course (and to a brief course on complex analysis written by Markushevich himself) is acknowledged by the credit line "based, in part, on material by A. I. Markushevich" appearing on the title page.

The fact that ICA is in part the offspring of the master course has a pleasant and useful side effect: An expanded discussion of material appearing in ICA can often be found in the master course *in precisely the same language*. The pedagogical assets of this "fringe benefit" should be obvious. Thus references to Markushevich's "Theory of Functions of a Complex Variable" (Prentice-Hall, Inc., Englewood Cliffs, New Jersey) are scattered throughout ICA, with "Volume I" abbreviated as M1 (M for "Markushevich"), "Volume II" as M2 and "Volume III" as M3. This feature also allows the student to pursue various advanced topics whose beginnings are to be found in ICA, without any problem of adjusting to new terminology and notation. References to M1, M2 and M3 are particularly prevalent in the problems. Thus the solutions of a number of problems can in effect be "looked up" in the master course, but it is hoped that the student will exercise restraint and try hard to solve such problems on his own.

Having disposed of the parentage of ICA, let us now examine the traits of the child. If complex analysis is regarded as differential and integral calculus in the complex domain, then Chapters 1 and 2 are essentially "precalculus." Differentiation appears in Chapter 3, together with the basic notion of an analytic function and a foretaste of conformal mapping. Chapters 4, 5 and 6 treat the elementary functions in considerably more detail than in most books on this level and, in my opinion, more in keeping with the needs of applied scientists.

Integration in the complex domain dominates Chapter 7 (on complex integrals and Cauchy's integral theorem) and Chapter 8 (on Cauchy's integral formula and its implications). Chapter 9 is devoted to the subject of complex series, as a prelude to Chapter 10 on the key topic of power series. Chapter 11 on Laurent series and Chapter 12 on the residue theorem and its implications (and manifold applications) lie deep in the home territory of complex analysis. Chapter 13 is devoted to harmonic functions, a subject too often slighted in first courses on complex analysis. A substantial chunk of the theory and technique of infinite product and partial fraction expansions is presented in Chapter 14.

The subject of conformal mapping is taken up again in Chapter 15, with considerable space devoted to the Schwarz-Christoffel transformation which figures so prominently in the applications. Finally, Chapter 16 deals with analytic continuation and related topics, notably the symmetry principle and Riemann surfaces.

The reference system is self-explanatory, with the prefix of each equation and figure number referring to the chapter in which the equation or figure occurs. A problem number unaccompanied by a section number always refers to a problem at the end of the section where the reference is made. Otherwise the problem number is preceded by the section number giving its proper address.

As in the case of all three volumes of the master course, ICA has benefited from a careful reading by Dr. T. J. Rivlin, who made many helpful suggestions.

R. A. S.

CONTENTS

CHAPTER **I**

COMPLEX NUMBERS, FUNCTIONS AND SEQUENCES

1. INTRODUCTORY REMARKS

Since the square of a real number is nonnegative, even the simple quadratic equation

$$x^2 + 1 = 0 \tag{1.1}$$

has no real solutions (roots). However, it seems perfectly reasonable to require that any number system suitable for computational purposes allow us to solve equation (1.1), or, for that matter, the general algebraic equation

$$a_0 + a_1 x + \cdots + a_n x^n = 0 \qquad (a_n \neq 0) \tag{1.2}$$

of degree n, where $a_0, a_1, \ldots, a_n$ are arbitrary real numbers. As already shown by (1.1), this goal can only be achieved if we somehow manage to extend the real number system by making it part of another, larger number system. Since (1.1) is in a certain sense the "simplest" algebraic equation with no real roots, an obvious first approach to our problem is to introduce an "imaginary unit" $i = \sqrt{-1}$, and then consider "complex numbers" of the form

$$a + bi \quad \text{or} \quad a + ib, \tag{1.3}$$

where a and b are arbitrary real numbers, and algebraic operations are defined in the natural way, i.e., expressions of the form (1.3) are treated exactly like binomials $a + bx$ in an unknown x, except that the rule

$$i^2 = -1, \quad i^3 = -i, \quad i^4 = 1, \quad i^5 = i, \ldots \tag{1.4}$$

is used to eliminate all powers of i higher than the first. If this is done, the roots of (1.1) are the particularly simple complex numbers i and $-i$. Moreover, setting $b = 0$, we see that the numbers of the form (1.3) include the real numbers as a special case (an essential feature).

Surprisingly enough, as we shall see later (Theorem 10.7, Corollary 2), it turns out that once we allow x to take complex values, equation (1.2) always has a root, even if the coefficients $a_0, a_1, \ldots, a_n$ are themselves complex numbers, a result known as the *fundamental theorem of algebra.*

2. COMPLEX NUMBERS AND THEIR GEOMETRIC REPRESENTATION

As already noted, by a *complex number* we mean an expression of the form $a + ib$, where a and b are real numbers and i is the "imaginary unit." If $c = a + ib$, a is called the *real part* of c, written $\operatorname{Re} c$, and b is called the *imaginary part* of c, written $\operatorname{Im} c$. By the complex number *zero,* we mean the number $0 = 0 + i0$, with zero real and imaginary parts. By definition, two complex numbers c_1 and c_2 are *equal* if and only if

$$\operatorname{Re} c_1 = \operatorname{Re} c_2, \qquad \operatorname{Im} c_1 = \operatorname{Im} c_2.$$

If $\operatorname{Im} c = 0$, $c = a + ib$ reduces to a real number, while if $\operatorname{Im} c \neq 0$, c is said to be *imaginary* and if $\operatorname{Re} c = 0$, $\operatorname{Im} c \neq 0$, c is said to be *purely imaginary.*[1]

Complex numbers can be represented geometrically as points in the plane, a fact which is not only useful but virtually indispensable. Introducing a rectangular coordinate system in the plane, we can identify the complex number $z = x + iy$ with the point $P = (x, y)$, as shown in Figure 1.1. In this way, we establish a one-to-one correspondence between the set of all complex numbers and the set of all points in the plane. Clearly, under this mapping, the set of all real numbers corresponds to the x-axis (hence called the *real axis*) and the set of all purely imaginary numbers corresponds to the y-axis (hence called the *imaginary axis*), while the set of all imaginary numbers corresponds to all points which do not lie on the real axis. Moreover, the complex number 0 corresponds to the point of intersection of the x-axis and y-axis, i.e., the origin of coordinates. The plane whose points

[1] In definitions, the locution "is said to be . . . if" will be used with the meaning "is . . . if and only if" (the dots indicate a missing predicate).

represent the complex numbers is called the *complex plane*, or the *z-plane*, *w-plane*, . . . , depending on the letter $z, w, \ldots$ used to denote a generic complex number. With the understanding that such a complex plane has been constructed, the terms "complex number $x + iy$" and "point $x + iy$" will be used interchangeably.

Another entirely equivalent way of representing the complex number $z = x + iy$ is to use the vector $\overrightarrow{OP}$ joining the origin O of the complex plane to the point $P = (x, y)$, instead of using the point P itself (see Figure 1.1). The length of the vector $\overrightarrow{OP}$ is called the *modulus* or *absolute value* of the

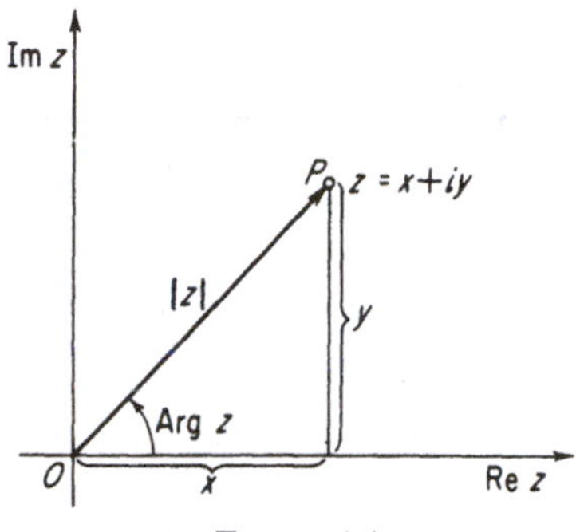

FIGURE 1.1

complex number z, and is denoted by $|z|$. The angle between the positive real axis and the vector $\overrightarrow{OP}$, more exactly, the angle through which the positive real axis must be rotated to cause it to have the same direction as $\overrightarrow{OP}$ (considered positive if the rotation is counterclockwise and negative otherwise), is called the *argument* of the complex number z, and is denoted by $\operatorname{Arg} z$.[2] In other words, $|z|$ and $\operatorname{Arg} z$ are the polar coordinates r and Φ of the point with rectangular coordinates x and y, i.e.,

$$x = \operatorname{Re} z = r \cos \Phi, \qquad y = \operatorname{Im} z = r \sin \Phi.$$

It follows at once that

$$z = x + iy = r(\cos \Phi + i \sin \Phi), \tag{1.5}$$

where (1.5) is called the *trigonometric form* of the complex number z.

Clearly, the quantity $\operatorname{Arg} z$ is defined only to within an integral multiple of 2π. However, there is one and only one value of $\operatorname{Arg} z$, say φ, which satisfies the inequality

$$-\pi < \varphi \leqslant \pi,$$

[2] In talking about the vector $\overrightarrow{OP}$, we have tacitly assumed that P is not the origin, i.e., that $z \neq 0$. If $z = 0$, then $|z| = 0$ and $\operatorname{Arg} z$ is indeterminate.

and we shall call φ the *principal value* of the argument z, written $\arg z$. The relation between $\operatorname{Arg} z$ and $\arg z$ is given by

$$\operatorname{Arg} z = \arg z + 2n\pi,$$

where n ranges over all the integers $0, \pm 1, \pm 2, \ldots$. It is an immediate consequence of Figure 1.1 and our definitions that[3]

$$|z| = \sqrt{x^2 + y^2}$$

and

$$\tan(\arg z) = \frac{y}{x}. \tag{1.6}$$

Some care is required in inverting (1.6), since the arc tangent of a real number x, written $\operatorname{Arc}\tan x$, is only defined to within an integral multiple of π. However, there is one and only one value of $\operatorname{Arc}\tan x$, say α, which satisfies the inequality

$$-\frac{\pi}{2} < \alpha < \frac{\pi}{2},$$

and we shall call α the *principal value* of the arc tangent of x, written $\arctan x$. We can now invert the relation (1.6), obtaining

$$\arg z = \begin{cases} \arctan \dfrac{y}{x} & \text{if} \quad x > 0, \\ \arctan \dfrac{y}{x} + \pi & \text{if} \quad x < 0,\ y \geqslant 0, \\ \arctan \dfrac{y}{x} - \pi & \text{if} \quad x < 0,\ y < 0. \end{cases}$$

Moreover, when y/x becomes infinite, we clearly have

$$\arg z = \begin{cases} \dfrac{\pi}{2} & \text{if} \quad x = 0,\ y > 0, \\ -\dfrac{\pi}{2} & \text{if} \quad x = 0,\ y < 0, \end{cases}$$

while the case $z = 0$ (i.e., $x = y = 0$) is indeterminate, as already noted.

The complex numbers $x + iy$ and $x - iy$ are said to be *conjugate complex numbers* or *complex conjugates* (of each other). If one of these numbers is denoted by z, the other is denoted by $\bar{z}$. Obviously, the points z and $\bar{z}$ are symmetric with respect to the real axis, and

$$|\bar{z}| = |z|, \qquad \overline{(\bar{z})} = z.$$

---

[3] By $\sqrt{a}$, where $a > 0$, we shall always mean the *positive* square root of a.

Moreover,

$$\arg z = -\arg \bar{z},$$

unless z is a negative real number, in which case

$$\arg z = \arg \bar{z} = \pi.$$

Problem. Represent the following complex numbers in trigonometric form: a) $1 + i$; b) $-1 + i$; c) $-1 - i$; d) $1 - i$; e) $1 + i\sqrt{3}$; f) $-1 + i\sqrt{3}$; g) $-1 - i\sqrt{3}$; h) $1 - i\sqrt{3}$; i) $\sqrt{3} - i$; j) $2 + \sqrt{3} + i$.

3. COMPLEX ALGEBRA

It follows from the prescription of Sec. 1 [cf. (1.4)] that the sum and product of two complex numbers $z_1 = x_1 + iy_1$, $z_2 = x_2 + iy_2$ are given by

$$z_1 + z_2 = (x_1 + x_2) + i(y_1 + y_2), \tag{1.7}$$

$$z_1 z_2 = (x_1 x_2 - y_1 y_2) + i(x_1 y_2 + x_2 y_1), \tag{1.8}$$

while the difference and quotient of z_1 and z_2 are given by

$$z_1 - z_2 = (x_1 - x_2) + i(y_1 - y_2),$$

$$\frac{z_1}{z_2} = \frac{x_1 x_2 + y_1 y_2}{x_2^2 + y_2^2} + i\,\frac{x_2 y_1 - x_1 y_2}{x_2^2 + y_2^2} \qquad (x_2^2 + y_2^2 \neq 0).$$

If z_1, z_2 and z_3 are any three complex numbers, the relations (1.7) and (1.8) imply that

$$z_1 + z_2 = z_2 + z_1,$$

$$z_1 z_2 = z_2 z_1$$

(addition and multiplication are *commutative*),

$$(z_1 + z_2) + z_3 = z_1 + (z_2 + z_3),$$

$$(z_1 z_2) z_3 = z_1 (z_2 z_3)$$

(addition and multiplication are *associative*), and

$$z_1(z_2 + z_3) = z_1 z_2 + z_1 z_3$$

(multiplication is *distributive* with respect to addition). Moreover, the product of two complex numbers is zero if and only if at least one of the factors is zero, any complex number z has a unique negative $-z$ such that $z + (-z) = 0$, and any nonzero complex number z has a unique inverse $z^{-1} = 1/z$ such that $zz^{-1} = 1$. (Verify these statements.) In other words, the complex numbers satisfy all the axioms of a *field*.[4]

---

[4] See e.g., G. Birkhoff and S. MacLane, *A Survey of Modern Algebra*, third edition, The Macmillan Co., New York, (1965) p. 33.

Guided by (1.7) and (1.8), one might define the complex numbers (without recourse to the "imaginary unit" i) as ordered pairs

$$(x_1, y_1), (x_2, y_2), \ldots$$

of real numbers $x_1, y_1, x_2, y_2, \ldots$, where two pairs (x_1, y_1) and (x_2, y_2) are equal if and only if $x_1 = x_2, y_1 = y_2$, while addition and multiplication of pairs are defined by

$$(x_1, y_1) + (x_2, y_2) = (x_1 + x_2, y_1 + y_2),$$
$$(x_1, y_1)(x_2, y_2) = (x_1x_2 - y_1y_2, x_1y_2 + x_2y_1).$$

With this approach, the relation $i^2 = -1$ has the analogue

$$(0, 1)(0, 1) = (-1, 0),$$

which is sometimes adduced as an "explanation" of the "imaginary unit" i. However, it should be noted that the extension of the real numbers to the

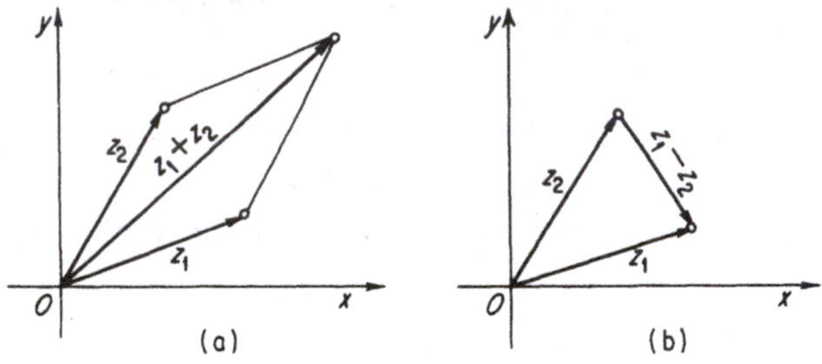

FIGURE 1.2

complex numbers is no more "mysterious" than the extension of the integers to the rational numbers, and less complicated than the extension of the rational numbers to the real numbers.[5]

Geometrically, addition of two complex numbers z_1 and z_2 corresponds to addition of the vectors representing them [see Figure 1.2(a)]. Similarly, the difference $z_1 - z_2$ is represented by the vector joining the point z_2 to the point z_1 [see Figure 1.2(b)]. It follows that $\rho(z_1, z_2)$, the distance between the two points z_1 and z_2, equals the modulus of the complex number $z_1 - z_2$:

$$\rho(z_1, z_2) = |z_1 - z_2|.$$

Then the familiar fact that the length of one side of a triangle cannot exceed the sum of the lengths of the other two sides implies the following two

[5] See e.g., G. Birkhoff and S. MacLane, *op. cit.*, Chaps. 2 and 4.

inequalities involving complex numbers:

$$|z_1 + z_2| \leqslant |z_1| + |z_2|, \tag{1.9}$$

$$|z_1 - z_2| \geqslant \big||z_1| - |z_2|\big|. \tag{1.10}$$

The relation (1.9) can be generalized immediately to the case of n complex numbers:

$$|z_1 + z_2 + \cdots + z_n| \leqslant |z_1| + |z_2| + \cdots + |z_n|. \tag{1.11}$$

In (1.9), (1.10) and (1.11), equality occurs if and only if all the (nonzero) complex numbers have the same argument. We note that according to (1.8), if $z = x + iy$, then

$$z\bar{z} = x^2 + y^2 = |z|^2,$$

and hence

$$|z| = \sqrt{x^2 + y^2}$$

and

$$|z_1 + \cdots + z_n| = \sqrt{(z_1 + \cdots + z_n)(\bar{z}_1 + \cdots + \bar{z}_n)}.$$

To give a geometric interpretation of multiplication and division of two complex numbers z_1 and z_2, we first write z_1 and z_2 in trigonometric form [cf. (1.5)], i.e.,

$$z_1 = r_1(\cos \Phi_1 + i \sin \Phi_1),$$
$$z_2 = r_2(\cos \Phi_2 + i \sin \Phi_2),$$

where

$$r_1 = |z_1|, \qquad \Phi_1 = \operatorname{Arg} z_1,$$
$$r_2 = |z_2|, \qquad \Phi_2 = \operatorname{Arg} z_2.$$

Then, according to (1.8),

$$\begin{aligned} z_1 z_2 &= r_1 r_2[(\cos \Phi_1 \cos \Phi_2 - \sin \Phi_1 \sin \Phi_2) \\ &\qquad + i(\sin \Phi_1 \cos \Phi_2 + \cos \Phi_1 \sin \Phi_2)] \\ &= r_1 r_2[\cos (\Phi_1 + \Phi_2) + i \sin (\Phi_1 + \Phi_2)], \end{aligned} \tag{1.12}$$

and therefore

$$|z_1 z_2| = |z_1|\,|z_2|,$$
$$\operatorname{Arg} z_1 z_2 = \operatorname{Arg} z_1 + \operatorname{Arg} z_2.$$

The last formula means that the set of values of $\operatorname{Arg} z_1 z_2$ is obtained by forming all possible sums of a value of $\operatorname{Arg} z_1$ and a value of $\operatorname{Arg} z_2$. Thus the geometric interpretation of multiplication of z_1 by z_2 ($z_1 \neq 0$, $z_2 \neq 0$) is the following: Draw the vector z_1 (with initial point at the origin), multiply its length by the factor $|z_2|$,[6] and rotate the resulting vector counterclockwise

[6] This can correspond to either stretching or shrinking (or neither, if $|z_2| = 1$).

through the angle Arg z_2. Similarly, if $z_2 \neq 0$, it is easily verified that

$$\left|\frac{z_1}{z_2}\right| = \frac{|z_1|}{|z_2|},$$
$$\operatorname{Arg}\frac{z_1}{z_2} = \operatorname{Arg} z_1 - \operatorname{Arg} z_2, \tag{1.13}$$

with a corresponding geometric interpretation of division. It follows from the second of the formulas (1.13) that $\operatorname{Arg}(z_1/z_2)$ is the angle between the vectors z_2 and z_1, measured from z_2 to z_1 in the counterclockwise direction.

Problem 1. Calculate the following expressions:

a) $\dfrac{1 + i\tan\alpha}{1 - i\tan\alpha}$; b) $\dfrac{(1 + 2i)^3 - (1 - i)^3}{(3 + 2i)^3 - (2 + i)^2}$; c) $\dfrac{(1 - i)^5 - 1}{(1 + i)^5 + 1}$;

d) $\dfrac{(1 + i)^9}{(1 - i)^7}$.

Problem 2. Calculate

a) $(a + b\omega + c\omega^2)(a + b\omega^2 + c\omega)$; b) $(a + b)(a + b\omega)(a + b\omega^2)$;
c) $(a + b\omega + c\omega^2)^3 + (a + b\omega^2 + c\omega)^3$; d) $(a\omega^2 + b\omega)(b\omega^2 + a\omega)$,
where $\omega = -\frac{1}{2} + \frac{1}{2}\sqrt{3}i$.

Ans. b) $a^3 + b^3$; d) $a^2 - ab + b^2$.

Problem 3. Find the complex numbers which are conjugate to a) their own squares; b) their own cubes.

Ans. b) $0, 1, i, -1, -i$.

Problem 4. Suppose a complex number u is obtained as a result of a finite number of rational operations (i.e., addition, subtraction, multiplication and division) applied to the complex numbers $z_1, z_2, \ldots, z_n$. Prove that the same operations applied to the complex conjugates $\bar{z}_1, \bar{z}_2, \ldots, \bar{z}_n$ leads to the number $\bar{u}$ which is the complex conjugate of u.

Problem 5. Defining rational numbers as ordered pairs of integers, write the corresponding rules for addition and multiplication of rational numbers. Verify that addition and multiplication are commutative and associative, and that multiplication is distributive with respect to addition.

Ans. $(x_1, y_1) + (x_2, y_2) = (x_1y_2 + x_2y_1, y_1y_2)$, $(x_1, y_1)(x_2, y_2) = (x_1x_2, y_1y_2)$, where $y_1 \neq 0$, $y_2 \neq 0$.

Problem 6. Prove the inequalities

$$|z_1 + z_2| \leqslant |z_1| + |z_2|,$$
$$|z_1 - z_2| \geqslant \big|\,|z_1| - |z_2|\,\big|$$

algebraically.

Problem 7. Prove the identity

$$|z_1 + z_2|^2 + |z_1 - z_2|^2 = 2(|z_1|^2 + |z_2|^2),$$

and interpret it geometrically.

Problem 8. By a purely geometric argument, prove that

$$|z - 1| \leqslant |\,|z| - 1| + |z|\,|\arg z|.$$

Problem 9. Solve the following equations:

a) $|z| - z = 1 + 2i$; b) $|z| + z = 2 + i$.

Ans. a) $z = \frac{3}{2} - 2i$.

Problem 10. Prove that every complex number (except -1) of unit modulus can be represented in the form

$$z = \frac{1 + it}{1 - it},$$

where t is a real number.

Problem 11. Prove that

$$|(1 + i)z^3 + iz| < \tfrac{3}{4},$$

if $|z| < \frac{1}{2}$.

Problem 12. Prove that

$$\arg \frac{z_3 - z_2}{z_3 - z_1} = \frac{1}{2} \arg \frac{z_2}{z_1},$$

if $|z_1| = |z_2| = |z_3|$.

Problem 13. Find a necessary and sufficient condition for three distinct points z_1, z_2 and z_3 to lie on the same line.

Problem 14. Suppose the points $z_1, z_2, \ldots, z_n$ all lie on one side of a line drawn through the origin of the complex plane. Prove that the same is true of the points $1/z_1, 1/z_2, \ldots, 1/z_n$. Moreover, show that

$$z_1 + z_2 + \cdots + z_n \neq 0, \qquad \frac{1}{z_1} + \frac{1}{z_2} + \cdots + \frac{1}{z_n} \neq 0.$$

Problem 15. Prove that if $z_1 + z_2 + z_3 = 0$ and $|z_1| = |z_2| = |z_3| = 1$, then the points z_1, z_2 and z_3 are the vertices of an equilateral triangle inscribed in the unit circle $|z| = 1$.

Problem 16. What are the loci of the points z which satisfy the following relations:

a) $|z - z_1| = |z - z_2|$; b) $|z - 2| + |z + 2| = 5$;
c) $|z - 2| - |z + 2| > 3$; d) $0 < \operatorname{Re}(iz) < 1$;
e) $|z| = \operatorname{Re} z + 1$; f) $\operatorname{Re} z + \operatorname{Im} z < 1$;

g) $\alpha < \arg z < \beta$, $\gamma < \operatorname{Re} z < \delta$, where $-\dfrac{\pi}{2} < \alpha, \beta < \dfrac{\pi}{2}$, $\gamma > 0$;

h) $\operatorname{Im} \dfrac{z - z_1}{z - z_2} = 0$; i) $\operatorname{Re} \dfrac{z - z_1}{z - z_2} = 0$?

Problem 17. Describe the families of curves in the z-plane with equations

$$\text{a) } \operatorname{Re}\frac{1}{z} = C; \quad \text{b) } \operatorname{Im}\frac{1}{z} = C; \quad \text{c) } \operatorname{Re} z^2 = C; \quad \text{d) } \operatorname{Im} z^2 = C,$$

where C is an arbitrary real number.

4. POWERS AND ROOTS OF COMPLEX NUMBERS

Let n be a positive integer and z a complex number. Then by the nth *power* of z, written z^n, we mean z multiplied by itself n times, just as in the case of real numbers. If $z = 0$, we obviously have $z^n = 0$. If $z \neq 0$, then, setting

$$z = r(\cos \Phi + i \sin \Phi) \tag{1.14}$$

and repeatedly applying (1.12), we find that

$$z^n = r^n(\cos n\Phi + i \sin n\Phi). \tag{1.15}$$

When $r = 1$, (1.14) and (1.15) together imply the following formula, known as *De Moivre's theorem*:

$$(\cos \Phi + i \sin \Phi)^n = \cos n\Phi + i \sin n\Phi. \tag{1.16}$$

Using the binomial theorem to expand the left-hand side of (1.16) and equating real and imaginary parts of the resulting equation, we can express $\cos n\Phi$ or $\sin n\Phi$ entirely in terms of powers of $\cos \Phi$ and $\sin \Phi$. For the same algebraic reasons as in the case of real numbers, we define

$$z^0 = 1, \qquad z^{-n} = \frac{1}{z^n}.$$

Then formulas (1.15) and (1.16) are valid for all integers $n = 0, \pm 1, \pm 2, \ldots$.

Next we consider the problem of extracting the root of a complex number. Again let n be a positive integer and z a complex number. Then by the nth *root* of z, written $\sqrt[n]{z}$ or $z^{1/n}$, we mean any complex number ζ satisfying the equation

$$\zeta^n = z.$$

If $z = 0$, we obviously have $\zeta = 0$. If $z \neq 0$, then, setting

$$\zeta = \rho(\cos \Theta + i \sin \Theta), \qquad z = r(\cos \Phi + i \sin \Phi),$$

we find that

$$\rho^n(\cos n\Theta + i \sin n\Theta) = r(\cos \Phi + i \sin \Phi).$$

It follows that[7]

$$\rho^n = r, \qquad \rho = \sqrt[n]{r} = \sqrt[n]{|z|}$$

and

$$n\Theta = \Phi + 2m\pi \qquad (m = 0, \pm 1, \pm 2, \ldots),$$

[7] By $\sqrt[n]{a} = a^{1/n}$, where $a > 0$ and n is a positive integer, we shall always mean the unique *positive* nth root of a (cf. footnote 3, p. 4).

i.e.,

$$n\Theta = \text{Arg } z, \qquad \Theta = \frac{\text{Arg } z}{n}.$$

Therefore we have

$$\sqrt[n]{z} = \sqrt[n]{|z|}\left(\cos\frac{\text{Arg } z}{n} + i\sin\frac{\text{Arg } z}{n}\right). \tag{1.17}$$

It is clear from (1.17) that values of Arg z which differ by $2k\pi$, where k is an integer not divisible by n, correspond to different values of $\sqrt[n]{z}$. Accordingly, when Arg z runs through the values

$$\arg z,\ \arg z + 2\pi, \ldots, \arg z + 2(n-1)\pi, \tag{1.18}$$

we obtain n different values of $\sqrt[n]{z}$. Since any other value of Arg z differs from one of the numbers (1.18) by a multiple of $2n\pi$, these n values of $\sqrt[n]{z}$ exhaust all the possibilities. Thus the nth root of z has n different values (if $z \neq 0$), which are all implicit in the formula (1.17). The value of $\sqrt[n]{z}$ equal to

$$\sqrt[n]{|z|}\left(\cos\frac{\arg z}{n} + i\sin\frac{\arg z}{n}\right)$$

is called the *principal value* of $\sqrt[n]{z}$, since when z is a positive real number, the principal value of $\sqrt[n]{z}$ coincides with the *positive* nth root of z (cf. footnote 7, p. 10). It follows from the formula

$$\text{Arg } z = \arg z + 2m\pi = \arg z + \text{Arg } 1$$

that

$$\sqrt[n]{z} = \sqrt[n]{|z|}\left(\cos\frac{\arg z + \text{Arg } 1}{n} + i\sin\frac{\arg z + \text{Arg } 1}{n}\right)$$

$$= \sqrt[n]{|z|}\left(\cos\frac{\arg z}{n} + i\sin\frac{\arg z}{n}\right)\left(\cos\frac{\text{Arg } 1}{n} + i\sin\frac{\text{Arg } 1}{n}\right),$$

i.e., all the values of $\sqrt[n]{z}$ can be obtained from the principal value by multiplying the latter by the different nth roots of unity.

Finally we define an arbitrary rational power of a complex number by the formula

$$z^{m/n} = (\sqrt[n]{z})^m,$$

where m and $n > 0$ are relatively prime integers. Then we have

$$\begin{aligned} z^{m/n} &= \left[\sqrt[n]{|z|}\cos\left(\frac{\text{Arg } z}{n} + i\sin\frac{\text{Arg } z}{n}\right)\right]^m \\ &= (\sqrt[n]{|z|})^m\left(\cos\frac{m\text{ Arg } z}{n} + i\sin\frac{m\text{ Arg } z}{n}\right) \\ &= |z|^{m/n}\left(\cos\frac{m\text{ Arg } z}{n} + i\sin\frac{m\text{ Arg } z}{n}\right), \end{aligned} \tag{1.19}$$

where $|z|^{m/n}$ is defined as the positive number $(\sqrt[n]{|z|})^m = \sqrt[n]{|z|^m}$.

Problem 1. Calculate the following quantities:

a) $(1 + i)^{25}$; b) $\left(\dfrac{1 + i\sqrt{3}}{1 - i}\right)^{30}$; c) $\left(1 - \dfrac{\sqrt{3} - i}{2}\right)^{24}$;

d) $\dfrac{(-1 + i\sqrt{3})^{15}}{(1 - i)^{20}} + \dfrac{(-1 - i\sqrt{3})^{15}}{(1 + i)^{20}}$.

Ans. a) $2^{12}(1 + i)$; c) $(2 - \sqrt{3})^{12}$.

Problem 2. Prove that

a) $(1 + i)^n = 2^{n/2}\left(\cos\dfrac{n\pi}{4} + i\sin\dfrac{n\pi}{4}\right)$;

b) $(\sqrt{3} - i)^n = 2^n\left(\cos\dfrac{n\pi}{6} - i\sin\dfrac{n\pi}{6}\right)$;

c) $(1 + \cos\alpha + i\sin\alpha)^n = 2^n\cos^n\dfrac{\alpha}{2}\left(\cos\dfrac{n\alpha}{2} + i\sin\dfrac{n\alpha}{2}\right)$.

Problem 3. Use De Moivre's theorem to express $\cos 5\Phi$, $\cos 8\Phi$, $\sin 6\Phi$ and $\sin 7\Phi$ in terms of $\cos\Phi$ and $\sin\Phi$.

Problem 4. Use De Moivre's theorem to express $\tan 6\Phi$ in terms of $\tan\Phi$.

Problem 5. Find all the values of the following roots:

a) $\sqrt[3]{1}$; b) $\sqrt[3]{i}$; c) $\sqrt[4]{-1}$; d) $\sqrt[6]{-8}$; e) $\sqrt[8]{1}$; f) $\sqrt{1 - i}$;

g) $\sqrt{3 + 4i}$; h) $\sqrt[3]{-2 + 2i}$; i) $\sqrt[5]{-4 + 3i}$; j) $\sqrt[6]{\dfrac{1 - i}{\sqrt{3} + i}}$;

k) $\sqrt[8]{\dfrac{1 + i}{\sqrt{3} - i}}$.

Problem 6. Prove that the sum of all the distinct nth roots of unity is zero. What geometric fact does this express?

Problem 7. Let m and $n \neq 0$ be any two integers. Show that $z^{m/n} = (z^{1/n})^m$ has $n/(m, n)$ distinct values, where (m, n) is the greatest common divisor of m and n. Prove that the sets of values $(z^{1/n})^m$ and $(z^m)^{1/n}$ coincide if and only if m and n are relatively prime [i.e., $(m, n) = 1$].

Problem 8. Calculate

a) $1 + 2\varepsilon + 3\varepsilon^2 + \cdots + n\varepsilon^{n-1}$; b) $1 + 4\varepsilon + 9\varepsilon^2 + \cdots + n^2\varepsilon^{n-1}$,

where ε is an nth root of unity.

Hint. Multiply by $1 - \varepsilon$.

Ans. a) $-\dfrac{n}{1 - \varepsilon}$ if $\varepsilon \neq 1$, $\dfrac{n(n + 1)}{2}$ if $\varepsilon = 1$.

Problem 9. Find all the complex numbers z satisfying the relation $\bar{z} = z^{n-1}$.

5. SET THEORY. COMPLEX FUNCTIONS

From now on, we shall use geometric language when talking about complex numbers, and therefore the study of various sets of complex numbers will be equivalent to the study of various sets of points in the plane. We shall employ standard set-theoretic terminology and notation. Thus $z \in E$ means that z *belongs* to the set E (synonymously, z is a *member* of E, or z is *contained* in E), and $z \notin E$ means that z does not belong to E. By $E \subset F$ (or $F \supset E$) we mean that the set E is a *subset* of F (synonymously, E is *included* in F, or E is *contained* in F), i.e., all the points of E are points of F. Two sets E and F are said to be *equal*, and we write $E = F$, if and only if $E \subset F$, $F \subset E$. If $E \subset F$ but $E \neq F$, we say that E is a *proper subset* of F, or that E is *strictly* included in F.

By the *empty set* we mean the set containing no points at all, and we denote the empty set by the same symbol 0 as used to denote the number zero. A set E is said to be *nonempty* if $E \neq 0$. By $\{a\}$ we mean the set whose only member is a, and more generally, by $\{a_1, \ldots, a_n\}$ we mean the set whose members are $a_1, \ldots, a_n$. By the *union* of two sets E and F, written $E \cup F$, we mean the set of points belonging to at least one of the sets E and F. By the *intersection* of two sets E and F, written $E \cap F$, we mean the set of points belonging to both E and F. Two sets E and F whose intersection is empty, i.e., which are such that $E \cap F = 0$, are said to be *disjoint*. By the *complement* of a set E, written E^c, we mean the set of all points (in the complex plane) which do not belong to E. By the *difference* $E - F$ between two sets E and F, we mean the set of points in E which do not belong to F, i.e.,

$$E - F = E \cap F^c.$$

Remark. One can also consider sets Σ whose members are themselves sets. If Σ is such a set, the union and intersection of all the sets in Σ are denoted by

$$\bigcup_{E \in \Sigma} E \quad \text{and} \quad \bigcap_{E \in \Sigma} E,$$

respectively. The sets in Σ are said to be *pairwise disjoint* (or simply *disjoint*) if $E \cap F = 0$ for any pair of distinct sets $E \in \Sigma$, $F \in \Sigma$.

Given two nonempty sets E and $\mathscr{E}$, by a *function* f we mean a set, written

$$f = \{(z, w_z) \mid z \in E\}, \tag{1.20}$$

whose members are distinct ordered pairs of the form (z, w_z), where $z \in E$, $w_z \in \mathscr{E}$. The subscript on w_z will often be omitted, and merely serves to emphasize that w_z is one of the (possibly several) complex numbers associated with z. The set E is called the *domain* (*of definition*) of the function f, and the set $\mathscr{E}$ is called the *range* of f. If f contains no two pairs (z, w) with the same first member z, then f is said to be *single-valued*, with the *value* w at the point z. Otherwise, f is said to be *multiple-valued*, and every w such that (z, w) is contained in f is called a *value* of f at the point z. In both cases, we use $f(z)$ to denote any value of f at the point z and $\mathscr{E}_z$ to denote the set of all values of f at z. We can also write (1.20) in the abbreviated form

$$f = \{w_z \mid z \in E\},$$

or, even more simply, as

$$w = f(z) \qquad (z \in E).$$

In fact, when there is no possibility of confusion, we often omit explicit reference to the domain E. Strictly speaking, one should distinguish between the function f and its values $f(z)$ at the point z, but within the framework of this simplified notation, it is customary to use $f(z)$ to denote both the function and its values at z.

By the *inverse* f^{-1} of the function f given by (1.20) we mean the set consisting of all the ordered pairs in f *written in reverse order*, i.e., $(w, z) \in f^{-1}$ if and only if $(z, w) \in f$. Obviously, f^{-1} is itself a function with domain $\mathscr{E}$ and range E, where E is the domain and $\mathscr{E}$ the range of the original function f. Paralleling our discussion of the function f, we can write f^{-1} as

$$f^{-1} = \{(w, z_w) \mid w \in \mathscr{E}\},$$

in the abbreviated form

$$f^{-1} = \{z_w \mid w \in \mathscr{E}\},$$

or simply as

$$z = f^{-1}(w) \qquad (w \in \mathscr{E}).$$

To avoid confusion, $f^{-1}(w)$ will never be used to denote the reciprocal $1/f(w)$.

Remark 1. Given a function $w = f(z)$ with domain E and range $\mathscr{E}$, the points of E are often called the values of the *independent variable*, and the points of $\mathscr{E}$ are often called the values of the *dependent variable*. If the domain E contains imaginary as well as real numbers, we sometimes emphasize this fact by referring to the function $w = f(z)$ as a *function of a complex variable*. Similarly, if the range $\mathscr{E}$ contains imaginary as well as real numbers, we sometimes emphasize this fact by calling $f(z)$ a *complex function*.

Remark 2. We have introduced the notion of a multiple-valued function of a complex variable in anticipation of future needs. However, from now on, unless the contrary is explicitly stated, all functions under discussion will be assumed to be single-valued.

Remark 3. Let $w = f(z)$, where $w = u + iv$ and $z = x + iy$, be a function of a complex variable. A special case of interest occurs when all the values of the function are real, i.e., $w = u$, $v = 0$. Then our function of a complex variable can be regarded as a real function of two real variables x and y. In fact, associating the real number u with the complex number $z = x + iy$ is equivalent to associating the real number u with the pair of real numbers x and y. In the general case where $w = u + iv$ is complex, defining a single complex function $w = f(z)$ on a set E is equivalent to defining two real functions

$$u = \text{Re } w = \varphi(x, y), \qquad v = \text{Im } w = \psi(x, y) \tag{1.21}$$

on E, each a function of two real variables x and y. Conversely, defining two real functions (1.21) on E is equivalent to defining a single complex function

$$w = \varphi(x, y) + i\psi(x, y) = f(z)$$

on E. Thus, it is clear that the whole theory of functions of a complex variable z can be interpreted in terms of pairs of real functions of two real variables x and y, and we shall sometimes make use of this interpretation.

Example 1. The single complex function

$$w = z^2 = (x + iy)^2 = x^2 - y^2 + 2ixy$$

corresponds to the two real functions

$$u = x^2 - y^2, \qquad v = 2xy.$$

Example 2. The two real functions

$$u = x^2 - y^2, \qquad v = e^{x^2+3y^2}$$

correspond to the single complex function

$$w = u + iv = (x^2 - y^2) + ie^{x^2+3y^2} = f(z).$$

If $f(z)$ is a single-valued function with domain E and range $\mathscr{E}$, the point $w = f(z) \in \mathscr{E}$ is called the *image* of the point $z \in E$ [under $f(z)$]. Similarly, if $D \subset E$, the set of all points $w = f(z)$ such that $z \in D$ is called the *image* of D, denoted by $f(D)$. Let $\mathscr{D}$ be any subset of $\mathscr{E}$ containing $f(D)$. Then we say that $f(z)$ is a *mapping* of D *into* $\mathscr{D}$. If $\mathscr{D} = f(D)$ and we want to emphasize this fact, we say that $f(z)$ is a *mapping* of D *onto* $\mathscr{D}$. Thus, every "onto" mapping is an "into" mapping, but not conversely. A mapping of a set into itself is often called a *transformation*.

If $f(z)$ is single-valued, the inverse function $f^{-1}(w)$ will in general be multiple-valued. If $f^{-1}(w)$ is single-valued, the original function or mapping $f(z)$ is said to be *one-to-one*. Then the point $z \in E$ which is the image of the point $w \in \mathscr{E}$ under $f^{-1}(w)$ is called the *inverse image* or *pre-image* of w [under $f(z)$]. Similarly, if $\mathscr{D} \subset \mathscr{E}$, the set of all points $z = f^{-1}(w)$ such that $w \in \mathscr{D}$ is called the *inverse image* or *pre-image* of $\mathscr{D}$, denoted by $f^{-1}(\mathscr{D})$. A one-to-one function $f(z)$ is said to establish a *one-to-one correspondence*, indicated by $z \leftrightarrow w$, between the points z in its domain and the points w in its range.

Remark. In the case of multiple-valued functions, images and inverse images are defined in the natural way. For example, if $f(z)$ takes several values at a given point $z \in E$, each of these values is called an *image* of the point z. Similarly, if there are several points $z \in E$ such that $f(z) = w$, where w is a given point of $\mathscr{E}$, then each of the points z is called an *inverse image* of w. In terms of the definition (1.20) of a function, the images of a given point $z \in E$ are all those points w appearing in some ordered pair (z, w) with z as its first member, and the inverse images of a given point $w \in \mathscr{E}$ are all those points z appearing in some ordered pair (z, w) with w as its second member.

Problem 1. Let Σ be a set whose members are themselves sets. Prove the relations

$$D \cap \Big(\bigcup_{E\in\Sigma} E\Big) = \bigcup_{E\in\Sigma} (D \cap E), \qquad D \cup \Big(\bigcap_{E\in\Sigma} E\Big) = \bigcap_{E\in\Sigma} (D \cup E),$$

where D is an arbitrary set. Prove the *De Morgan formulas*

$$D - \bigcup_{E\in\Sigma} E = \bigcap_{E\in\Sigma} (D - E), \qquad D - \bigcap_{E\in\Sigma} E = \bigcup_{E\in\Sigma} (D - E),$$

and specialize these formulas to the case where D is the whole complex plane.

Problem 2. By a *finite* set is meant a set consisting of a finite number of elements, and by an *infinite* set is meant a set which is not finite. (The empty set is regarded as finite.) Two sets are said to be *equivalent* if a one-to-one correspondence can be established between them. A set is said to be *countably infinite* if it is equivalent to the set of all positive integers. By a *countable* set is meant either a finite set or a countably infinite set. A set which is not countable is said to be *uncountable.*

Prove that

a) The set of all rational numbers is countable;
b) The set of all real numbers is uncountable;
c) Any subset of a countable set is countable;
d) The union of countably many countable sets is countable.

6. COMPLEX SEQUENCES

Let I be the set of all positive integers $1, 2, \ldots, n, \ldots$. Then any single-valued complex function $f(z)$ with domain I is called an *infinite sequence* (of complex numbers), or simply a (complex) *sequence*, and the values $z_n = f(n)$ are called the *terms* of the sequence. The sequence itself is usually denoted by listing its terms in order of increasing indices, i.e.,

$$z_1, z_2, \ldots, z_n, \ldots,$$

or by the more concise notation $\{z_n\}$.[8]

If $k_1, k_2, \ldots, k_n, \ldots$ is a sequence of positive integers such that $k_n < k_{n+1}$ for all $n = 1, 2, \ldots$, then the corresponding terms of the sequence $\{z_n\}$, i.e., $z_{k_1}, z_{k_2}, \ldots, z_{k_n}, \ldots$ form a new sequence $\{z_{k_n}\}$, called a *subsequence* of the

[8] Not to be confused with the set whose only member is z_n.

sequence $\{z_n\}$. For example, the sequences

$$z_1, z_3, z_5, \ldots, z_{2n-1}, \ldots,$$

$$z_2, z_4, z_6, \ldots, z_{2n}, \ldots,$$

$$z_1, z_4, z_9, \ldots, z_{n^2}, \ldots$$

are all subsequences of the sequence $\{z_n\}$.

Remark. The subsequences of $\{z_n\}$ can also be written in the form

$$z_{n_1}, z_{n_2}, \ldots, z_{n_k}, \ldots,$$

or briefly $\{z_{n_k}\}$, where $n_1, n_2, \ldots, n_k, \ldots$ is a sequence of positive integers such that $n_k < n_{k+1}$ for all $k = 1, 2, \ldots$, so that k rather than n runs through the positive integers. In other words, the running index is always the "lowest subscript," regardless of the particular symbol used.

Let ζ be a point of the complex plane. Then the interior of the circle of radius ε with ζ as its center, i.e., the set of all points z satisfying the inequality $|z - \zeta| < \varepsilon$, is called an *$\varepsilon$-neighborhood* of ζ, or simply a *neighborhood* of ζ, and is denoted by $\mathscr{N}(\zeta; \varepsilon)$, or simply by $\mathscr{N}(\zeta)$ if ε is unimportant. A sequence $\{z_n\}$ is said to be *convergent*, with *limit* ζ, if any given neighborhood of ζ contains all the terms of $\{z_n\}$ starting from some value of n, i.e., if given any $\varepsilon > 0$, there exists an integer $N(\varepsilon) > 0$ such that $|z_n - \zeta| < \varepsilon$ for all $n > N(\varepsilon)$. We then say that $\{z_n\}$ *converges to* ζ, and write $z_n \to \zeta$ as $n \to \infty$ or

$$\lim_{n\to\infty} z_n = \zeta.$$

Given a complex sequence $\{z_n\}$, let $z_n = x_n + iy_n$, i.e., let x_n and y_n be the real and imaginary parts of the term z_n. Then every sequence of complex numbers $\{z_n\}$ corresponds to two sequences of real numbers $\{x_n\}$ and $\{y_n\}$.

THEOREM 1.1. *The complex sequence $\{z_n\}$, where $z_n = x_n + iy_n$, converges to the limit $\zeta = \xi + i\eta$ if and only if the real sequences $\{x_n\}$ and $\{y_n\}$ converge to the limits ξ and η, respectively.*

Proof. If $z_n \to \zeta$ as $n \to \infty$, then for any $\varepsilon > 0$ there exists an integer $N(\varepsilon) > 0$ such that $|z_n - \zeta| < \varepsilon$ if $n > N(\varepsilon)$. It follows that

$$|x_n - \xi| = |\mathrm{Re}\,(z_n - \zeta)| \leqslant |z_n - \zeta| < \varepsilon,$$

$$|y_n - \eta| = |\mathrm{Im}\,(z_n - \zeta)| \leqslant |z_n - \zeta| < \varepsilon$$

if $n > N(\varepsilon)$, and hence $x_n \to \xi$, $y_n \to \eta$ as $n \to \infty$.

Conversely, if $x_n \to \xi$, $y_n \to \eta$ as $n \to \infty$, then for any $\varepsilon > 0$ there exists an integer $N(\varepsilon) > 0$ such that $|x_n - \xi| < \varepsilon$, $|y_n - \eta| < \varepsilon$ if $n > N(\varepsilon)$. It follows that

$$|z_n - \zeta| = |(x_n - \xi) + i(y_n - \eta)| \leqslant |x_n - \xi| + |y_n - \eta| < 2\varepsilon$$

if $n > N(\varepsilon)$, and hence $z \to \zeta$ as $n \to \infty$, since ε is arbitrary.

According to Theorem 1.1, convergence of a complex sequence $\{z_n\}$ is equivalent to convergence of two real sequences, i.e., those formed from the real and imaginary parts of $\{z_n\}$. This allows us to generalize the whole theory of real sequences to the case of complex sequences. For example, we have

THEOREM 1.2 (*Cauchy convergence criterion*; M1, p. 31).[9] *The sequence $\{z_n\}$ is convergent if and only if given any $\varepsilon > 0$, there exists an integer $N(\varepsilon) > 0$ such that*

$$|z_{n+p} - z_n| < \varepsilon$$

for all $n > N(\varepsilon)$ and all $p > 0$.

Similarly, the familiar results concerning algebraic operations on convergent real sequences can be carried over without change to the case of complex sequences. Thus, given any two complex sequences $\{z_n\}$ and $\{z'_n\}$, suppose that $z_n \to \zeta$, $z'_n \to \zeta'$ as $n \to \infty$. Then

$$z_n \pm z'_n \to \zeta \pm \zeta', \qquad z_n z'_n \to \zeta\zeta'$$

and

$$\frac{z_n}{z'_n} \to \frac{\zeta}{\zeta'}$$

as $n \to \infty$, where in the last formula it is assumed that $z'_n \neq 0$ $(n = 1, 2, \ldots)$ and $\zeta' \neq 0$.

Problem 1. Find the limit, if any, of each of the following sequences:

a) $\left\{\dfrac{2^n}{n!} + \dfrac{in}{2^n}\right\}$; b) $\{\sqrt[n]{n} + inq^n\}$ $(|q| < 1)$; c) $\left\{\sqrt[n]{a} + i \sin \dfrac{1}{n}\right\}$ $(a > 0)$.

Problem 2. Prove that if $\{z_n\}$ converges to the limit ζ, then so does every subsequence $\{z_{n_k}\}$.

Problem 3 (M1, p. 35). Prove that if $z_n \to \zeta$ as $n \to \infty$, then $|z_n| \to |\zeta|$ as $n \to \infty$, and moreover, if $\zeta \neq 0$ and Φ is any value of $\operatorname{Arg} \zeta$, then there is a sequence $\{\Phi_n\}$, where each Φ_n is a value of $\operatorname{Arg} z_n$, such that $\Phi_n \to \Phi$ as $n \to \infty$ (ignore the finite number of terms z_n which may vanish).

Comment. This last fact is indicated by writing

$$\lim_{n\to\infty} \operatorname{Arg} z_n = \operatorname{Arg} \zeta.$$

Problem 4 (M1, p. 34). Give an example where $\{z_n\}$ converges to a nonzero limit ζ, but $\{\arg z_n\}$ fails to converge. When can we choose the sequence $\{\Phi_n\}$ in Prob. 3 to be $\{\arg z_n\}$?

---

[9] The meaning of the reference "M1, p. 31" (and similar references to M2 and M3) is explained in the preface. An easy proof of Theorem 1.2 is suggested in Sec. 9, Prob. 9.

Problem 5. Given a sequence $\{z_n\}$, let ζ be a nonzero complex number such that

$$\lim_{n\to\infty} |z_n| = |\zeta|, \qquad \lim_{n\to\infty} \operatorname{Arg} z_n = \operatorname{Arg} \zeta.$$

Prove that $z_n \to \zeta$ as $n \to \infty$.

7. PROPER AND IMPROPER COMPLEX NUMBERS

In the theory of functions of a complex variable, an important role is played by an "improper" complex number called *infinity* and denoted by the symbol ∞. In order to define this concept suitably, we first introduce a new interpretation of complex numbers, which comes up quite naturally in problems involving limits.

Two convergent sequences of complex numbers $\{u_n\}$ and $\{v_n\}$ are said to be *coterminal* if they have the same limit. Let E be the set of all convergent sequences of complex numbers, and divide E into all classes of coterminal sequences. In other words, $\{u_n\}$ and $\{v_n\}$ belong to the same class if and only if $\{u_n\}$ and $\{v_n\}$ converge to the same limit. Obviously, these classes are disjoint and their union is E. If the common limit of all the sequences in a given class is denoted by a small letter, say a, then we denote the class itself by the corresponding capital letter, say A, i.e., $u_n \to a$ as $n \to \infty$ if and only if $\{u_n\} \in A$. In this way, we establish a one-to-one correspondence

$$a \leftrightarrow A, \qquad b \leftrightarrow B, \ldots \tag{1.22}$$

between the complex numbers and classes of coterminal sequences.

If A and B are two classes of coterminal sequences, we define $A \pm B$ as the class of all sequences of the form $\{u_n \pm v_n\}$, AB as the class of all sequences of the form $\{u_n v_n\}$, and A/B as the class of all sequences of the form $\{u_n/v_n\}$, where $\{u_n\} \in A$, $\{v_n\} \in B$ but are otherwise arbitrary.[10] Since $\{u_n\} \in A$, $\{v_n\} \in B$, we have

$$\lim_{n\to\infty} u_n = a, \qquad \lim_{n\to\infty} v_n = b$$

and hence

$$\lim_{n\to\infty} (u_n \pm v_n) = a \pm b, \qquad \lim_{n\to\infty} u_n v_n = ab, \qquad \lim_{n\to\infty} \frac{u_n}{v_n} = \frac{a}{b}. \tag{1.23}$$

It follows that $A \pm B$, AB and A/B are again classes of coterminal sequences. Moreover (1.23) implies that

$$a \pm b \leftrightarrow A \pm B, \quad ab \leftrightarrow AB, \quad \frac{a}{b} \leftrightarrow \frac{A}{B},$$

i.e., the one-to-one correspondence (1.22) is an *isomorphism*[11] between the

---

[10] Whenever we write u_n/v_n or a/b, it is assumed that $v_n \neq 0$ or $b \neq 0$.

[11] See e.g., G. Birkhoff and S. MacLane, *op. cit.*, p. 31.

field of complex numbers and the field of coterminal sequences. In this sense, we are justified in identifying a and A, b and B, etc., and then every class of coterminal sequences is called a *proper complex number*. In particular the class A can be represented geometrically by the same point in the complex plane as that representing the complex number a.

We now adjoin a single *improper complex number* to the set of all proper complex numbers (i.e., the set of all classes of coterminal sequences). This improper complex number, which we denote by ∞, is the class of all sequences $\{z_n\}$ with the property that given any $\rho > 0$, there exists an integer $n_0 > 0$ (depending on ρ and $\{z_n\}$) such that $|z_n| > \rho$ whenever $n > n_0$. If a sequence belongs to the class ∞, we say that $\{z_n\}$ *converges to infinity*, and we write $z_n \to \infty$ as $n \to \infty$ or

$$\lim_{n\to\infty} z_n = \infty.$$

The union of the set of all proper complex numbers and the improper complex number ∞ is called the *extended complex number system*.

Problem 1. Prove that if $\{u_n\}$ and $\{v_n\}$ are coterminal sequences, then

$$\lim_{n\to\infty} |u_n - v_n| = 0.$$

Problem 2. Suppose $z_n \to \infty$ as $n \to \infty$. What does this imply about the behavior of Re z_n, Im z_n, $|z_n|$ and Arg z_n?

Problem 3 (M1, p. 79). Prove that if a is a proper complex number, then

a) $a \pm \infty = \infty \pm a = \infty$, but $\infty \pm \infty$ is meaningless (i.e., does not lead to a class of coterminal sequences);
b) $a \cdot \infty = \infty \cdot a = \infty$ if $a \neq 0$ and moreover $\infty \cdot \infty = \infty$, but $0 \cdot \infty$ is meaningless;
c) $a/\infty = 0$ and $\infty/a = \infty$, but ∞/∞ is meaningless;
d) $a/0 = \infty$ if $a \neq 0$, but $0/0$ is meaningless.

8. INFINITY AND STEREOGRAPHIC PROJECTION

In order to represent the extended complex number system geometrically, it is convenient to use the following construction, due to Riemann. Consider a sphere Σ of unit radius and center O, and let Π be a plane passing through O (see Figure 1.3). Introducing a rectangular coordinate system in the plane Π, with origin at O, we can represent any proper complex number $z = x + iy$ by a point (x, y) in the plane Π. To associate a point on the sphere Σ with a given point $P \in \Pi$, we first draw the diameter NS of the sphere which is perpendicular to Π and intersects Π at O. Then we draw the line segment joining one end of this diameter, say N, to the point P. The line segment NP (or its prolongation) intersects the sphere Σ in some point P^* different from N. It is clear that this construction establishes a one-to-one correspondence

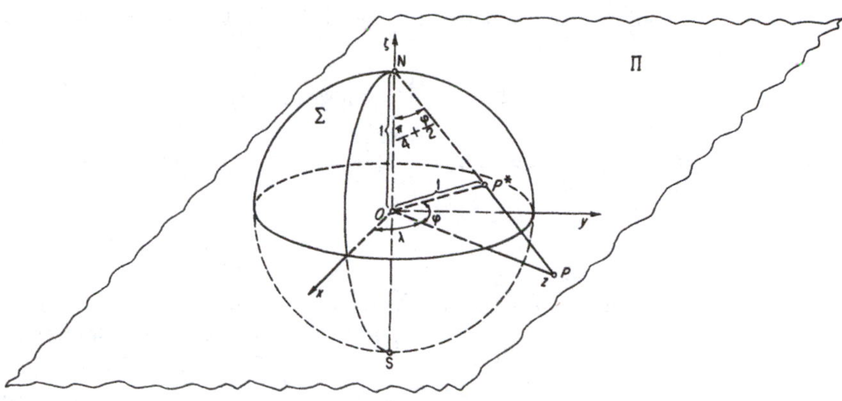

FIGURE 1.3

between the points of the sphere Σ (except for the point N itself) and the points of the plane Π. This mapping of the sphere into the plane (or of the plane into the sphere) is called *stereographic projection*, and the sphere Σ is called the *Riemann sphere*. If the point $P \in \Pi$ represents the complex number z, we also regard the point $P^* \in \Sigma$ as representing z.

To be as descriptive as possible, we use geographic terminology. Thus the circle in which the sphere Σ intersects the plane is called the *equator*, the points N and S are called the *north pole* and *south pole*, respectively, the great circles going through N and S are called *meridians*, and in particular, the meridian lying in the plane NOx is called the *prime* (or *initial*) *meridian*. Then it is easy to see that the points of the plane Π lying inside the unit circle $|z| = 1$ (which coincides with the equator) are mapped into points of the *southern hemisphere* (containing S), while the points of Π lying outside the unit circle are mapped into points of the *northern hemisphere* (containing N). Similarly, the upper half-plane $y > 0$ is mapped into the *eastern hemisphere* (which is intersected by the positive y-axis), while the lower half-plane $y < 0$ is mapped into the *western hemisphere*, and so on.

We now introduce spherical (or geographic) coordinates on Σ, i.e., the *latitude* φ, measured from the equator and ranging from 0 to $\pi/2$ in the northern hemisphere and from 0 to $-\pi/2$ in the southern hemisphere, and the *longitude* λ, measured from the prime meridian (more exactly, from the point of intersection of the prime meridian with the positive x-axis) and ranging from 0 to π (including π) in the eastern hemisphere and from 0 to $-\pi$ (excluding $-\pi$) in the western hemisphere. As shown by Figure 1.3, under stereographic projection we have

$$\arg z = \lambda, \qquad |z| = \tan\left(\frac{\pi}{4} + \frac{\varphi}{2}\right), \tag{1.24}$$

and hence the point of the sphere with coordinates λ and φ is the image of the complex number

$$z = \tan\left(\frac{\pi}{4} + \frac{\varphi}{2}\right)(\cos\lambda + i\sin\lambda). \tag{1.25}$$

Conversely, the image on the Riemann sphere of the complex number $z \neq 0$ has spherical coordinates

$$\lambda = \arg z, \qquad \varphi = 2 \operatorname{arc\,tan} |z| - \frac{\pi}{2}.$$

Every circle γ on the Riemann sphere Σ which does not go through a given point $P^* \in \Sigma$ divides Σ into two parts, such that one part contains P^* and the other does not. The part of Σ containing P^* (and not including γ) will be called a *neighborhood* of P^*. We are now in a position to introduce the concept of a convergent sequence of points $\{P_n^*\}$ on Σ, i.e., P_n^* is said to be *convergent*, with *limit* P^*, if any given neighborhood of P^* contains all the terms of $\{P_n\}$ starting from some value of n. We then say that $\{P_n^*\}$ *converges to* P^*, and write $P_n^* \to P^*$ as $n \to \infty$ or

$$\lim_{n\to\infty} P_n^* = P^*.$$

Now consider a sequence $\{a_n\}$ belonging to the improper class ∞ (see p. 20). Then, given any $\rho > 0$, there exists an integer $n_0 > 0$ such that the points P_n representing the numbers a_n in the complex plane lie outside the circle $|z| = \rho$ if $n > n_0$. The corresponding points P_n^* of the Riemann sphere lie in the neighborhood of the north pole N bounded by the parallel of latitude φ whose projection onto the plane is the circle $|z| = \rho$, and moreover, according to the second of the formulas (1.24), $|z| = \rho$ approaches $+\infty$ if and only if φ approaches $\pi/2$.[12] Therefore $\{P_n^*\}$ converges to the north pole N if and only if $\{a_n\}$ converges to ∞, and in this sense, N is the geometric image of the improper complex number ∞.

Let Σ denote the Riemann sphere, let $\Sigma - \{N\}$ denote Σ with the single point N (the north pole) deleted, and let Π be the ordinary or *finite* (complex) plane. Then, as we have just seen, stereographic projection is a one-to-one mapping of $\Sigma - \{N\}$ onto Π. Therefore $\Sigma - \{N\}$ is just as suitable as Π itself for representing the proper complex numbers. On the other hand, the presence on the sphere Σ of the extra point N, which it is natural to regard as the image of the complex number ∞, suggests that we think of Σ as the

[12] Note the distinction between the improper real number $+\infty$ and the improper complex number ∞ ($+\infty$ is the class of all real sequences $\{a_n\}$ with the property that given any $\rho > 0$, there exists an integer $n_0 > 0$ such that $a_n > \rho$ whenever $n > n_0$). However, we usually write ∞ instead of $+\infty$ whenever $+\infty$ is a "value" of a real quantity which is inherently nonnegative, or whenever the context precludes any possibility of confusion.

image of an *extended* (*complex*) *plane* $\Pi \cup \{\infty\}$, obtained by adjoining to the finite complex plane a single "improper" or "ideal" point, called the *point at infinity* and denoted by ∞, the same symbol as used to denote the corresponding improper complex number. In other words, by imagining that stereographic projection has been carried out, we can think of the extended plane as the union of the finite plane and a single extra point ∞, the point at infinity, which is the image under stereographic projection of the north pole of the Riemann sphere. By the same token, both Σ and $\Pi \cup \{\infty\}$ can be regarded as representing the extended complex number system (see p. 20).

Under stereographic projection, transformations of the Riemann sphere Σ into itself correspond to transformations of the extended plane into itself. For example, suppose we rotate Σ through the angle π about the diameter of Σ directed along the x-axis. Then the northern hemisphere goes into the southern hemisphere, and vice versa. In particular, the north and south poles (i.e., the points representing ∞ and 0) are interchanged. Moreover, the eastern and western hemispheres change places, while the equator and the prime meridian are transformed into themselves. It follows that the given transformation of the sphere into itself is equivalent to the transformation

$$\varphi \to -\varphi, \qquad \lambda \to -\lambda \tag{1.26}$$

of the spherical coordinates φ, λ of an arbitrary point of Σ.

To find the corresponding transformation of the extended plane into itself, we carry out the transformation (1.26) in formula (1.25). As a result, the point z goes into the new point

$$\zeta = \tan\left(\frac{\pi}{4} - \frac{\varphi}{2}\right)(\cos\lambda - i\sin\lambda) = \frac{1}{\tan\left(\dfrac{\pi}{4} + \dfrac{\varphi}{2}\right)(\cos\lambda + i\sin\lambda)} = \frac{1}{z},$$

i.e., rotation of the sphere through the angle π about the real axis corresponds to the transformation of the plane given by

$$\zeta = \frac{1}{z}.$$

Under this transformation, the exterior of the unit circle (corresponding to the northern hemisphere) goes into the interior of the unit circle (corresponding to the southern hemisphere), and vice versa. Moreover, the upper and lower half-planes (corresponding to the eastern and western hemispheres) also change places, but, on the other hand, the unit circle (the image of the equator) and the real axis (the image of the prime meridian) transform into themselves.

It is particularly important to note that under the transformation $\zeta = 1/z$, the point at infinity and the origin of coordinates (the images of the north and south poles) change places, and furthermore, neighborhoods of these points change places. This is completely consistent with the relations

$$\infty = \frac{1}{0}, \qquad 0 = \frac{1}{\infty}$$

satisfied by the improper complex number ∞. The transformation $\zeta = 1/z$ is habitually used in problems involving the point at infinity, since it reduces problems involving ∞ and its neighborhoods to problems involving 0 and its neighborhoods.

Problem 1 (M1, p. 81). Derive the formulas relating the rectangular coordinates (ξ, η, ζ) of the point $P^* \in \Sigma$ and the rectangular coordinates $(x, y, 0)$ of the point $P \in \Pi$ which corresponds to P^* under stereographic projection.

Ans. $$\xi = \frac{2x}{x^2 + y^2 + 1}, \quad \eta = \frac{2y}{x^2 + y^2 + 1}, \quad \zeta = \frac{x^2 + y^2 - 1}{x^2 + y^2 + 1};$$

$$x = \frac{\xi}{1 - \zeta}, \quad y = \frac{\eta}{1 - \zeta}.$$

Problem 2 (M1, p. 82). Use the preceding problem to show that stereographic projection maps circles on the Riemann sphere into circles or straight lines in the plane, and conversely.

Problem 3. What is the relation satisfied by two points z_1 and z_2 which are the images under stereographic projection of a pair of diametrically opposite points of the Riemann sphere?

Ans. $z_1 \bar{z}_2 = -1$.

Problem 4. Find the images of the following curves under the transformation $\zeta = 1/z$:

a) The family of circles $x^2 + y^2 = ax$;
b) The family of circles $x^2 + y^2 = by$;
c) The family of parallel lines $y = x + b$;
d) The family of lines $y = kx$ passing through the origin.

CHAPTER 2

LIMITS AND CONTINUITY

9. MORE SET THEORY. THE HEINE-BOREL THEOREM

Let E be an arbitrary set of points in the complex plane. A point ζ (not necessarily in E) is said to be a *limit point* of the set E if every neighborhood of ζ contains infinitely many (distinct) points of E. Obviously, a set consisting of a finite number of points has no limit points. A set E is said to be *bounded* if there exists a neighborhood of the point $z = 0$ containing all the points of E, i.e., if there exists a $\rho > 0$ such that $|z| < \rho$ for all $z \in E$; otherwise E is said to be *unbounded.*

A set F is said to be *closed* if it contains all its limit points.[1] Sets with no limit points (e.g., finite sets, the empty set, the set of all positive integers) are also regarded as closed. The reason for this is apparent from the following negative form of the definition of a closed set: A set F is said to be closed if no point which is not a member of F can be a limit point of F. A set F which is both closed and bounded is said to be *compact.*

Example 1. Any finite set is compact.

Example 2. The set of all points in the plane and the set of all positive integers are closed, but not compact.

---

[1] A closed set is often denoted by the letter F (from the French *fermé*).

Example 3. The set F of all points belonging to a given straight line is closed, since any point $z_0 \notin F$ has a neighborhood consisting entirely of points not in F and hence z_0 cannot be a limit point of F. However, F is unbounded and hence not compact.

Example 4. The set F of all points belonging to a given circle is closed, for the same reason as given in Example 3. Moreover, F is obviously bounded and hence compact.

By an *interior point* of a set E, we mean a point of E which has a neighborhood contained in E. A set O is said to be *open* if all its points are interior points.[2] The empty set is regarded as open.

Example 1. Any neighborhood $\mathcal{N}(z)$ is open.

Example 2. The set of all points in the plane is open.

Example 3. The set of all points which do not belong to a given finite family of lines or circles is open.

A key proposition of set theory is

THEOREM 2.1 (*Heine-Borel theorem*). *Let F be a compact set, and suppose a neighborhood $\mathcal{N}(z)$ is associated with each point $z \in F$.*[3] *Then there exists a finite number of points $z_1, \ldots, z_n$ in F such that*

$$F \subset \mathcal{N}(z_1) \cup \cdots \cup \mathcal{N}(z_n),$$

i.e., F can be covered by a finite number of neighborhoods $\mathcal{N}(z)$.

Proof. Let Q_0 be a square with sides parallel to the coordinate axes and with center at the origin, which contains the set F (see Figure 2.1).

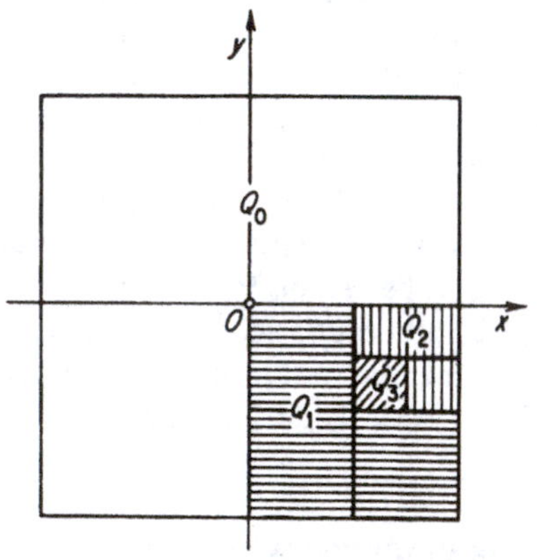

FIGURE 2.1

[2] An open set is often denoted by the letter O (from the French *ouvert*).

[3] The set of neighborhoods $\mathcal{N}(z)$, $z \in F$ is in general uncountable (see Sec. 5, Prob. 2).

Suppose the theorem is false, i.e., suppose F cannot be covered by a finite number of neighborhoods $\mathcal{N}(z)$. The coordinate axes divide Q_0 into four squares, and correspondingly, the set F is divided into four subsets, i.e., the intersections of F with each of these squares. At least one of the four subsets of F cannot be covered by a finite number of neighborhoods $\mathcal{N}(z)$. Let Q_1 be the square containing this subset. Similarly, dividing Q_1 into four congruent subsquares, we can find a new square, say Q_2, whose intersection with F cannot be covered by a finite number of neighborhoods $\mathcal{N}(z)$. Continuing this construction, we arrive at a sequence of squares

$$Q_0 \supset Q_1 \supset Q_2 \supset \cdots \supset Q_n \cdots,$$

each strictly including the next, such that $Q_n \cap F$ cannot be covered by a finite number of neighborhoods. The projections of these squares onto the x and y-axes, respectively, form two "nested" sequences of closed intervals

$$X_0 \supset X_1 \supset X_2 \supset \cdots \supset X_n \supset \cdots,$$

$$Y_0 \supset Y_1 \supset Y_2 \supset \cdots \supset Y_n \supset \cdots,$$

where each of the intervals X_n is twice as long as the next, and similarly for the intervals Y_n. As is familiar from elementary analysis,[4] there is a unique point ξ in all the intervals X_n and a unique point η in all the intervals Y_n. Hence there is a unique point $\zeta = \xi + i\eta$ in all the squares Q_n $(n = 0, 1, 2, \ldots)$. Every Q_n contains an infinite number of points of F, for otherwise we could immediately find a trivial finite covering of $Q_n \cap F$. Therefore ζ must be a limit point of F, since any neighborhood of ζ contains all the squares Q_n starting from some value of n, and hence contains infinitely many points of F. Since F is closed, ζ must belong to F, and hence there is a neighborhood containing ζ in the original set of neighborhoods, i.e., $\mathcal{N}(\zeta)$. But as just noted, starting from some value of n we have $Q_n \subset \mathcal{N}(\zeta)$ and hence $Q_n \cap F \subset \mathcal{N}(\zeta)$, i.e., $Q_n \cap F$ is covered by the single neighborhood $\mathcal{N}(\zeta)$, contrary to hypothesis. This contradiction proves the theorem.

Remark. Obviously, the theorem remains true if $\mathcal{N}(z)$ is replaced by any neighborhood containing z (not necessarily one with z as its center), or more generally by any open set containing z.

The next theorem, of importance in its own right, illustrates the use of

---

[4] See e.g., K. Knopp, *Infinite Sequences and Series* (translated by F. Bagemihl), Dover Publications, Inc., New York (1956), p. 39.

the Heine-Borel theorem:

THEOREM 2.2 (*Bolzano-Weierstrass theorem*). *Every bounded infinite set E has at least one limit point.*

Proof. Let F be a compact set containing E. Consider the set of all neighborhoods containing only finitely many points of E. Every point of F which is not a limit point of E has such a neighborhood. Therefore, if F does not contain a limit point of E, F can be covered by a family of neighborhoods, each containing only a finite number of points of E, and hence, by the Heine-Borel theorem, F can be covered by a finite number of such neighborhoods. But then F contains only finitely many points of E, and hence E is finite since $E \subset F$. This contradiction shows that F must contain a limit point of E, thereby proving the theorem.

Next let E be a nonempty set, and let z be any point in the complex plane. By the *distance between z and E*, we mean the quantity

$$\rho(z, E) = \inf_{z' \in E} |z - z'|, \tag{2.1}$$

i.e., the greatest lower bound (denoted by the symbol inf) of the distances between z and all the points of E. If $\rho(z, E) > 0$, then obviously z cannot belong to E, but if $\rho(z, E) = 0$, z may or may not belong to E. In the latter case, E contains points arbitrarily close to z, i.e., z is a limit point of E. Since a closed set contains all its limit points, we have proved

THEOREM 2.3. *If E is a closed set and z is a point not in E, then $\rho(z, E) > 0$.*

Now let D and E be two nonempty sets. By the *distance between D and E* we mean the quantity

$$\rho(D, E) = \inf_{\substack{z \in D \\ z' \in E}} |z - z'|, \tag{2.2}$$

i.e., the greatest lower bound of the distances between all pairs of points z and z', with $z \in D$, $z' \in E$. If D consists of a single point z, then (2.2) reduces to (2.1). Moreover, it is obvious that $\rho(D, E) = \rho(E, D)$. If $\rho(D, E) > 0$, then D and E must be disjoint, but if $\rho(D, E) = 0$, D and E may or may not have a point in common. In the latter case, there are pairs of points z and z', with $z \in D$, $z' \in E$, such that $|z - z'|$ is arbitrarily small. Clearly, this is possible when one of the sets is not closed,[5] but it is even possible when D and E are both closed. For example, let D consist of the points belonging to a hyperbola, and let E consist of the points belonging to the asymptotes of the hyperbola. In this case, D and E are both closed and $D \cap E = 0$, but

[5] If D is not closed, let E consist of the limit points of D which do not belong to D. Then $D \cap E = 0$ and $\rho(D, E) = 0$.

$\rho(D, E) = 0$. Actually, the appropriate generalization of Theorem 2.3 is another easy consequence of the Heine-Borel theorem:

THEOREM 2.4. *If D and E are two disjoint closed sets, and if one of the sets, say D, is bounded, then $\rho(D, E) > 0$.*

Proof. If $z \in D$, then $z \notin E$ and hence $\rho(z, E) > 0$, by Theorem 2.3. With each point $z \in D$ we associate the "half-size" neighborhood $\mathcal{N}(z; \frac{1}{2}\rho(z, E))$. According to Theorem 2.1, since D is compact, there exists a finite number of points $z_1, \ldots, z_n$ in D such that

$$D \subset \mathcal{N}(z_1; \tfrac{1}{2}\rho_1) \cup \cdots \cup \mathcal{N}(z_n; \tfrac{1}{2}\rho_n), \tag{2.3}$$

where $\rho_1 = \rho(z_1, E), \ldots, \rho_n = \rho(z_n, E)$. Let

$$\delta = \tfrac{1}{2} \min (\rho_1, \ldots, \rho_n),$$

where $\min (\rho_1, \ldots, \rho_n)$ denotes any of the numbers $\rho_1, \ldots, \rho_n$ which does not exceed the others. Consider any point $z \in D$. According to (2.3), z belongs to some neighborhood $\mathcal{N}(z_k; \frac{1}{2}\rho_k)$, $1 \leqslant k \leqslant n$, and since $\mathcal{N}(z_k; \rho_k)$ contains no points of E, if z' is any point of E we have

$$|z - z'| \geqslant \tfrac{1}{2}\rho_k \geqslant \delta.$$

Therefore

$$\rho(D, E) = \inf_{\substack{z \in D \\ z' \in E}} |z - z'| \geqslant \delta > 0,$$

as asserted.

Relative to any open set O, all the points of the plane fall into one of the following three categories:

1. Points which belong to O, called *interior points* of O;
2. Points which do not belong to O and are limit points of O, called *boundary points* of O;
3. Points which do not belong to O and are not limit points of O, called *exterior points of O*.

The set of all boundary points of O is called the *boundary* of O. By the *closure* $\bar{O}$ of an open set O, we mean the union of O and its boundary. These concepts are all easily generalized to the case of arbitrary sets (see M1, pp. 75–76), but the present definitions are enough for our purposes.

Problem 1. Find the limit points of the set of all points z such that

a) $z = 1 + (-1)^n \dfrac{n}{n+1}$ $\quad (n = 1, 2, \ldots)$;

b) $z = \dfrac{1}{m} + \dfrac{i}{n}$ $\quad (m, n = \pm 1, \pm 2, \ldots)$;

c) $z = \dfrac{p}{m} + i\dfrac{q}{n}$ $\quad (m, n, p, q = \pm 1, \pm 2, \ldots)$;

d) $|z| < 1$.

Problem 2. Given an arbitrary set E, a point $z \in E$ is said to be *isolated* if z is not a limit point of E, i.e., if there is a neighborhood of z containing no point of E other than z itself. Prove that the set of all isolated points of E is countable (see Sec. 5, Prob. 2).

Problem 3. Prove that
a) A set is closed if and only if its complement is open;
b) A set is open if and only if its complement is closed.

Problem 4. Prove that
a) The intersection of an arbitrary number of closed sets is closed;
b) The union of a finite number of closed sets is closed.

Problem 5. Prove that
a) The union of an arbitrary number of open sets is open;
b) The intersection of a finite number of open sets is open.

Problem 6. Give examples showing why the word "finite" is needed in Probs. 4b and 5b.

Problem 7. Let E be an unbounded infinite set. Defining a limit point at ∞ in the obvious way, prove that E has at least one limit point in the extended plane.

Problem 8 (M1, pp. 43–44). Prove that ζ is a limit point of a set E if and only if there exists a sequence $\{z_n\}$ of distinct points of E converging to ζ. In particular, prove that every bounded infinite set contains a convergent sequence consisting of distinct points of E.

Problem 9. Use the Bolzano-Weierstrass theorem (Theorem 2.2) to prove the Cauchy convergence criterion (Theorem 1.2).

Problem 10. Let O be an open set. Prove that
a) The boundary of O is closed;
b) $\bar{O}$ is closed;
c) If O has an exterior point, it has infinitely many exterior points;
d) If O has an exterior point, it has infinitely many boundary points.

Hint. Concerning d, see M1, p. 66.

10. THE LIMIT OF A FUNCTION AT A POINT

Let E be an infinite set, let $f(z)$ be a complex function defined on E, and let z_0 $(\neq \infty)$ be a limit point of E. Suppose that given any $\varepsilon > 0$, there exists a number $\delta = \delta(\varepsilon) > 0$ such that

$$|f(z) - A| < \varepsilon \qquad (A \neq \infty) \tag{2.4}$$

whenever

$$|z - z_0| < \delta \qquad (z \in E,\ z \neq z_0)$$

(here A is a fixed complex number). Then $f(z)$ is said to have the *limit A at the point* z_0 (relative to the set E), and we write

$$\lim_{\substack{z\to z_0\\ z\in E}} f(z) = A,$$

or simply[6]

$$\lim_{z\to z_0} f(z) = A \tag{2.5}$$

in cases where the exact nature of E does not matter (or is known from the context).

Remark. Our definition of the limit of a function of a complex variable includes as a special case the familiar definition of the limit of a real function of one or two real variables. In fact, if $f(z)$ is real, we can write

$$f(z) = f(x + iy) = u(x, y),$$

where u takes only real values. Let $z_0 = x_0 + iy_0$, and suppose that

$$\lim_{z\to z_0} f(z) = A.$$

Then, given any $\varepsilon > 0$, there exists a $\delta(\varepsilon) > 0$ such that

$$|f(z) - A| = |u(x, y) - A| < \varepsilon \tag{2.6}$$

whenever

$$0 < |z - z_0| = \sqrt{(x - x_0)^2 + (y - y_0)^2} < \delta(\varepsilon).$$

But (2.6) holds whenever

$$0 < |x - x_0| < \delta'(\varepsilon), \qquad 0 < |y - y_0| < \delta'(\varepsilon),$$

where $\delta'(\varepsilon) = \delta(\varepsilon)/\sqrt{2}$, and this is precisely what is meant by writing

$$\lim_{\substack{x\to x_0\\ y\to y_0}} u(x, y) = A.$$

THEOREM 2.5. *A necessary and sufficient condition for*

$$\lim_{z\to z_0} f(z) = A,$$

where

$$f(z) = u(x, y) + iv(x, y), \qquad A = a + ib,$$

is that

$$\lim_{\substack{x\to x_0\\ y\to y_0}} u(x, y) = a, \qquad \lim_{\substack{x\to x_0\\ y\to y_0}} v(x, y) = b.$$

[6] Instead of (2.5), we sometimes write $f(z) \to A$ as $z \to z_0$.

Proof. Use the above remark and the same argument as in the proof of Theorem 1.1.

It follows from Theorem 2.5 that the results concerning algebraic operations on limits of real functions (familiar from elementary calculus) carry over to the case of functions of a complex variable. Thus, if

$$\lim_{z\to z_0} f(z) = A, \qquad \lim_{z\to z_0} g(z) = B,$$

then

$$\begin{aligned} \lim_{z\to z_0} [f(z) \pm g(z)] &= A \pm B, \\ \lim_{z\to z_0} f(z)g(z) &= AB, \\ \lim_{z\to z_0} \frac{f(z)}{g(z)} &= \frac{A}{B}, \end{aligned} \tag{2.7}$$

where in the last formula it is assumed that $B \neq 0$.

The concept of a limit at infinity is introduced in the natural way. Thus if $f(z)$ is defined on a set E with the point at infinity as a limit point and if given any $\varepsilon > 0$, there exists a number $\rho = \rho(\varepsilon) > 0$ such that (2.4) holds whenever

$$|z| > \rho \qquad (z \in E, z \neq \infty),$$

then $f(z)$ is said to have the *limit A at the point* ∞ (relative to the set E), and we write

$$\lim_{\substack{z\to\infty \\ z\in E}} f(z) = A \quad \text{or simply} \quad \lim_{z\to\infty} f(z) = A.$$

Similarly, we can drop the requirement that the limit A be finite. Thus if $f(z)$ is defined on a set E with z_0 as a limit point (where z_0 may be finite or infinite) and if given any $M > 0$, there exists a neighborhood $\mathcal{N}(z_0)$ [of the form $|z - z_0| < \delta$ or $|z| > \rho$, depending on whether z_0 is finite or infinite] such that

$$|f(z)| > M$$

whenever

$$z \in \mathcal{N}(z_0) \qquad (z \in E, z \neq z_0),$$

then $f(z)$ is said to have the *limit* ∞ *at the point* z_0 (relative to the set E), and we write

$$\lim_{\substack{z\to z_0 \\ z\in E}} f(z) = \infty \quad \text{or simply} \quad \lim_{z\to z_0} f(z) = \infty.$$

Remark. In the case where $A = \infty$, we must be careful in applying the rules (2.7), since the expressions $\infty \pm \infty$, $0 \cdot \infty$ and ∞/∞ are meaningless (see Sec. 7, Prob. 3).

Problem 1. Summarize all the above definitions of the limit of a function at a point in a single general definition.

Hint. Consider a neighborhood of the limit A, as well as a neighborhood of the point z_0.

Problem 2 (M1, pp. 44–45). Prove that $f(z) \to A$ as $z \to z_0$ if and only if the sequence $\{f(z_n)\}$ converges to A for every sequence $\{z_n\}$ converging to z_0.

Problem 3. Give an example of a function $f(z)$ which has a limit A at a point z_0 relative to every straight line passing through z_0, but not relative to a neighborhood of z_0.

11. CONTINUOUS FUNCTIONS

Next we introduce the concept of continuity for functions of a complex variable. Let $f(z)$ be a function defined on a set E, and let $z_0 \in E$ be a (finite or infinite) limit point of E. Then the function $f(z)$ is said to be *continuous at the point* z_0 if $f(z_0) \neq \infty$ and

$$\lim_{z \to z_0} f(z) = f(z_0). \tag{2.8}$$

If $z_0 \in E$ is not a limit point of E, then $f(z)$ is automatically regarded as continuous at z_0. A function which is continuous at every point of a set E is said to be *continuous on E*.[7] Since, when $f(z)$ takes only real values, our definition of a limit reduces to the definition familiar from elementary calculus, the same is true of our definition of continuity. In fact, according to Theorem 2.5, if

$$f(z) = u(x, y) + iv(x, y),$$

we can replace (2.8) by the two equivalent relations

$$\lim_{\substack{x \to x_0 \\ y \to y_0}} u(x, y) = u(x_0, y_0), \qquad \lim_{\substack{x \to x_0 \\ y \to y_0}} v(x, y) = v(x_0, y_0),$$

where $z = x + iy$, $z_0 = x_0 + iy_0$. In other words, the complex function $f(z)$ is continuous at the point $z_0 = x_0 + iy_0$ if and only if the real and imaginary parts of $f(z)$, regarded as functions of the two real variables x and y, are continuous at the point (x_0, y_0).

The following properties of continuous functions of a complex variable are immediate consequences of our definition of continuity:

1. If $f(z)$ and $g(z)$ are continuous at the point $z_0 \in E$, then

$$f(z) \pm g(z), \quad f(z)g(z), \quad \frac{f(z)}{g(z)}$$

are also continuous at z_0 [in the last case, it is assumed that $g(z_0) \neq 0$].

---

[7] A property is said to hold "on" a set of points E if it holds "at all points of" E, but when talking about intervals, neighborhoods, planes or half-planes, the word "in" is usually preferred to "on."

2. Suppose the function $f(z)$ with domain E and range $\mathscr{E}$ is continuous at the point $z_0 \in E$, and suppose $\mathscr{E}$ is an infinite set with $w_0 = f(z_0)$ as a limit point. Moreover, suppose the function $\varphi(w)$ has domain $\mathscr{E}$ and is continuous at the point $w_0 \in \mathscr{E}$. Then the *composite function*

$$\varphi[f(z)]$$

is continuous at the point z_0 (supply the details).

THEOREM 2.6. *Let F be a compact set, and let $f(z)$ be a continuous function on F. Then the set $f(F)$ is compact, i.e., the continuous image of a compact set is compact.*

Proof. First we prove that $\mathscr{F} = f(F)$ is bounded. Since $f(z)$ is continuous on F, given any $\varepsilon > 0$, we can associate a neighborhood $\mathscr{N}(z)$ with each $z \in F$ such that

$$|f(z) - f(z')| < \varepsilon$$

whenever $z' \in \mathscr{N}(z) \cap F$. In other words,

$$f(z') \in \mathscr{N}(f(z); \varepsilon)$$

whenever $z' \in \mathscr{N}(z) \cap F$. According to Theorem 2.1, there exists a finite number of points $z_1, \ldots, z_n$ in F such that

$$F \subset \mathscr{N}(z_1) \cup \cdots \cup \mathscr{N}(z_n),$$

and hence

$$\mathscr{F} \subset \mathscr{N}(f(z_1); \varepsilon) \cup \cdots \cup \mathscr{N}(f(z_n); \varepsilon),$$

i.e., $\mathscr{F}$ can be covered by a finite number of neighborhoods and hence is bounded.

Next we prove that $\mathscr{F}$ is closed. Let ω be a limit point of $\mathscr{F}$. Then, according to Sec. 9, Prob. 8, there is a sequence $\{w_n\}$ of distinct points of $\mathscr{F}$ converging to ω. For each n, let z_n be a point in F such that $f(z_n) = w_n$, and consider the sequence $\{z_n\}$. Since $\{z_n\}$ is bounded[8] (being a subset of the compact set F), $\{z_n\}$ has at least one limit point ζ, according to Theorem 2.2, and since F is closed, ζ must belong to F. By Sec. 9, Prob. 8 again, $\{z_n\}$ contains a subsequence $\{z_{k_n}\}$ converging to $\zeta \in F$. Moreover, since $f(z)$ is continuous on F (and hence at ζ),

$$\lim_{n \to \infty} f(z_{k_n}) = f(\zeta) \tag{2.9}$$

(see Sec. 10, Prob. 2). On the other hand, $\{f(z_{k_n})\}$ is a subsequence of $\{w_n\}$, and hence itself converges to ω (see Sec. 6, Prob. 2):

$$\lim_{n \to \infty} f(z_{k_n}) = \omega. \tag{2.10}$$

[8] A sequence is said to be *bounded* if the set consisting of its (distinct) terms is bounded.

It follows by comparing (2.9) and (2.10) that $\omega = f(\zeta)$, i.e., that $\omega \in \mathscr{F}$ and $\mathscr{F}$ is closed, as asserted.

COROLLARY. *Let F be a compact set, and let f(z) be a continuous function on F. Then there exists a constant $M > 0$ such that*

$$|f(z)| \leqslant M$$

for all $z \in F$, i.e., f(z) is bounded on F.[9]

Remark. If $f(z)$ is continuous on F, so is $|f(z)|$, since

$$\big||f(z)| - |f(z_0)|\big| \leqslant |f(z) - f(z_0)|$$

[cf. (1.10)].

THEOREM 2.7. *Let F be a compact set, and let f(z) be a continuous function on F. Then there exist points z_0, Z_0 in F such that*

$$|f(z_0)| \leqslant |f(z)| \leqslant |f(Z_0)|$$

for all $z \in F$, i.e., $|f(z)|$ achieves its greatest lower bound and its least upper bound on F.[10]

Proof. Since $|f(z)|$ is continuous on F, the set $\mathscr{F}$ of all real numbers

$$w = |f(z)|, \qquad z \in F$$

is compact, according to Theorem 2.6. Let m and M be the greatest lower bound and the least upper bound, respectively, of $\mathscr{F}$ (these exist since $\mathscr{F}$ is bounded). By the very definition of m, given any $\varepsilon > 0$, the neighborhood $\mathscr{N}(m; \varepsilon)$ contains at least one point of $\mathscr{F}$ (which may be m itself). Therefore m is either already known to be in $\mathscr{F}$, or else m is a limit point of $\mathscr{F}$ and hence an element of $\mathscr{F}$ (since $\mathscr{F}$ is closed). By a similar argument, $M \in \mathscr{F}$. In other words, there exist points z_0, Z_0 in F such that

$$f(z_0) = m, \qquad f(Z_0) = M.$$

THEOREM 2.8. *Let F be a compact set, and let f(z) be a continuous function on F. Then, given any $\varepsilon > 0$, there exists a number $\delta = \delta(\varepsilon) > 0$ such that*

$$|f(z) - f(z')| < \varepsilon$$

---

[9] Equivalently, a function $f(z)$ is said to be *bounded* on a set E if $f(E)$ is a bounded set.

[10] An equivalent version of Theorem 2.7 is the following: *Let F be a compact set, and let f(z) be a continuous function on F. Then there exist points z_0, Z_0 in F such that*

$$|f(z_0)| = \min_{z \in F} |f(z)|, \qquad |f(Z_0)| = \max_{z \in F} |f(z)|.$$

[The right-hand sides of these formulas denote the minimum and maximum values taken by $|f(z)|$ at a point of F.]

for every pair of points z, z' in F such that

$$|z - z'| < \delta, \tag{2.11}$$

i.e., $f(z)$ is uniformly continuous on F.

Proof. The important point here is that one choice of $\delta(\varepsilon)$ suffices for *all* $z, z' \in F$. Since $f(z)$ is continuous on F, given any $\varepsilon > 0$, we can associate a neighborhood $\mathcal{N}(z; \delta(z))$ with each $z \in F$ such that

$$|f(z) - f(z')| < \tfrac{1}{2}\varepsilon$$

whenever $z' \in \mathcal{N}(z; \delta(z))$. Now consider the set of "half-size" neighborhoods $\mathcal{N}(z; \frac{1}{2}\delta(z))$, $z \in F$. According to Theorem 2.1, there exists a finite number of points $z_1, \ldots, z_n$ in F such that

$$F \subset \mathcal{N}(z_1; \tfrac{1}{2}\delta_1) \cup \cdots \cup \mathcal{N}(z_n; \tfrac{1}{2}\delta_n), \tag{2.12}$$

where $\delta_1 = \delta(z_1), \ldots, \delta_n = \delta(z_n)$. Then a suitable δ for use in the inequality (2.11) is given by

$$\delta = \tfrac{1}{2}\min(\delta_1, \ldots, \delta_n).$$

To see this, let z and z' be any two points in F such that $|z - z'| < \delta$. According to (2.12), z belongs to some neighborhood $\mathcal{N}(z_k; \frac{1}{2}\delta_k)$, $1 \leqslant k \leqslant n$, and hence

$$|f(z) - f(z_k)| < \tfrac{1}{2}\varepsilon,$$

since obviously $\mathcal{N}(z_k; \frac{1}{2}\delta_k) \subset \mathcal{N}(z_k; \delta_k)$. Moreover, z' belongs to $\mathcal{N}(z_k; \delta_k)$, since

$$|z' - z_k| = |(z' - z) + (z - z_k)| \leqslant |z' - z| + |z - z_k| < \delta + \tfrac{1}{2}\delta_k \leqslant \delta_k.$$

It follows that

$$|f(z') - f(z_k)| < \tfrac{1}{2}\varepsilon,$$

and hence

$$\begin{aligned} |f(z) - f(z')| &= |[f(z) - f(z_k)] + [f(z_k) - f(z')]| \\ &\leqslant |f(z) - f(z_k)| + |f(z') - f(z_k)| < \varepsilon, \end{aligned}$$

as required.

Finally, we relax the requirement that $f(z_0)$ be finite in (2.8). Again let E be an infinite set and let $z_0 \in E$ be a (finite or infinite) limit point of E, but this time suppose $f(z_0) = \infty$. Then we say that $f(z)$ is *wide-sense continuous at z_0* if

$$\lim_{z \to z_0} f(z) = \infty.$$

A function which is continuous or wide-sense continuous at every point of a set E is said to be *wide-sense continuous on E*.

Example. The function

$$f(z) = \begin{cases} \dfrac{1}{z} & \text{if} \quad z \neq 0,\ z \neq \infty, \\ 0 & \text{if} \quad z = \infty, \\ \infty & \text{if} \quad z = 0 \end{cases}$$

is wide-sense continuous in the extended plane. In particular,

$$\lim_{z\to\infty} f(z) = 0 = f(\infty), \qquad \lim_{z\to 0} f(z) = \infty = f(0).$$

Problem 1. Let K be the open unit disk $|z| < 1$. Are the functions

$$\frac{1}{1-z}, \qquad \frac{1}{1+z^2}$$

continuous on K, and if so, are they uniformly continuous on K?

Problem 2. The functions

$$\frac{\operatorname{Re} z}{|z|}, \quad \frac{z}{|z|}, \quad \frac{\operatorname{Re} z^2}{|z|^2}, \quad \frac{z \operatorname{Re} z}{|z|}$$

are all defined for $z \neq 0$. Which of them can be defined at the point $z = 0$ in such a way that the extended functions are continuous at $z = 0$?

Ans. Only $f(z) = \dfrac{z \operatorname{Re} z}{|z|}$, $f(0) = 0$.

Problem 3. Which, if any, of the theorems of this section generalize to the case of wide-sense continuous functions?

Problem 4 (M1, p. 53). Let $\rho(z, E)$ be the same as in (2.1). Prove that if E is a compact set and z is an arbitrary point, then there exists a point ζ in E such that $\rho(z, E) = |z - \zeta|$.

Problem 5 (M1, p. 53). Let $\rho(D, E)$ be the same as in (2.2). Prove that if D and E are disjoint compact sets, then there exists a pair of points z_1 and z_2, with $z_1 \in D$, $z_2 \in E$, such that $\rho(D, E) = |z_1 - z_2|$.

Problem 6. Prove that in the preceding two problems, E need only be closed.

12. CURVES AND DOMAINS

Let $z = f(t)$ be a complex function of a real variable t, defined and continuous in a closed interval $a \leqslant t \leqslant b$. Then $z = f(t)$ is said to define a (*continuous*) *curve* L. The values of the function are called the *points* of the curve, and the set of points in the curve, i.e., the range of $f(t)$, is often referred to simply as the *curve* (when no ambiguity can arise). In particular, the points

$z_0 = f(a)$ and $z_1 = f(b)$ are called the *initial* and *final points* of the curve, respectively, and z_0, z_1 are called the *end points* of the curve. The initial and final points of a curve may coincide, in which case the curve is said to be *closed* (not to be confused with the concept of a closed set). The real variable t is called the *parameter* of the curve, and the equation $z = f(t)$, mapping the values of the parameter onto the points of the curve, is called the (*parametric*) *equation* of the curve.

Remark. We can also think of L as an *oriented curve*, in the sense that a point $z' = f(t') \in L$ is regarded as *distinct* from a point $z'' = f(t'') \in L$ if $t' \neq t''$ and as *preceding* z'' if $t' < t''$. The oriented curve L is then said to be "traversed in the direction of increasing t." It will always be clear from the context whether L is a curve in the set-theoretic sense, i.e., the continuous image of a closed interval, or an oriented curve as just described.

Two curves L and L', with equations $z = f(t)$, $a \leqslant t \leqslant b$ and $z = \varphi(\tau)$, $\alpha \leqslant \tau \leqslant \beta$, respectively, are regarded as identical (except possibly for direction) if the equation of one curve can be transformed into the equation of the other by means of a *continuous strictly monotonic* change of parameter, i.e., if there exists a continuous strictly monotonic function $\tau = \alpha(t)$, $a \leqslant t \leqslant b$ (see Prob. 1) such that the function $\varphi[\alpha(t)]$ coincides with $f(t)$ on the interval $[a, b]$.[11] Then we say that the two curves have the *same* direction if the function $\alpha(t)$ is *increasing*, or *opposite* directions if $\alpha(t)$ is *decreasing*. In the latter case, the initial point of L is the same as the final point of L', and vice versa. The curve differing from L only by the direction in which it is traversed will be denoted by $-L$.

If the same point z corresponds to more than one parameter value in the half-open interval $a \leqslant t < b$, we say that z is a *multiple point* of the curve $z = f(t)$, $a \leqslant t \leqslant b$. A curve with no multiple points is called a *Jordan curve*. It is to allow for the possibility of *closed* Jordan curves, i.e., Jordan curves whose end points coincide, that we consider parameter values in the *half-open* interval $a \leqslant t < b$.

Example 1. The functions

$$z = t, \quad z = t^2, \quad z = 1 - t \qquad (0 \leqslant t \leqslant 1) \tag{2.13}$$

all define the same continuous curve (except for direction), corresponding to the segment of the real axis lying between the points $z = 0$ and $z = 1$. The first and second representations of the curve have the same direction (with initial point $z = 0$ and final point $z = 1$), but the third representation has the opposite direction (with initial point $z = 1$ and final point $z = 0$).

[11] By $[a, b]$ we mean the *closed* interval $a \leqslant t \leqslant b$, and by (a, b) the *open* interval $a < t < b$. Similarly $[a, b)$ and $(a, b]$ denote the *half-open* intervals $a \leqslant t < b$ and $a < t \leqslant b$, respectively.

Obviously, the curve (2.13) has no multiple points and hence is a Jordan curve. Since its initial and final points do not coincide, (2.13) is not a closed Jordan curve.

Example 2. The functions

$$z = t, \quad z = \sin \pi t \qquad (0 \leqslant t \leqslant 1) \tag{2.14}$$

define different continuous curves L and L', since there is no continuous strictly monotonic change of parameter carrying one of these curves into the other. In fact, such a transformation would have to carry the monotonic function $f(t) = t$ into another monotonic function, but the function $\varphi(t) = \sin \pi t$ is not monotonic in the interval [0, 1]. We note that although the two curves (2.14) are different, they consist of the same set of points, i.e., the points in the interval [0, 1]. Therefore the fact that two continuous curves consist of the same points is not a sufficient condition (but is obviously a necessary condition) for the curves to be identical. We also note the following facts:

1. The curve $z = t$ is a Jordan curve, but not a closed curve;
2. The curve $z = \sin \pi t$ is not a Jordan curve, since it has multiple points
$$\sin \pi t = \sin \pi(1 - t) \qquad (0 < t < \tfrac{1}{2});$$
3. The curve $z = \sin \pi t$ is a closed curve, since its initial and final points coincide, i.e., $\sin \pi \cdot 0 = \sin \pi \cdot 1 = 0$.

Example 3. Let $\Delta_1, \ldots, \Delta_n$ be a family of (straight) line segments in the plane, with definite directions, such that the final point of each segment Δ_k $(k < n)$ coincides with the initial point of the next segment Δ_{k+1}. Let $a_1, \ldots, a_n$ be the complex numbers represented by the vectors $\Delta_1, \ldots, \Delta_n$, respectively, and let z_0 be the initial point of the segment Δ_1. Then with the given family of segments we can associate the following continuous curve, defined on the interval $0 \leqslant t \leqslant n$:

$$z = z_0 + a_1 + \cdots + a_{k-1} + a_k(t - k + 1) \qquad (k - 1 \leqslant t \leqslant k; k = 1, \ldots, n).$$

A curve of this type is called a *polygonal curve*, with segments $\Delta_1, \ldots, \Delta_n$. Whether or not a polygonal curve is closed depends on whether or not the final point of Δ_n coincides with the initial point of Δ_1. A polygonal curve is a Jordan curve if and only if it does not intersect itself, which means that its segments have no points in common except those permitted by the construction, i.e., the final point of Δ_k $(k < n)$ coincides with the initial point of Δ_{k+1}, and the final point of Δ_n may coincide with the initial point of Δ_1 (if the curve is closed).

Example 4. The circle

$$z = z_0 + r(\cos t + i \sin t) \qquad (0 \leqslant t \leqslant 2\pi),$$

with center z_0 and radius r, is a closed Jordan curve.

A set E is said to be (*arcwise*) *connected* if every pair of points $z, z' \in E$ can be joined by a continuous curve consisting only of points of E, with one of the points z, z' as its initial point and the other as its final point. A nonempty open connected set G is called a *domain*.[12] If G is a domain, the set $\bar{G}$ (the closure of G) is called a *closed domain.* For example, the open disk $|z - z_0| < \rho$ is a domain, and the closed disk $|z - z_0| \leqslant \rho$ is a closed domain. Each of these domains has the circle $|z - z_0| = \rho$ as its boundary.

Let O be any nonempty open set, and let z_0 be any point of O. By the (*connected*) *component* of O containing the point z_0, we mean the set $\mathscr{C}(z_0)$ of all points in O which can be joined to z_0 by a continuous curve contained in O.

THEOREM 2.9. *Every component $\mathscr{C}(z_0)$ of a nonempty open set O is a domain.*

Proof. $\mathscr{C}(z_0)$ is nonempty, since it contains z_0 and some neighborhood of z_0. $\mathscr{C}(z_0)$ is open, since if z_1 can be joined to z_0 by a continuous curve $L \subset O$, then any point z in a neighborhood $\mathscr{N}(z_1)$ of z_1 such that $\mathscr{N}(z_1) \subset O$ can be joined to z_0 by a continuous curve contained in O, i.e., the union of L and the line segment with end points z and z_1. Finally, $\mathscr{C}(z_0)$ is connected, by definition.

Given a closed set F, let F^c be its complement. Noting that F^c is open (see Sec. 9, Prob. 3) we consider the connected components of F^c. For example, if F consists of the points belonging to the two infinite families of lines $x = 0, \pm 1, \pm 2, \ldots$ and $y = 0, \pm 1, \pm 2, \ldots$ parallel to the coordinate axes, F^c has infinitely many components, i.e., all open squares of the form

$$m < x < m + 1, \qquad n < y < n + 1 \qquad (m, n = 0, \pm 1, \pm 2, \ldots).$$

However, if F is a closed Jordan curve, F^c can only have two components, as shown by the following basic result:

THEOREM 2.10 (*Jordan curve theorem*).[13] *The complement γ^c of any closed Jordan curve γ has exactly two components, with γ as their common boundary. One of these components $I(\gamma)$, called the interior of γ, is bounded, and the other component $E(\gamma)$, called the exterior of γ, is unbounded.*

---

[12] A domain is often denoted by the letter G (from the German *Gebiet*).

[13] The proof of Theorem 2.10 is beyond the scope of this book, but can be found in many textbooks on topology. See e.g., P. S. Aleksandrov, *Combinatorial Topology, Vol. 1* (translated by H. Komm), Graylock Press, Rochester, N.Y. (1956), Chap. 2.

Example. The interior of the circle $|z - z_0| = \rho$ is the disk $|z - z_0| < \rho$,[14] and the exterior is the set of points z satisfying $|z - z_0| > \rho$.

A domain G is said to be *simply connected* if whenever G contains a closed Jordan curve γ, G also contains $I(\gamma)$. Otherwise G is said to be *multiply connected.*

Example 1. The interior of any circle is a simply connected domain, while the set of points lying between two concentric circles (called an *annulus*) is a multiply connected domain.

Example 2. The whole plane is simply connected, and so are the domains a, b, c, and d shown in Figure 2.2, i.e., the half-plane, the wedge, the strip and the half-strip.

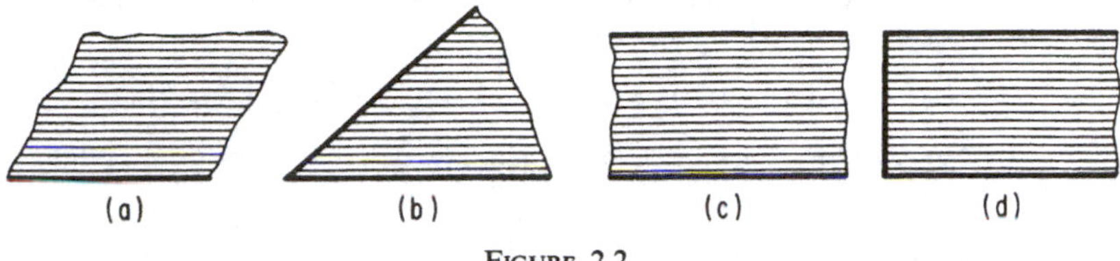

FIGURE 2.2

Example 3. The exterior of a circle is multiply connected.

Problem 1. A real function $\lambda(t)$ defined on a subset E of the real line is said to be *increasing* (or *nondecreasing*) on E if for every pair of points t' and t'' in E, $t' < t''$ implies $\lambda(t') \leqslant \lambda(t'')$. If $t' < t''$ implies $\lambda(t') < \lambda(t'')$, then $\lambda(t)$ is said to be *strictly increasing* on E. *Decreasing* (or *nonincreasing*) functions and *strictly decreasing* functions are defined similarly by writing $\lambda(t') \geqslant \lambda(t'')$ instead of $\lambda(t') \leqslant \lambda(t'')$ and $\lambda(t') > \lambda(t'')$ instead of $\lambda(t') < \lambda(t'')$. A function is said to be (*strictly*) *monotonic* on E if it is (strictly) increasing or (strictly) decreasing on E.

Prove that if $\lambda(t)$ is monotonic in an interval $[a, b]$, then $\lambda(t)$ has a left-hand limit $\lambda(t_0-)$ and a right-hand limit $\lambda(t_0+)$ at every point $t_0 \in (a, b)$, and

$$\lambda(t_0-) \leqslant \lambda(t_0) \leqslant \lambda(t_0+).$$

Moreover, prove that $\lambda(a+)$ and $\lambda(b-)$ exist, and

$$\lambda(a) \leqslant \lambda(a+), \qquad \lambda(b-) \leqslant \lambda(b).$$

Problem 2. A complex function $z = f(t)$ defined and *wide-sense* continuous in an interval $[a, b]$, possibly infinite, is said to define an *unbounded* curve if $f(t_0) = \infty$ for at least one $t_0 \in [a, b]$. Jordan curves and closed Jordan curves are defined in the same way as for ordinary curves. Which of the following

[14] Note the distinction between a circle and a disk.

unbounded curves are Jordan curves:

a) The straight line

$$z = (\alpha t + \beta) + i(\gamma t + \delta) \qquad (\alpha^2 + \gamma^2 \neq 0;\ -\infty < t < +\infty);$$

b) The parabola

$$z = (\alpha t^2 + \beta t + \gamma) + i(\delta t + \varepsilon) \qquad (\alpha, \delta \neq 0;\ -\infty < t < +\infty);$$

c) The hyperbola

$$z = \frac{\alpha(1 + t^2) + 2i\beta t}{1 - t^2} \qquad (\alpha, \beta \neq 0;\ -\infty < t < +\infty)?$$

Visualize these curves on the Riemann sphere.

Ans. The straight line and parabola are closed Jordan curves. The hyperbola is not a Jordan curve (∞ is a multiple point corresponding to $t = \pm 1$).

Problem 3. A set E is said to be *convex* if whenever E contains two points z_1 and z_2, E also contains the line segment joining z_1 and z_2. Prove that every convex set is connected.

Problem 4. Is the set of points z such that $0 < |z| < 1$ a domain?

Problem 5. Give an example of a domain G such that G and $\bar{G}$ have different boúndaries.

Problem 6 (M1, p. 68). Prove that any two points of a domain G can be joined by a Jordan curve contained in G, which can always be chosen to be a polygonal curve.

Problem 7 (M1, p. 65). Prove that every nonempty open set O is the union of countably many disjoint domains, i.e., the connected components of O.

Problem 8. Let E be the set consisting of all points of the upper half-plane $\operatorname{Im} z > 0$ which do not lie on the family of line segments

$$z = \frac{1}{n} + it \qquad (0 < t < 1;\, n = \pm 1, \pm 2, \ldots).$$

What is the boundary of E? Is E a domain, and if so, is E simply connected?

Problem 9. Let G be a domain *in the extended plane*, i.e., the image under stereographic projection of an open connected set on the Riemann sphere (G is now allowed to contain the point at infinity). Then G is said to be simply connected if whenever G contains a closed Jordan curve, G also contains either $I(\gamma)$ or $E(\gamma)$.[15] Discuss Examples 1, 2 and 3, p. 41 from the standpoint of this new definition.

Ans. The exterior of a circle is now simply connected.

Comment. Whenever we have in mind domains of this type, the phrase "in the extended plane" will be included.

---

[15] This is the natural definition on the Riemann sphere, where the "inside" and "outside" domains are not distinguishable. Note that $E(\gamma)$ must now be regarded as containing the point at infinity.

CHAPTER 3

DIFFERENTIATION. ANALYTIC FUNCTIONS

13. DERIVATIVES. RULES FOR DIFFERENTIATING COMPLEX FUNCTIONS

Let $f(z)$ be a function of a complex variable which is defined and single-valued on a set E, and let z_0 be any point of E which is a limit point of E. Then the *difference quotient*

$$\frac{f(z) - f(z_0)}{z - z_0} \tag{3.1}$$

is a function of z defined for any point $z \neq z_0$ of the set E. The limit of (3.1) as $z \to z_0$, $z \in E$, provided it exists, is called the *derivative* of the function $f(z)$ at the point z_0 relative to the set E, denoted by $f'_E(z_0)$ or simply $f'(z_0)$, and the function $f(z)$ itself is said to be *differentiable* at the point z_0 relative to the set E.

In the special case where E is an interval (finite or infinite) of the real axis, $f(z)$ is a function of the real variable $z = x$ and takes values which are in general complex:

$$f(z) = f(x) = \varphi(x) + i\psi(x).$$

If $\psi(x) \equiv 0$, if the values of $f(z)$ are real, our definitions of a derivative and of differentiability reduce to the usual definitions given in elementary calculus.

However, if $\psi(x) \not\equiv 0$, then, writing

$$\frac{f(x) - f(x_0)}{x - x_0} = \frac{\varphi(x) - \varphi(x_0)}{x - x_0} + i\,\frac{\psi(x) - \psi(x_0)}{x - x_0},$$

we conclude that $f'(x_0)$ exists if and only if the derivatives $\varphi'(x_0)$ and $\psi'(x_0)$ exist, and that

$$f'(x_0) = \varphi'(x_0) + i\psi'(x_0)$$

(see Sec. 10). For example, if

$$f(x) = a\cos x + ib\sin x$$

(where x is any real number), then

$$f'(x) = -a\sin x + ib\cos x.$$

Introducing the notation $\Delta z = z - z_0$ for the increment of the independent variable and $\Delta_E f(z) = f(z) - f(z_0)$ for the increment of the function $f(z)$ (relative to the set E), we have the following necessary and sufficient condition for differentiability:

THEOREM 3.1. *The function $f(z)$ is differentiable at the point $z_0 \in E$ (relative to the set E) if and only if $f(z)$ can be written in the form*

$$\Delta_E f(z) = A\,\Delta z + \varepsilon(z, z_0)\,\Delta z, \tag{3.2}$$

where $\varepsilon(z, z_0) \to 0$ as $\Delta z \to 0$ ($z \in E$), and A is a constant independent of Δz and ε.

Proof. If $f(z)$ has a derivative $f_E'(z_0)$ at z_0, then, by definition

$$\frac{\Delta_E f(z)}{\Delta z} = f_E'(z_0) + \varepsilon(z, z_0), \tag{3.3}$$

where $\varepsilon(z, z_0) \to 0$ as $\Delta z \to 0$. Multiplying (3.3) by Δz, we find that $\Delta_E f(z)$ can be written in the form (3.2) with $A = f_E'(z_0)$. Conversely, if $\Delta_E f(z)$ can be written in the form (3.2), then dividing by Δz and taking the limit as $\Delta z \to 0$, we find that $f_E'(z_0)$ exists and equals A.

Remark. It is an immediate consequence of (3.2) that if $f(z)$ is differentiable at the point $z_0 \in E$, it is also continuous at z_0.[1]

By the *differential* of a function $f(z)$ which is differentiable at the point $z_0 \in E$, we mean the principal linear part of the increment $\Delta_E f(z)$, i.e., the quantity

$$d_E f(z_0) = f_E'(z_0)\,\Delta z = f_E'(z_0)\,dz, \tag{3.4}$$

[1] For brevity, we shall often omit the phrase "relative to E," which is tacitly assumed in any context like this.

where, as in elementary calculus, we set $dz = \Delta z$ (see Prob. 1). We can then write

$$f'_E(z_0) = \frac{d_E f(z_0)}{dz}. \tag{3.5}$$

For simplicity, we usually omit the subscript E in expressions like (3.2) through (3.5), unless we want to emphasize the role of the set E.

Example. To clarify the role of the set E relative to which the derivative is taken, let E be the real axis and consider the function

$$f(z) = f(x) = x. \tag{3.6}$$

Obviously, the derivative $f'_E(z)$ exists for all $z = x \in E$, since

$$\frac{f(z) - f(z_0)}{z - z_0} = \frac{x - x_0}{x - x_0} = 1 \qquad (z \in E).$$

We now extend the function $f(z)$ to the whole complex plane $\mathscr{E}$, by writing (3.6) for all $z = x + iy \in \mathscr{E}$. Then $f(z)$ is continuous for any $z \in \mathscr{E}$ and coincides with the original function when $z \in E$. However, the difference quotient is now

$$\frac{f(z) - f(z_0)}{z - z_0} = \frac{x - x_0}{(x - x_0) + i(y - y_0)}, \tag{3.7}$$

which has no limit as $z \to z_0 \in \mathscr{E}$. In fact, (3.7) approaches 0 if $z \to z_0$ along the line $x = x_0$, whereas (3.7) approaches 1 if $z \to z_0$ along the line $y = y_0$. In other words, $f'_{\mathscr{E}}(z_0)$ does not exist for any $z \in \mathscr{E}$. Thus it is already apparent from this simple example that the requirement that $f(z)$ be differentiable regarded as a function of a complex variable $z = x + iy$ is much more stringent than the requirement that $f(z)$ be differentiable only for real values of z.

It follows from our definition of a derivative and the properties of limits (see Sec. 10) that the basic differentiation rules familiar from elementary calculus can be extended to the case of functions of a complex variable. We now list some of these rules. In each case, $f(z)$, $f_1(z)$, $f_2(z)$, etc., is assumed to be differentiable at a given point $z \in E$ (relative to E).

Rule 1. If $f(z) \equiv z$, then

$$\frac{df(z)}{dz} \equiv 1,$$

whereas if $f(z) \equiv$ const, then

$$\frac{df(z)}{dz} \equiv 0.$$

Loosely speaking, the derivative of the independent variable is 1, and the derivative of any constant is 0.

Rule 2 (*Differentiation of sums, products and powers*). If c is a constant, then

$$\frac{d[cf(z)]}{dz} = c\,\frac{df(z)}{dz}.$$

Moreover we have

$$\frac{d}{dz}[f_1(z) + f_2(z) + \cdots + f_n(z)] = \frac{df_1(z)}{dz} + \frac{df_2(z)}{dz} + \cdots + \frac{df_n(z)}{dz},$$

$$\frac{d}{dz}[f_1(z)f_2(z)\cdots f_n(z)] = \frac{df_1(z)}{dz} f_2(z)f_3(z)\cdots f_n(z)$$

$$+ f_1(z)\frac{df_2(z)}{dz} f_3(z)\cdots f_n(z) + \cdots + f_1(z)f_2(z)\cdots f_{n-1}(z)\frac{df_n(z)}{dz},$$

$$\frac{d}{dz}[f(z)]^n = n[f(z)]^{n-1}f'(z).$$

In particular,

$$\frac{d}{dz} z^n = nz^{n-1}$$

and

$$\frac{d}{dz}(a_0 + a_1 z + a_2 z^2 + \cdots + a_n z^n) = a_1 + 2a_2 z + \cdots + na_n z^{n-1}.$$

Rule 3 (*Differentiation of a quotient*). If $f_2(z) \neq 0$, then

$$\frac{d}{dz}\left[\frac{f_1(z)}{f_2(z)}\right] = \frac{f_2(z)\dfrac{df_1(z)}{dz} - f_1(z)\dfrac{df_2(z)}{dz}}{[f_2(z)]^2}.$$

Rule 4 (*Differentiation of a composite function*). Suppose the function $f(z)$ with domain E and range $\mathscr{E}$ is differentiable at the point $z_0 \in E$, and suppose $\mathscr{E}$ is an infinite set with $w_0 = f(z_0)$ as a limit point. Moreover, suppose the function $\varphi(w)$ has domain $\mathscr{E}$ and is differentiable at the point $w_0 \in \mathscr{E}$. Then the *composite function*

$$\varphi[f(z)]$$

is differentiable at the point z_0, and

$$\frac{d_E\varphi[f(z_0)]}{dz} = \frac{d_{\mathscr{E}}\varphi(w_0)}{dw}\,\frac{d_E f(z_0)}{dz}. \tag{3.8}$$

The proof is trivial if there exists a neighborhood $\mathcal{N}(z_0)$ of the point z_0 such that $f(z) = w \neq w_0$ for all $z \in \mathcal{N}(z_0) \cap E$, for then

$$\begin{aligned}\frac{\varphi[f(z)] - \varphi[f(z_0)]}{z - z_0} &= \frac{\varphi(w) - \varphi(w_0)}{w - w_0}\frac{w - w_0}{z - z_0} \\ &= \frac{\varphi(w) - \varphi(w_0)}{w - w_0}\frac{f(z) - f(z_0)}{z - z_0}.\end{aligned} \tag{3.9}$$

Since $w \to w_0$ as $z \to z_0$ (see the remark on p. 44), taking the limit of (3.9) as $z \to z_0$ we immediately obtain (3.8).

Now suppose every neighborhood of z_0 contains a point $z \in E$, $z \neq z_0$ such that $f(z) = f(z_0) = w_0$. Then there exists a sequence $\{z_n\}$ of distinct points in E converging to z_0 such that

$$f(z_n) = f(z_0) \qquad (n = 1, 2, \ldots).$$

For this sequence

$$\frac{f(z_n) - f(z_0)}{z_n - z_0} = 0 \qquad (n = 1, 2, \ldots),$$

and hence the derivative

$$\frac{d_E f(z_0)}{dz} = \lim_{z \to z_0} \frac{f(z) - f(z_0)}{z - z_0}$$

(which by hypothesis exists) must vanish. Therefore the right-hand side of (3.8) vanishes, and to prove (3.8) we have to show that its left-hand side also vanishes. Since the difference quotient

$$\frac{\varphi[f(z)] - \varphi[f(z_0)]}{z - z_0}$$

is obviously zero for every $z \neq z_0$ such that $f(z) = f(z_0)$, the proof reduces to showing that

$$\lim_{n \to \infty} \frac{\varphi[f(z_n^*)] - \varphi[f(z_0)]}{z_n^* - z_0} \tag{3.10}$$

vanishes for every sequence $\{z_n^*\}$ converging to z_0 such that $f(z_n^*) = w_n^* \neq w_0 = f(z_0)$. However, for such a sequence, (3.10) equals

$$\frac{\varphi(w_n^*) - \varphi(w_0)}{w_n^* - w_0}\frac{w_n^* - w_0}{z_n^* - z_0} = \frac{\varphi(w_n^*) - \varphi(w_0)}{w_n^* - w_0}\frac{f(z_n^*) - f(z_0)}{z_n^* - z_0}$$

[cf. (3.9)], which approaches

$$\frac{d_{\mathscr{E}}\varphi(w_0)}{dw}\frac{d_E f(z_0)}{dz}$$

as $n \to \infty$. But as just shown, $d_E f(z_0)/dz$ vanishes, and hence (3.8) holds.

Rule 5 (*Differentiation of an inverse function*). Suppose $w = f(z)$ is a one-to-one function defined on E, and suppose the inverse function $z = f^{-1}(w) = \varphi(w)$ is continuous on $\mathscr{E}$, the range of $f(z)$. Then if $f(z)$ is differentiable at the point $z_0 \in E$ and if $f'_E(z_0) \neq 0$, the function $\varphi(w)$ is differentiable at the point $w_0 = f(z_0) \in \mathscr{E}$, and

$$\varphi'_{\mathscr{E}}(w_0) = \frac{1}{f'_E(z_0)}.$$

The proof goes as follows: Since the function $w = f(z)$ is one-to-one, $w \neq w_0$ implies $z \neq z_0$, and hence the difference quotient of the function $\varphi(w)$ can be written in the form

$$\frac{\varphi(w) - \varphi(w_0)}{w - w_0} = \frac{z - z_0}{w - w_0} = \frac{1}{\dfrac{w - w_0}{z - z_0}}.$$

Moreover, since $\varphi(w)$ is continuous on $\mathscr{E}$, $\varphi(w) \to \varphi(w_0)$ as $w \to w_0$, i.e., $z \to z_0$ as $w \to w_0$. Therefore

$$\varphi'_{\mathscr{E}}(w_0) = \lim_{w \to w_0} \frac{\varphi(w) - \varphi(w_0)}{w - w_0} = \frac{1}{\lim\limits_{z \to z_0} \dfrac{w - w_0}{z - z_0}} = \frac{1}{\lim\limits_{z \to z_0} \dfrac{f(z) - f(z_0)}{z - z_0}} = \frac{1}{f'_E(z_0)},$$

as asserted.

Problem 1. Justify the formula $dz = \Delta z$ ("the differential and the increment of the independent variable are equal") and hence formula (3.4).

Problem 2. Show that the function $f(z) = \bar{z}$ is not differentiable at any point z_0 relative to any neighborhood $\mathscr{N}(z_0)$.

14. THE CAUCHY-RIEMANN EQUATIONS. ANALYTIC FUNCTIONS

From now on, we shall be concerned mainly with functions defined on some domain $E = G$, and we shall drop the subscript E in $f'_E(z)$ and $d_E f(z)/dz$. Suppose $u(x, y)$ is a real function of two real variables x and y defined on G. Then $u(x, y)$ is said to be *differentiable* at the point $(x_0, y_0) \in G$ if $u(x, y)$ can be written in the form

$$\begin{aligned} u(x, y) - u(x_0, y_0) &= A(x_0, y_0)(x - x_0) + B(x_0, y_0)(y - y_0) \\ &\quad + \varepsilon_1(x, y; x_0, y_0)(x - x_0) + \varepsilon_2(x, y; x_0, y_0)(y - y_0), \end{aligned} \tag{3.11}$$

where

$$\varepsilon_1(x, y; x_0, y_0) \to 0, \qquad \varepsilon_2(x, y; x_0, y_0) \to 0$$

as $(x, y) \to (x_0, y_0)$. The coefficients $A(x_0, y_0)$ and $B(x_0, y_0)$ in the right-hand side of (3.11) are just the partial derivatives of the function $u(x, y)$ at the point (x_0, y_0):

$$A(x_0, y_0) = \left.\frac{\partial u(x, y)}{\partial x}\right|_{\substack{x=x_0,\\ y=y_0}} \qquad B(x_0, y_0) = \left.\frac{\partial u(x, y)}{\partial y}\right|_{\substack{x=x_0.\\ y=y_0}}$$

THEOREM 3.2 (*Cauchy-Riemann equations*). *Let*

$$f(z) = u(x, y) + iv(x, y)$$

be a function of a complex variable defined on a domain G. Then a necessary and sufficient condition for $f(z)$ to be differentiable (as a function of a complex variable) at the point $z_0 = x_0 + iy_0 \in G$ is that the functions $u(x, y)$ and $v(x, y)$ be differentiable (as functions of the two real variables x and y) at the point (x_0, y_0) and satisfy the Cauchy-Riemann equations

$$\frac{\partial u}{\partial x} = \frac{\partial v}{\partial y}, \qquad \frac{\partial u}{\partial y} = -\frac{\partial v}{\partial x} \tag{3.12}$$

at (x_0, y_0). If these conditions are satisfied, $f'(z_0)$ can be represented in any of the forms

$$f'(z_0) = \frac{\partial u}{\partial x} + i\frac{\partial v}{\partial x} = \frac{\partial v}{\partial y} - i\frac{\partial u}{\partial y} = \frac{\partial u}{\partial x} - i\frac{\partial u}{\partial y} = \frac{\partial v}{\partial y} + i\frac{\partial v}{\partial x}, \tag{3.13}$$

where the partial derivatives are all evaluated at (x_0, y_0).

Proof. First we prove that the conditions are necessary. Consider the increments

$$\Delta z = z - z_0 = (x - x_0) + i(y - y_0) = \Delta x + i\,\Delta y,$$
$$\Delta f(z) = f(z) - f(z_0) = [u(x, y) - u(x_0, y_0)] + i[v(x, y) - v(x_0, y_0)]$$
$$= \Delta u + i\,\Delta v$$

of the independent variable z and of the function $f(z)$. According to Theorem 3.1, if $f(z)$ is differentiable at z_0, then

$$\Delta f(z) = f'(z_0)\,\Delta z + \varepsilon\,\Delta z, \tag{3.14}$$

where $\varepsilon \to 0$ as $\Delta z \to 0$. Writing

$$f'(z_0) = a + ib, \qquad \varepsilon = \varepsilon_1 + i\varepsilon_2,$$

and taking the real and imaginary parts of (3.14), we find that

$$\Delta u = a\,\Delta x - b\,\Delta y + \varepsilon_1\,\Delta x - \varepsilon_2\,\Delta y,$$
$$\Delta v = b\,\Delta x + a\,\Delta y + \varepsilon_2\,\Delta x + \varepsilon_1\,\Delta y,$$

where $\varepsilon_1, \varepsilon_2 \to 0$ as $\Delta x, \Delta y \to 0$, since

$$|\Delta z| = \sqrt{(\Delta x)^2 + (\Delta y)^2}, \quad |\varepsilon_1| \leqslant |\varepsilon|, \quad |\varepsilon_2| \leqslant |\varepsilon|.$$

It follows that the functions $u(x, y)$ and $v(x, y)$ are differentiable at (x_0, y_0) and

$$\frac{\partial u}{\partial x} = a, \quad \frac{\partial u}{\partial y} = -b, \quad \frac{\partial v}{\partial x} = b, \quad \frac{\partial v}{\partial y} = a. \tag{3.15}$$

But (3.15) immediately implies (3.12) and (3.13).

To prove that the conditions of the theorem are sufficient, we reverse the preceding argument. Thus, suppose $u(x, y)$ and $v(x, y)$ are differentiable at the point (x_0, y_0), and suppose (3.12) holds. Then

$$\begin{aligned}\Delta u &= \frac{\partial u}{\partial x}\Delta x + \frac{\partial u}{\partial y}\Delta y + \alpha_1 \Delta x + \alpha_2 \Delta y \\ &= a\,\Delta x - b\,\Delta y + \alpha_1 \Delta x + \alpha_2 \Delta y, \\ \Delta v &= \frac{\partial v}{\partial x}\Delta x + \frac{\partial v}{\partial y}\Delta y + \beta_1 \Delta x + \beta_2 \Delta y \\ &= b\,\Delta x + a\,\Delta y + \beta_1 \Delta x + \beta_2 \Delta y,\end{aligned} \tag{3.16}$$

where $\alpha_1, \alpha_2, \beta_1, \beta_2 \to 0$ as $\Delta x, \Delta y \to 0$ and we have written

$$a = \frac{\partial u}{\partial x} = \frac{\partial v}{\partial y}, \qquad b = -\frac{\partial u}{\partial y} = \frac{\partial v}{\partial x}. \tag{3.17}$$

Substituting (3.16) into the formula $\Delta f(z) = \Delta u + i\,\Delta v$, we find that

$$\begin{aligned}\Delta f(z) &= a(\Delta x + i\,\Delta y) + ib(\Delta x + i\,\Delta y) + (\alpha_1 + i\beta_1)\Delta x + (\alpha_2 + i\beta_2)\Delta y \\ &= (a + ib)\,\Delta z + \left[(\alpha_1 + i\beta_1)\frac{\Delta x}{\Delta z} + (\alpha_2 + i\beta_2)\frac{\Delta y}{\Delta z}\right]\Delta z\end{aligned}$$

or

$$\Delta f(z) = (a + ib)\,\Delta z + \varepsilon\,\Delta z, \tag{3.18}$$

where

$$\begin{aligned}|\varepsilon| &= \left|(\alpha_1 + i\beta_1)\frac{\Delta x}{\Delta z} + (\alpha_2 + i\beta_2)\frac{\Delta y}{\Delta z}\right| \\ &\leqslant |\alpha_1 + i\beta_1|\left|\frac{\Delta x}{\Delta z}\right| + |\alpha_2 + i\beta_2|\left|\frac{\Delta y}{\Delta z}\right| \\ &\leqslant |\alpha_1 + i\beta_1| + |\alpha_2 + i\beta_2| \leqslant |\alpha_1| + |\beta_1| + |\alpha_2| + |\beta_2|.\end{aligned}$$

Since $\alpha_1, \alpha_2, \beta_1, \beta_2 \to 0$ as $\Delta x, \Delta y \to 0$, it follows that $\varepsilon \to 0$ as $\Delta z \to 0$, and hence, according to (3.18) and Theorem 3.1, $f(z)$ is differentiable at z_0, with derivative

$$f'(z_0) = a + ib. \tag{3.19}$$

The various representations (3.13) of $f'(z_0)$ are now immediate consequences of (3.19) and (3.17).

A function $f(z)$ which is differentiable on a domain G, i.e., at every point of G, is said to be *analytic* (synonymously, *holomorphic* or *regular*) *on* G. If $f(z)$ is analytic in a neighborhood of z_0, $f(z)$ is said to be *analytic at* z_0.

Remark. As we know from calculus,[2] a sufficient condition for the differentiability of the functions $u(x, y)$ and $v(x, y)$ on a domain G is that the partial derivatives

$$\frac{\partial u}{\partial x}, \quad \frac{\partial u}{\partial y}, \quad \frac{\partial v}{\partial x}, \quad \frac{\partial v}{\partial y} \tag{3.20}$$

exist and be continuous on G. Therefore a sufficient condition for the function $f(z) = u + iv$ to be analytic on G is that the partial derivatives (3.20) exist, be continuous and satisfy the equations (3.12) on G.

Example 1. For the function

$$f(z) = e^x(\cos y + i \sin y), \tag{3.21}$$

defined in the whole plane, we have

$$u = e^x \cos y, \qquad v = e^x \sin y,$$

with continuous partial derivatives

$$\frac{\partial u}{\partial x} = e^x \cos y = \frac{\partial v}{\partial y}, \qquad \frac{\partial u}{\partial y} = -e^x \sin y = -\frac{\partial v}{\partial x}.$$

Therefore the Cauchy-Riemann equations (3.12) are satisfied, and the function (3.21) is analytic in the whole plane, with derivative

$$f'(z) = \frac{\partial u}{\partial x} + i\frac{\partial v}{\partial x} = e^x \cos y + ie^x \sin y = f(z).$$

Example 2. For the function $f(z) = x$, considered in the example on p. 45, we have

$$u = x, \qquad v = 0$$

and

$$\frac{\partial u}{\partial x} = 1, \quad \frac{\partial u}{\partial y} = 0, \quad \frac{\partial v}{\partial x} = 0, \quad \frac{\partial v}{\partial y} = 0.$$

Since

$$\frac{\partial u}{\partial x} \neq \frac{\partial v}{\partial y},$$

the Cauchy-Riemann equations are not satisfied, and hence this function is not differentiable anywhere in the plane.

---

[2] See e.g., D. V. Widder, *Advanced Calculus*, second edition, Prentice-Hall, Inc., Englewood Cliffs, N.J. (1961), p. 17.

In many cases, it is important to express the differentiability conditions for a function $f(z) = u + iv$ at a point $z \neq 0$ in terms of the polar coordinates

$$r = |z|, \qquad \Phi = \operatorname{Arg} z.$$

The appropriate necessary and sufficient conditions for differentiability are that $u(r, \Phi)$ and $v(r, \Phi)$ be differentiable (as functions of the two real variables r and Φ) and satisfy the polar form of the Cauchy-Riemann equations, i.e.,

$$\frac{\partial u}{\partial r} = \frac{1}{r}\frac{\partial v}{\partial \Phi}, \qquad \frac{\partial v}{\partial r} = -\frac{1}{r}\frac{\partial u}{\partial \Phi} \tag{3.22}$$

(at a given nonzero point P). To verify these conditions, we must prove that 1) u and v are differentiable as functions of r and Φ at P if and only if they are differentiable as functions of x and y at P, and 2) under these conditions, the equations (3.22) are equivalent to the equations (3.12). The first assertion follows from the familiar fact that a differentiable function ($u = u(x, y)$, say) of differentiable functions ($x = r\cos\Phi$, $y = r\sin\Phi$, say) is also differentiable (with respect to the new variables r and Φ).[3] The second assertion can be verified by direct calculation. For example, if u and v are differentiable functions of x and y, and if the equations (3.12) hold, then

$$\begin{aligned}
\frac{\partial u}{\partial r} &= \frac{\partial u}{\partial x}\cos\Phi + \frac{\partial u}{\partial y}\sin\Phi = \frac{\partial v}{\partial y}\cos\Phi - \frac{\partial v}{\partial x}\sin\Phi = \frac{1}{r}\frac{\partial v}{\partial \Phi}, \\
\frac{\partial v}{\partial r} &= \frac{\partial v}{\partial x}\cos\Phi + \frac{\partial v}{\partial y}\sin\Phi = -\frac{\partial u}{\partial y}\cos\Phi + \frac{\partial u}{\partial x}\sin\Phi = -\frac{1}{r}\frac{\partial u}{\partial \Phi}.
\end{aligned} \tag{3.23}$$

Writing (3.23) in the form

$$\frac{\partial u}{\partial r} = \frac{\partial u}{\partial x}\cos\Phi - \frac{\partial v}{\partial x}\sin\Phi,$$

$$\frac{\partial v}{\partial r} = \frac{\partial u}{\partial x}\sin\Phi + \frac{\partial v}{\partial x}\cos\Phi,$$

and solving for $\partial u/\partial x$ and $\partial u/\partial y$, we obtain

$$\frac{\partial u}{\partial x} = \frac{\partial u}{\partial r}\cos\Phi + \frac{\partial v}{\partial r}\sin\Phi,$$

$$\frac{\partial v}{\partial x} = -\frac{\partial u}{\partial r}\sin\Phi + \frac{\partial v}{\partial r}\cos\Phi,$$

[3] The reader should verify this assertion, guided by (3.11).

and hence

$$\begin{aligned} f'(z) &= \frac{\partial u}{\partial x} + i\,\frac{\partial v}{\partial x} = \frac{\partial u}{\partial r}(\cos\Phi - i\sin\Phi) + i\,\frac{\partial v}{\partial r}(\cos\Phi - i\sin\Phi) \\ &= \left(\frac{\partial u}{\partial r} + i\,\frac{\partial v}{\partial r}\right)(\cos\Phi - i\sin\Phi) = \frac{r}{z}\left(\frac{\partial u}{\partial r} + i\,\frac{\partial v}{\partial r}\right). \end{aligned} \tag{3.24}$$

This formula is convenient for calculating $f'(z)$ with the help of polar coordinates. Using (3.22), we can write $f'(z)$ in the form

$$f'(z) = \frac{1}{z}\left(\frac{\partial v}{\partial\Phi} - i\,\frac{\partial u}{\partial\Phi}\right).$$

Example. Consider the function

$$\begin{aligned} f(z) = z^{m/n} &= |z|^{m/n}\left(\cos\frac{m\,\mathrm{Arg}\,z}{n} + i\sin\frac{m\,\mathrm{Arg}\,z}{n}\right) \\ &= r^{m/n}\left(\cos\frac{m\Phi}{n} + i\sin\frac{m\Phi}{n}\right), \end{aligned} \tag{3.25}$$

where m and $n > 0$ are integers (see Sec. 4). This function is defined on the domain G consisting of all nonzero points of the complex plane, and is multiple-valued (unless m/n is an integer), since $\Phi = \mathrm{Arg}\,z$ is multiple-valued. Before we can talk about the derivative of $z^{m/n}$, we must first make $z^{m/n}$ single-valued in the following sense: Let $z_0 \in G$, and choose a neighborhood $\mathscr{N}(z_0)$ which does not contain the origin. Fix one of the values of $\Phi_0 = \mathrm{Arg}\,z_0$, and for the argument of any other point $z \in \mathscr{N}(z_0)$ choose the unique value Φ satisfying the condition

$$|\Phi - \Phi_0| < \frac{\pi}{2}$$

(see Figure 3.1). Using this value of Φ in (3.25), we obtain a single-valued

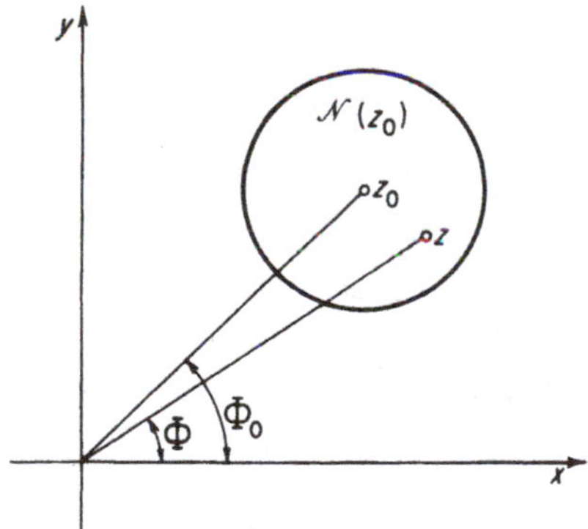

FIGURE 3.1

function defined on $\mathcal{N}(z_0)$, which we call a *single-valued branch* of the function (3.25), denoted by $z^{m/n}$ as before. Thus at any point $z \in \mathcal{N}(z_0)$, including z_0 itself, we can now write

$$u = r^{m/n} \cos \frac{m\Phi}{n}, \qquad v = r^{m/n} \sin \frac{m\Phi}{n}$$

and

$$\frac{\partial u}{\partial r} = \frac{m}{n} r^{(m/n)-1} \cos \frac{m\Phi}{n} = \frac{1}{r} \frac{\partial v}{\partial \Phi},$$

$$\frac{\partial v}{\partial r} = \frac{m}{n} r^{(m/n)-1} \sin \frac{m\Phi}{n} = -\frac{1}{r} \frac{\partial u}{\partial \Phi},$$

i.e., $f(z)$ satisfies the equations (3.22), and is therefore differentiable on $\mathcal{N}(z_0)$. According to formula (3.24),

$$f'(z) = \frac{r}{z}\left(\frac{m}{n} r^{(m/n)-1} \cos \frac{m\Phi}{n} + i \frac{m}{n} r^{(m/n)-1} \sin \frac{m\Phi}{n}\right)$$

$$= \frac{m}{n} r^{m/n}\left(\cos \frac{m\Phi}{n} + i \sin \frac{m\Phi}{n}\right)\frac{1}{z} = \frac{m}{n}\frac{f(z)}{z},$$

so that the rule for differentiating a fractional power $z^{m/n}$ is formally the same as the rule for differentiating the corresponding function $x^{m/n}$ of a real variable. It should be kept in mind that our calculation is subject to the condition $z \neq 0$, which can only be dropped if m/n is a nonnegative integer.

Problem 1. Show that the function $f(z) = z \operatorname{Re} z$ is differentiable only at the point $z = 0$, and find $f'(0)$.

Comment. Thus $f(z) = z \operatorname{Re} z$ is differentiable but not analytic at $z = 0$.

Problem 2. Show that the function

$$u(x, y) = \begin{cases} x & \text{if } |y| > |x|, \\ -x & \text{if } |y| \leqslant |x| \end{cases}$$

has partial derivatives $\partial u/\partial x$ and $\partial u/\partial y$ at the origin, but is not differentiable there.

Problem 3. Show that the function $f(z) = \sqrt{|xy|}$ satisfies the Cauchy-Riemann equations at the point $z = 0$, but is not differentiable there.

Problem 4. Establish the following generalization of the Cauchy-Riemann equations: If $f(z) = u + iv$ is differentiable at a point $z_0 = x_0 + iy_0$ of a domain G, then

$$\frac{\partial u}{\partial s} = \frac{\partial v}{\partial n}, \qquad \frac{\partial u}{\partial n} = -\frac{\partial v}{\partial s}$$

at (x_0, y_0), where $\partial/\partial s$ and $\partial/\partial n$ denote directional differentiation in any two orthogonal directions s and n at (x_0, y_0) such that n is obtained from s by making a counterclockwise rotation.

Problem 5. Use the preceding problem to deduce the equations (3.22).

15. GEOMETRIC INTERPRETATION OF Arg $f'(z)$ AND $|f'(z)|$. CONFORMAL MAPPING

Let l be a continuous curve with equation $z = \lambda(t)$, $t \in [a, b]$, and suppose $\lambda(t)$ is differentiable at a point $t_0 \in [a, b]$ (relative to the set $[a, b]$). Let $\{t_n\}$ be an arbitrary sequence of points in $[a, b]$ converging to t_0 $(t_n \neq t_0)$, and consider the difference quotient

$$r_n = \frac{\lambda(t_n) - \lambda(t_0)}{t_n - t_0}. \tag{3.26}$$

Obviously $r_n \to r_0 = \lambda'(t_0)$ as $n \to \infty$.

DEFINITION. *The curve l is said to have a tangent at the point $z_0 = \lambda(t_0)$ if the limit*

$$\theta = \lim_{n\to\infty} \operatorname{Arg} r_n \tag{3.27}$$

exists, and then the tangent is said to have inclination θ. Geometrically, the tangent to l at z_0 is represented by the ray τ emanating from z_0 which makes the angle θ with the positive real axis.[4]

Remark 1. Clearly, if $\lambda'(t_0) \neq 0$, l has a tangent at z_0, since

$$\theta = \lim_{n\to\infty} \operatorname{Arg} r_n = \operatorname{Arg} r_0 = \operatorname{Arg} \lambda'(t_0)$$

(see Sec. 6, Prob. 3), and then the inclination of the tangent is just the argument of the complex number $\lambda'(t_0)$. On the other hand, if $\lambda'(t_0) = 0$, l may or may not have a tangent at z_0, since the fact that $r_n \to 0$ implies only that $|r_n| \to 0$ and says nothing about the behavior of Arg r_n. However, if l has a tangent at z_0, then $\lambda(t_n) \neq \lambda(t_0)$ for all t_n sufficiently close to t_0, since Arg 0 is meaningless.[5]

---

[4] Formula (3.27) means that given any $\varepsilon > 0$, there is an integer $N(\varepsilon) > 0$ and a sequence $\{\theta_n\}$, where each θ_n is a value of Arg r_n, such that $|\theta_n - \theta| < \varepsilon$ for all $n > N(\varepsilon)$. Clearly, θ is only defined to within a multiple of 2π. The angle θ will always be measured from the positive real axis to the tangent τ (in the counterclockwise direction for a positive value of θ).

[5] This condition is automatically satisfied if $\lambda'(t_0) \neq 0$, since otherwise $r_n = 0$ for t arbitrarily close to t_0, which implies that $\lambda'(t_0) = 0$, contrary to hypothesis.

Remark 2. As just defined, the tangent is a ray, not a vector. If $\lambda'(t_0) \neq 0$, we can also introduce a *tangent vector* to l at z_0, defined as the vector of length $|\lambda'(t_0)|$ which makes the angle θ with the positive real axis.

THEOREM 3.3. *Let G be a domain, and let $f(z)$ be a continuous function of a complex variable defined on G. Suppose $f(z)$ has a nonzero derivative $f'(z_0)$ at a point $z_0 \in G$, and let l be a curve which passes through z_0 and has a tangent τ at z_0. Then $w = f(z)$ maps l into a curve L in the w-plane which passes through the point $w_0 = f(z_0)$ and has a tangent T at w_0. Moreover, the inclination of T exceeds the inclination of τ by the angle* $\operatorname{Arg} f'(z_0)$.

Proof. Suppose l has the equation $z = \lambda(t)$, $t \in [a, b]$, and let $z_0 = \lambda(t_0)$. By hypothesis,

$$\theta = \lim_{n\to\infty} \operatorname{Arg} r_n$$

exists, where r_n is given by (3.26). The function $w = f(z)$ maps l into a curve L in the w-plane with equation

$$w = f[\lambda(t)] = \Lambda(t), \qquad t \in [a, b], \tag{3.28}$$

where $w_0 = f(z_0) = \Lambda(t_0)$. Let $\{t_n\}$ be an arbitrary sequence of points in $[a, b]$ converging to t_0, and let

$$R_n = \frac{\Lambda(t_n) - \Lambda(t_0)}{t_n - t_0}.$$

Then the tangent to L at w_0 has inclination

$$\Theta = \lim_{n\to\infty} \operatorname{Arg} R_n,$$

provided this limit exists. Clearly we have

$$R_n = \frac{\Lambda(t_n) - \Lambda(t_0)}{\lambda(t_n) - \lambda(t_0)} \frac{\lambda(t_n) - \lambda(t_0)}{t_n - t_0},$$

where the first factor in the right-hand side is well defined, since $\lambda(t_n) \neq \lambda(t_0)$ for all t_n sufficiently close to t_0 (l is assumed to have a tangent at z_0). Therefore

$$\Theta = \lim_{n\to\infty} \operatorname{Arg} R_n = \lim_{n\to\infty} \operatorname{Arg} \frac{\Lambda(t_n) - \Lambda(t_0)}{\lambda(t_n) - \lambda(t_0)} \frac{\lambda(t_n) - \lambda(t_0)}{t_n - t_0}$$

$$= \lim_{n\to\infty} \operatorname{Arg} \frac{w_n - w_0}{z_n - z_0} r_n = \lim_{n\to\infty} \operatorname{Arg} \frac{w_n - w_0}{z_n - z_0} + \lim_{n\to\infty} \operatorname{Arg} r_n \tag{3.29}$$

$$= \operatorname{Arg} \lim_{n\to\infty} \frac{w_n - w_0}{z_n - z_0} + \theta = \operatorname{Arg} f'(z_0) + \theta,$$

where $w_n = \Lambda(t_n)$, $z_n = \lambda(t_n)$ and θ is the inclination of τ at z_0.[6] It follows from (3.29) that Θ exists and that

$$\Theta - \theta = \operatorname{Arg} f'(z_0),$$

as asserted. We note that things are particularly simple in the case where $\lambda'(t_0) \neq 0$, since then

$$\begin{aligned}\Theta &= \operatorname{Arg} \Lambda'(t_0) = \operatorname{Arg} [f'(z_0)\lambda'(t_0)] \\ &= \operatorname{Arg} f'(z_0) + \operatorname{Arg} \lambda'(t_0) = \operatorname{Arg} f'(z_0) + \theta,\end{aligned}$$

by the rule for differentiating the composite function (3.28).

Now let l_1 and l_2 be two curves with a common initial point z_0, which have tangents τ_1 and τ_2 at z_0, and suppose the angle between τ_1 and τ_2 is measured from τ_1 to τ_2. Suppose l_1 and l_2 have images L_1 and L_2 under $f(z)$. Then, according to Theorem 3.3, if $f'(z_0) \neq 0$, L_1 and L_2 have tangents T_1 and T_2 at the point $w_0 = f(z_0)$, where T_1 and T_2 are obtained by rotating τ_1 and τ_2 through the same angle $\operatorname{Arg} f'(z_0)$. Therefore the angle between L_1 and L_2 equals the angle between l_1 and l_2, and is measured in the same direction, i.e., from L_1 to L_2. In other words, a continuous function $w = f(z)$ with a nonzero derivative $f'(z_0)$ maps all curves in the z-plane which pass through z_0 and have tangents at z_0 into curves in the w-plane which pass through $w_0 = f(z_0)$ and have tangents at w_0, and moreover, *the mapping preserves angles between curves.* A mapping by a continuous function which preserves angles between curves passing through a given point z_0 is said to be *conformal at* z_0. If a conformal mapping preserves the directions in which angles are measured (as well as their magnitudes), it is called a *conformal mapping of the first kind*, but if it reverses the directions in which angles are measured, it is called a *conformal mapping of the second kind.* Thus Theorem 3.3 has the following consequence:

THEOREM 3.4. *Let G be a domain, and let $f(z)$ be an analytic function on G. Then $f(z)$ is a conformal mapping of the first kind at every point of G where $f'(z) \neq 0$.*

Example 1. Reflection in the real axis, i.e., the transformation $w = \bar{z}$, is a conformal mapping of the second kind. A more general example is the complex conjugate

$$w = \overline{f(z)}$$

of an analytic function $f(z)$, where $f'(z) \neq 0$.

---

[6] In reversing the order of the operations Arg and $\lim\limits_{n\to\infty}$, we have used the fact that $f'(z_0) \neq 0$.

Example 2. At a point where the derivative vanishes, angles may or may not be preserved, as can be seen by comparing the mappings

$$f_1(z) = r^2(\cos \Phi + i \sin \Phi) = rz,$$
$$f_2(z) = r^2(\cos 2\Phi + i \sin 2\Phi) = z^2$$

at the point $z = 0$.

As we have just seen, $\operatorname{Arg} f'(z_0)$ represents the rotation undergone by the tangent to a curve l at the point $z_0 \in l$ when transforming to the new curve $L = f(l)$ and the new point $w_0 = f(z_0)$. In particular, if $f'(z_0)$ is a positive real number, the tangents to l at z_0 and to L at w_0 are parallel and point in the same direction.

To explain the geometric meaning of the quantity $|f'(z_0)|$, i.e., the absolute value of the derivative at z_0, we note that

$$|f'(z_0)| = \lim_{z \to z_0} \frac{|f(z) - f(z_0)|}{|z - z_0|}.$$

The numbers $|z - z_0|$ and $|f(z) - f(z_0)|$ are the distance between the points z and z_0 in the z-plane, and the distance between their images $f(z)$ and $f(z_0)$ in the w-plane, respectively. Thus, interpreting

$$\frac{|f(z) - f(z_0)|}{|z - z_0|}$$

as the *linear magnification ratio* (or simply the *magnification*) of the vector $z - z_0$ under the mapping $w = f(z)$,[7] we can regard $|f'(z_0)|$ as the *magnification at the point* z_0 under $w = f(z)$.

Remark. The size of the magnification at the point z_0 does not depend on the choice of the finite vector $z - z_0$ drawn from z_0, since $|f'(z_0)|$ is not the actual magnification of any such vector, but rather the limiting magnification as $z \to z_0$.

Problem 1. With the same notation as on p. 55, a curve l is said to have a *left-hand tangent* (of inclination θ) at the point $z_0 = \lambda(t_0)$ if the limit (3.27) exists, subject to the extra condition that every point of the sequence $\{t_n\}$ converging to t_0 be less than t_0. The *right-hand tangent* is defined similarly by requiring that $t_n > t_0$ for every n. Give an example of a (continuous) curve l which has a left-hand tangent but no right-hand tangent (and hence no tangent) at a point $z_0 \in l$.

Problem 2. Verify that the function

$$f_1(z) = r^2(\cos \Phi + i \sin \Phi) = rz$$

used in Example 2 above is differentiable at $z = 0$.

---

[7] Here the word *magnification* is used in a general sense, and can correspond to *stretching* if $|f'(z_0)| > 1$ or *shrinking* if $|f'(z_0)| < 1$ [or neither if $|f'(z_0)| = 1$].

Problem 3. Find the angle through which a curve drawn from the point z_0 is rotated under the mapping $w = z^2$ if

a) $z_0 = i$; b) $z_0 = -\frac{1}{4}$; c) $z_0 = 1 + i$; d) $z_0 = -3 + 4i$.

Also find the corresponding values of the magnification.

Problem 4. Carry out the same calculations as in the preceding problem, this time applied to the function $w = z^3$.

Problem 5. Which part of the plane is shrunk and which part stretched under the following mappings:

a) $w = z^2$; b) $w = z^2 + 2z$; c) $w = \dfrac{1}{z}$?

16. THE MAPPING $w = \dfrac{az + b}{cz + d}$

To illustrate the above considerations, we now examine the *fractional linear transformation* or *Möbius transformation*

$$L(z) = \frac{az + b}{cz + d}, \tag{3.30}$$

where a, b, c, d are arbitrary complex numbers (except that c and d are not both zero). First suppose that $c = 0$. Then $L(z)$ reduces to

$$L(z) = \alpha z + \beta \qquad (\alpha = a/d, \quad \beta = b/d), \tag{3.31}$$

and is sometimes called the *entire linear transformation*. The transformation (3.31) is defined for all values of z, and if $\alpha \neq 0$, its derivative $L'(z)$ is a nonzero constant, so that (3.31) is conformal at every point of the (finite) z-plane. Under this transformation, the tangents to all curves in the z-plane are rotated through the same angle, equal to Arg α, and the magnification at every point equals $|\alpha|$. If $\alpha = 1$, then

$$\text{Arg}\,\alpha = 2k\pi, \qquad |\alpha| = 1,$$

where k is an integer, and then both the rotation and expansion produce no effect. In this case, the transformation takes the form

$$w = z + \beta,$$

which obviously corresponds to displacing the whole plane by the vector β. On the other hand, if $\alpha \neq 1$ (and $\alpha \neq 0$), the transformation (3.31) can be written in the form

$$w - z_0 = \alpha(z - z_0),$$

where z_0 is determined from the equation[8]

$$z_0 = \alpha z_0 + \beta.$$

Then it is immediately clear that the transformation (3.31) is equivalent to a rotation of the whole plane through the angle Arg α about the point $z_0 = \beta/(1 - \alpha)$, together with a uniform magnification by the factor $|\alpha|$ relative to the point z_0. This magnification is sometimes called a *homothetic transformation* (or *transformation of similitude*) with *ray center* z_0 and *ray ratio* $|\alpha|$.

Next suppose that $c \neq 0$ in (3.30). Then the derivative

$$L'(z) = \frac{ad - bc}{(cz + d)^2} = \frac{ad - bc}{c^2}\frac{1}{(z - \delta)^2}$$

exists, if $z \neq \delta$, where $\delta = -d/c$. If the determinant

$$ad - bc = \begin{vmatrix} a & b \\ c & d \end{vmatrix}$$

vanishes, then its rows are proportional, i.e.,

$$\frac{a}{c} = \frac{b}{d} = \mu \quad \text{or} \quad a = \mu c, \quad b = \mu d,$$

where μ is a constant, so that (3.30) reduces to the trivial transformation

$$L(z) = \frac{az + b}{cz + d} = \frac{\mu cz + \mu d}{cz + d} \equiv \mu.$$

If $ad - bc \neq 0$, then $L'(z) \neq 0$ for all $z \neq \delta$, and hence the mapping $w = L(z)$ is conformal at all finite points except possibly at $z = \delta$. Under the mapping, the tangents to curves passing through any point $z \neq \delta$ are rotated through an angle equal to

$$\operatorname{Arg} L'(z) = \operatorname{Arg}\frac{ad - bc}{c^2} - 2\operatorname{Arg}(z - \delta),$$

while the magnification at z equals

$$|L'(z)| = \left|\frac{ad - bc}{c^2}\right|\frac{1}{|z - \delta|^2}.$$

The angle through which tangents are rotated has the same value for all points with equal values of Arg $(z - \delta)$, i.e., along any ray drawn from δ, but otherwise varies from point to point. Similarly, in general the magnification varies with z, but it has the same value for all points with equal values of $|z - \delta|$, i.e., along any circle with center δ. In particular, the magnification

---

[8] Obviously, z_0 is invariant under the transformation (3.31), i.e., z_0 is a *fixed point* of the transformation. If $\alpha \neq 0$, the point at infinity is also a fixed point (see Sec. 25).

is equal to 1 at every point of the circle C with equation

$$|z - \delta| = \frac{1}{|c|}\sqrt{|ad - bc|}$$

(called the *isometric circle* of the Möbius transformation), is greater than 1 inside C (approaching ∞ as $z \to \delta$), and is less than 1 outside C (approaching 0 as $z \to \infty$). The situation is shown schematically in Figure 3.2.

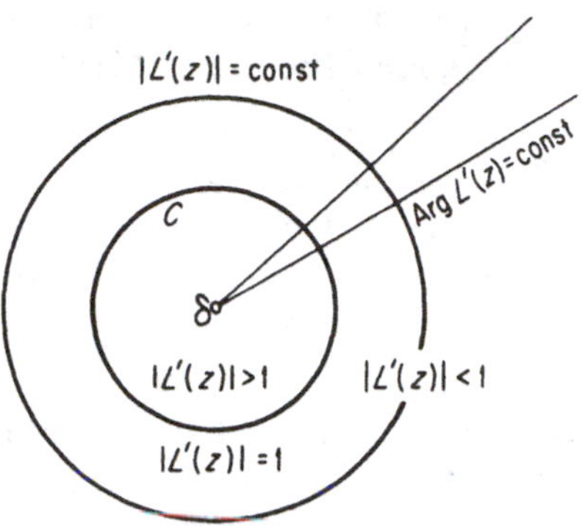

FIGURE 3.2

Problem 1. As shown on p. 60, the entire linear transformation $w = \alpha z + \beta$ is equivalent to a rotation and a magnification relative to the fixed point $z_0 = \beta/(1 - \alpha)$, provided that $\alpha \neq 0$. Find the rotation, magnification and (finite) fixed point, if such exists, corresponding to each of the following transformations, and write each in the canonical form $w - z_0 = \alpha(z - z_0)$:

a) $w = 2z + 1 - 3i$; b) $w = iz + 4$;

c) $w = z + 1 - 2i$; d) $w - w_1 = a(z - z_1)$ $(a \neq 0)$.

Problem 2. Find the entire linear transformation with fixed point $1 + 2i$ carrying the point i into the point $-i$.

Ans. $w = (2 + i)z + 1 - 3i$.

Problem 3. Find the entire linear transformation carrying the triangle with vertices at the points $0, 1, i$ into the similar triangle with vertices at the points $0, 2, 1 + i$.

17. CONFORMAL MAPPING OF THE EXTENDED PLANE

As in the preceding section, let $c \neq 0$ and $ad - bc \neq 0$. Then it is clear that

$$\lim_{z\to\delta} \frac{az + b}{cz + d} = \infty, \qquad \lim_{z\to\infty} \frac{az + b}{cz + d} = \frac{a}{c} = A,$$

where $\delta = -d/c$. Suppose we complete the definition of $L(z)$ by setting

$$L(\delta) = \infty, \qquad L(\infty) = A. \tag{3.32}$$

Then $L(z)$ is defined in the whole extended plane, and maps the finite point δ into ∞ (the point at infinity) and ∞ into the finite point A. We now show that the mapping $w = L(z)$ is conformal at the points δ and A (it has already been shown that the mapping is conformal everywhere else). First we must suitably define the concept of an angle with its vertex at infinity (for the justification of this definition, see Probs. 3 and 4; also recall the last paragraph of Sec. 8):

DEFINITION.[9] *Two continuous curves γ_1 and γ_2 in the extended plane form an angle of α radians with its vertex at infinity if and only if their images $\tilde{\gamma}_1$ and $\tilde{\gamma}_2$ in the extended plane under the transformation $\zeta = 1/z$ form an angle of α radians with its vertex at the origin.*

Example. The real and imaginary axes form an angle of $\pi/2$ radians with its vertex at infinity. In fact, under the transformation $\zeta = 1/z$, the real and imaginary axes are carried into themselves, and they obviously form an angle of $\pi/2$ radians with its vertex at the origin.

Returning to the mapping $w = L(z)$, let γ_1 and γ_2 be two curves forming an angle θ with its vertex at the point δ, and let Γ_1 and Γ_2 be their images in the w-plane. To prove that Γ_1 and Γ_2 form an angle θ with its vertex at infinity, we subject the w-plane to the transformation

$$\eta = \frac{1}{w}.$$

Then the curves Γ_1 and Γ_2 go into two curves Γ_1^* and Γ_2^*, and the point at infinity goes into the origin of coordinates (see Figure 3.3). Obviously, we can go from γ_1 and γ_2 in the z-plane to Γ_1^* and Γ_2^* in the η-plane by making the Möbius transformation

$$\eta = \frac{cz + d}{az + b},$$

which is conformal at the point $z = \delta = -d/c$. It follows that the curves Γ_1^* and Γ_2^* form an angle θ with its vertex at the origin. Therefore the curves Γ_1 and Γ_2 also form an angle θ with its vertex at infinity. This proves that the mapping $w = L(z)$ is conformal at the point $z = \delta$.

The fact that $w = L(z)$ is conformal at ∞ is proved similarly. In fact, if the curves γ_1 and γ_2 go through the point at infinity in the z-plane, their images Γ_1 and Γ_2 in the w-plane go through the point A. Suppose γ_1 and γ_2

[9] The curves γ_1 and γ_2 are unbounded, in the sense of Sec. 12, Prob. 2.

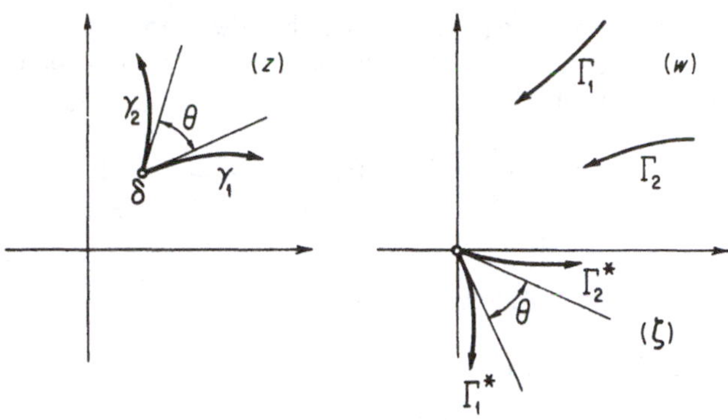

FIGURE 3.3

form an angle θ with its vertex at infinity. This means that their images γ_1^* and γ_2^* under the transformation $\zeta = 1/z$ form an angle θ with its vertex at the origin. But we can obviously go from γ_1^* and γ_2^* to Γ_1 and Γ_2 by making the Möbius transformation

$$w = \frac{az + b}{cz + d} = \frac{a\dfrac{1}{\zeta} + b}{c\dfrac{1}{\zeta} + d} = \frac{a + b\zeta}{c + d\zeta},$$

which is conformal at the point $\zeta = 0$. It follows that Γ_1 and Γ_2 form the same angle θ at the point $A = a/c$. This proves that the mapping $w = L(z)$ is conformal at ∞. The situation can be summarized by saying that *the transformation* $w = L(z)$ *is a conformal mapping of the extended plane onto itself.*

Remark 1. These considerations suggest the following definition: A function $f(z)$ is said to be *analytic at* $z = \infty$ if the function $f^*(\zeta) = f(1/\zeta)$ is analytic at $\zeta = 0$. In particular, if $f(z)$ is analytic at $z = \infty$, the limit

$$\lim_{z\to\infty} f(z) = \lim_{\zeta\to 0} f^*(\zeta) = f(\infty)$$

always exists and is finite. We define the derivative of $f(z)$ at $z = \infty$ to be the quantity

$$f'(\infty) = f^{*\prime}(0),$$

where it should be noted that in general

$$f'(\infty) \neq \lim_{z\to\infty} f'(z)$$

(see Prob. 2). Then, by the argument given above for the special case of the

Möbius transformation, it is easy to see that the mapping $w = f(z)$ is conformal at ∞ if $f'(\infty) \neq 0$. With this approach, the conformality at ∞ of the Möbius transformation

$$L(z) = \frac{az + b}{cz + d} \qquad (ad - bc \neq 0, c \neq 0)$$

follows from the fact that

$$L'(\infty) = -\frac{ad - bc}{c^2} \neq 0.$$

Remark 2. Similarly, if

$$\lim_{z \to a} f(z) = \infty,$$

but if

$$\varphi(z) = \frac{1}{f(z)}$$

is analytic at $z = a$, with derivative $\varphi'(a) \neq 0$, then, just as in the case of the Möbius transformation, the mapping $w = f(z)$ is conformal at $z = a$.

Problem 1. Prove that the transformation $L(z) = \alpha z + \beta$ is conformal at infinity if $\alpha \neq 0$.

Problem 2. Prove that if $f(z)$ is analytic at infinity, then

$$\lim_{z \to \infty} f'(z) = 0.$$

Problem 3 (M1, p. 87). Prove that stereographic projection is conformal, i.e., that under stereographic projection the angle between any two curves on the Riemann sphere (with its vertex at any point except the north pole) equals the angle between the images of the curves in the finite plane.

Problem 4 (M1, Sec. 25). Prove that two curves γ_1 and γ_2 in the extended plane form an angle of α radians with its vertex at infinity if and only if their images γ_1^* and γ_2^* on the Riemann sphere under stereographic projection form an angle of α radians with its vertex at the north pole.

CHAPTER 4

POLYNOMIALS AND RATIONAL FUNCTIONS

18. POLYNOMIALS. THE MAPPING $w = P_n(z)$

The simplest class of differentiable functions of a complex variable is the class of single-valued functions which are analytic everywhere in the finite plane.[1] Such functions are known as *entire functions*. An important (but very special) subclass of the class of entire functions consists of *polynomials*, i.e., functions of the form

$$P_n(z) = a_0 + a_1 z + \cdots + a_n z^n \qquad (a_n \neq 0 \text{ for } n > 0),$$

where n is a nonnegative integer, called the *degree* of $P_n(z)$, and $a_0, a_1, \ldots, a_n$ are finite complex numbers, called the *coefficients* of $P_n(z)$. If $n = 0$, $P_n(z)$ reduces to a constant, but if $n > 0$,

$$\lim_{z \to \infty} P_n(z) = \infty,$$

and we set $P(\infty) = \infty$ by definition, thereby making $P_n(z)$ wide-sense continuous.

According to the fundamental theorem of algebra (see Theorem 10.7, Corollary 2), the equation

$$P_n(z) = 0 \qquad (n > 0)$$

[1] It will be shown later (see Theorem 10.7, Corollary 1) that any function which is analytic in the *extended plane* must be the trivial function $f(z) \equiv$ const.

always has a complex root, say α_1. It follows that

$$P_n(z) = (z - \alpha_1)P_{n-1}(z),$$

where $P_{n-1}(z)$ is a polynomial of degree $n - 1$ (see Probs. 1 and 2). If $P_{n-1}(\alpha_1) = 0$, then by the same argument,

$$P_{n-1}(z) = (z - \alpha_1)P_{n-2}(z),$$

where $P_{n-2}(z)$ is a polynomial of degree $n - 2$, and hence

$$P_n(z) = (z - \alpha_1)^2 P_{n-2}(z).$$

Continuing this process k_1 times ($1 \leqslant k_1 \leqslant n$), we eventually find that

$$P_n(z) = (z - \alpha_1)^{k_1} P_{n-k_1}(z),$$

where $P_{n-k_1}(z)$ is a polynomial of degree $n - k_1$ and $P_{n-k_1}(\alpha_1) \neq 0$. The number α_1 is called a *root of multiplicity* k_1 (or a k_1*-fold root*) of the equation $P_n(z) = 0$; it is also called a *zero* (or *zero-point*) *of order* k_1 of the polynomial $P_n(z)$. If $k_1 < n$, then, applying the fundamental theorem of algebra again, we find that $P_{n-k_1}(z)$ has a zero, say α_2, where $\alpha_2 \neq \alpha_1$. Therefore, by the previous argument,

$$P_n(z) = (z - \alpha_1)^{k_1}(z - \alpha_2)^{k_2} P_{n-k_1-k_2}(z),$$

where $1 \leqslant k_2 \leqslant n - k_1$ and $P_{n-k_1-k_2}(\alpha_2) \neq 0$. Repeating this process as many times as necessary (but no more than n times), we finally arrive at the unique factorization

$$P_n(z) = a_n(z - \alpha_1)^{k_1} \cdots (z - \alpha_p)^{k_p}, \tag{4.1}$$

where $1 \leqslant p \leqslant n$ and $k_1 + \cdots + k_p = n$ (see Probs. 3 and 4).

Now let A be an arbitrary finite complex number. Then, since $P_n(z) - A$ is obviously a polynomial of degree n, the equation

$$P_n(z) = A$$

also has n solutions (which are not necessarily distinct). As for the equation

$$P_n(z) = \infty, \tag{4.2}$$

it has only the solution $z = \infty$, since $P_n(z) = \infty$ if and only if $z = \infty$. For consistency, we formally regard $z = \infty$ as an n-fold root of the equation (4.2).

It follows that *every polynomial* $w = P_n(z)$ *of degree* $n > 0$ *maps the extended plane onto itself in such a way that every point* w *has at most* n *distinct inverse images*. Moreover, only certain exceptional values of w (always including ∞ if $n > 1$) can have fewer than n distinct inverse images:

THEOREM 4.1. *If* $P_n(z)$ *is a polynomial of degree* $n > 1$, *there are at most* n *points in the extended* w*-plane with fewer than* n *distinct inverse images under the mapping* $w = P_n(z)$.

Proof. Since $n > 1$, the point ∞ has fewer than n inverse images (in fact, just one). If $A \neq \infty$ has fewer than n distinct inverse images, the equation

$$P_n(z) = A$$

must have a multiple root. As is well known (see Prob. 5), any multiple root of this equation satisfies the equation

$$P'_n(z) = 0,$$

where $P'_n(z)$ is the derivative of $P_n(z)$. But this equation, being of degree $n - 1$, has no more than $n - 1$ distinct roots $\gamma_1, \ldots, \gamma_r$ $(1 \leqslant r \leqslant n - 1)$. Therefore the numbers

$$P_n(\gamma_1), \ldots, P_n(\gamma_r), \infty \qquad (1 \leqslant r \leqslant n - 1)$$

are the only values of A for which the equation $P_n(z) = A$ can have a multiple root. Since at most n of these numbers are distinct, the theorem is proved. (If $n = 0$ or $n = 1$, the theorem is meaningless.)

According to the general theory of Chap. 3, the mapping $w = P_n(z)$ is conformal at all points of the z-plane, with the possible exception of the point at infinity and the points $\gamma_1, \ldots, \gamma_r$ $(1 \leqslant r \leqslant n - 1)$ at which the derivative $P'_n(z)$ vanishes. In the case $n = 1$, $w = P_n(z)$ reduces to the entire linear transformation (3.31), and the mapping is one-to-one and conformal everywhere in the extended plane (cf. Sec. 19, Prob. 1). However, for $n > 1$ the mapping actually fails to be conformal at the points $\gamma_1, \ldots, \gamma_r, \infty$, as we now show:

THEOREM 4.2. *Let z_0 be a root of multiplicity $k > 1$ of the equation $P_n(z) = P_n(z_0)$. Then, under the mapping $w = P_n(z)$, every angle with its vertex at z_0 is enlarged k times. Moreover, every angle with its vertex at infinity is enlarged n times.*

Proof. If z_0 is a root of multiplicity $k > 1$ of the equation

$$P_n(z) = P_n(z_0),$$

and hence a root of multiplicity $k - 1$ of the equation

$$P'_n(z) = 0,$$

then

$$P_n(z) - P_n(z_0) = (z - z_0)^k Q(z), \tag{4.3}$$

where $Q(z)$ is a polynomial which does not vanish for $z = z_0$. It follows from (4.3) that

$$\operatorname{Arg} \frac{w - w_0}{(z - z_0)^k} = \operatorname{Arg} Q(z), \tag{4.4}$$

where $w = P_n(z)$, $w_0 = P_n(z_0)$. Now let l be a curve in the z-plane, with equation $z = \lambda(t)$, $t \in [a, b]$, which has a tangent τ with inclination θ at the point $z_0 = \lambda(t_0)$, $t_0 \in [a, b]$. Then the image of l under $f(z)$ is a curve L in the w-plane, with equation

$$w = P_n[\lambda(t)] = \Lambda(t), \qquad t \in [a, b].$$

Let $\{t_n\}$ be an arbitrary sequence of points in $[a, b]$ converging to t_0 ($t_n \neq t_0$). By a familiar argument (see Sec. 15), the inclination of the tangent to L at w_0 (if such exists) equals

$$\Theta = \lim_{n\to\infty} \operatorname{Arg} \frac{\Lambda(t_n) - \Lambda(t_0)}{t_n - t_0} = \lim_{n\to\infty} \operatorname{Arg} \frac{w_n - w_0}{(z_n - z_0)^k} \frac{(z_n - z_0)^k}{t_n - t_0}, \tag{4.5}$$

where $w_n = \Lambda(t_n)$, $w_0 = \Lambda(t_0)$, $z_n = \lambda(t_n)$, $z_0 = \lambda(t_0)$. Moreover $z_n \to z_0$ as $n \to \infty$, but $z_n \neq z_0$ for t_n sufficiently close to t_0 (since τ exists); this last fact has been anticipated in writing (4.5). Since $Q(z_0) \neq 0$, it follows from (4.5), (4.4) and the definition of θ that Θ exists and equals

$$\Theta = \lim_{n\to\infty} \operatorname{Arg} Q(z_n) + k \lim_{n\to\infty} \operatorname{Arg} \frac{\lambda(t_n) - \lambda(t_0)}{t_n - t_0} = \operatorname{Arg} Q(z_0) + k\theta \tag{4.6}$$

(cf. the proof of Theorem 3.3).[2]

Next let l_1 and l_2 be two curves in the z-plane with a common initial point z_0, which have tangents τ_1 and τ_2 at z_0. Then the angle with vertex z_0 formed by l_1 and l_2 (measured from l_1 to l_2) equals

$$\delta = \theta_2 - \theta_1,$$

where θ_1 and θ_2 are the inclinations of τ_1 and τ_2, respectively. According to (4.6), the images L_1 and L_2 of l_1 and l_2 under the mapping $w = P_n(z)$ form an angle

$$\begin{aligned}\Delta = \Theta_2 - \Theta_1 = [\operatorname{Arg} Q(z_0) + k\theta_2] - [\operatorname{Arg} Q(z_0) + k\theta_1]\\ = k(\theta_2 - \theta_1) = k\delta,\end{aligned}$$

with its vertex at the point $w_0 = P_n(z_0)$. This proves the first part of the theorem.

To prove the second part of the theorem, we make use of the transformation $\zeta = 1/z$, obtaining the new function

$$\eta = f(\zeta) = \frac{1}{P_n(1/\zeta)} = \frac{\zeta^n}{a_n + a_{n-1}\zeta + \cdots + a_0\zeta^n} \qquad (a_n \neq 0).$$

Then the point $\zeta = 0$ corresponds to the point $z = \infty$, and we have to prove that under the mapping $\eta = f(\zeta)$, every angle in the ζ-plane with

[2] Note that even if $\lambda'(t_0) \neq 0$, we still cannot write $\Theta = \operatorname{Arg} \Lambda'(t_0)$, since $\Lambda'(t_0) = P_n'(z_0)\lambda'(t_0) = 0$.

its vertex at the origin is enlarged n times. Let l be a curve in the ζ-plane, with equation $\zeta = \lambda(t)$, $t \in [a, b]$, which goes through the point $\zeta = 0$ and has a tangent τ with inclination θ at $\zeta = 0$. Then, by a slight modification of the previous argument, we find that the curve L in the η-plane with equation $\eta = f[\lambda(t)] = \Lambda(t)$ goes through the point $\eta = 0$ and has a tangent with inclination

$$\Theta = n\theta - \operatorname{Arg} a_n$$

at $\eta = 0$. The rest of the proof follows as before (note that $w = 1/\eta$).

Problem 1. Given two polynomials $f(z)$ and $g(z)$, prove that there exist uniquely defined polynomials $q(z)$ and $r(z)$ such that

$$f(z) = q(z)g(z) + r(z),$$

where the degree of $r(z)$ is less than the degree of $g(z)$.

Hint. Suppose $f(z) = a_0 + a_1 z + \cdots + a_m z^m$, $g(z) = b_0 + b_1 z + \cdots + b_n z^n$, where $a_m \neq 0$, $b_n \neq 0$. If $m < n$, the result is trivial. If $m \geqslant n$, form the new polynomial

$$f_1(z) = f(z) - \frac{a_m}{b_n} z^{m-n} g(z).$$

Repeat this argument if necessary.

Problem 2. Use the result of the preceding problem to prove that a necessary and sufficient condition for a polynomial $f(z)$ to be divisible by $z - a$ without a remainder is that $f(a) = 0$, i.e., that $z = a$ be a zero of $f(z)$.

Problem 3. Prove that if a polynomial of degree no higher than n has more than n distinct zeros, then the polynomial is identically zero. Prove that if the values of two polynomials of degree no higher than n are the same for more than n different values of z, then the two polynomials are identically equal.

Problem 4. Prove that the factorization (4.1) is unique.

Hint. Suppose that besides (4.1),

$$P(z) = a_n'(z - \alpha_1')^{k_1'} \cdots (z - \alpha_q')^{k_q'}.$$

Equate the two expressions for $P(z)$, and observe that at least one of the numbers $\alpha_1', \ldots, \alpha_q'$ must equal α_1. Repeat this argument if necessary.

Problem 5. Prove that a necessary and sufficient condition for $z = a$ to be a zero of order k of the polynomial $f(z)$ is that

$$f(a) = f'(a) = \cdots = f^{(k-1)}(a) = 0, \qquad f^{(k)}(a) \neq 0.$$

Hint. Consider (4.1).

Problem 6. Prove that if $z = a + ib$ is a zero of order k of the polynomial $f(z)$ with real coefficients, then $\bar{z} = a - ib$ is also a zero of order k of $f(z)$.

19. THE MAPPING $w = (z - a)^n$

We now make a detailed study of the mapping

$$w = (z - a)^n \qquad (n > 1). \tag{4.7}$$

This function maps the extended z-plane onto the extended w-plane in such a way that every point w has n distinct inverse images, with the exception of the two points $w = 0$ and $w = \infty$, for which the n inverse images "coalesce" into the single points $z = a$ and $z = \infty$, respectively. To find the n inverse images of w when $w \neq 0$, $w \neq \infty$, we solve (4.7) for z, obtaining

$$z = a + \sqrt[n]{w} = a + \sqrt[n]{|w|}\left(\cos\frac{\operatorname{Arg} w}{n} + i\sin\frac{\operatorname{Arg} w}{n}\right). \tag{4.8}$$

Obviously, the n distinct points (4.8) lie at the vertices of a regular n–gon with its center at the point $z = a$. The mapping (4.7) is conformal at all points except $z = a$, $z = \infty$, and every angle with its vertex at one of these two points is enlarged n times.

To get a clearer picture of the mapping (4.7), we observe that

$$|w| = |z - a|^n, \qquad \operatorname{Arg} w = n \operatorname{Arg}(z - a),$$

which implies that every circle of radius r with its center at the point $z = a$ is mapped into a circle of radius r^n with its center at the point $w = 0$. Moreover, as the point z goes around the circle $|z - a| = r$ once in the positive direction [so that $\operatorname{Arg}(z - a)$ increases continuously by 2π], the image point w goes around the circle $|w| = r^n$ n times in the same direction [since $\operatorname{Arg} w$ increases continuously by $2n\pi$]. We also note that as the point z sweeps out the ray

$$\operatorname{Arg}(z - a) = \varphi_0 + 2k\pi \qquad (k = 0, \pm 1, \pm 2, \ldots)$$

going from a to ∞, the image point w sweeps out the ray

$$\operatorname{Arg} w = n\varphi_0 + 2m\pi \qquad (m = 0, \pm 1, \pm 2, \ldots)$$

going from 0 to ∞.

Next consider the domain G consisting of all points z such that

$$\varphi_0 + 2k\pi < \operatorname{Arg}(z - a) < \varphi_1 + 2k\pi \qquad (k = 0, \pm 1, \pm 2, \ldots),$$

where $0 < \varphi_1 - \varphi_0 \leqslant 2\pi/n$. Such a domain will be called the *interior* of the angle of $\varphi_1 - \varphi_0$ radians formed by the rays

$$\operatorname{Arg}(z - a) = \varphi_0 + 2k\pi, \quad \operatorname{Arg}(z - a) = \varphi_1 + 2k\pi \quad (k = 0, \pm 1, \pm 2, \ldots),$$

the term *angle* itself being reserved either for the figure formed by these two rays or for the quantity $\varphi_1 - \varphi_0$. Then the image of G under the mapping (4.7) is the domain

$$n\varphi_0 + 2m\pi < \operatorname{Arg} w < n\varphi_1 + 2m\pi \qquad (m = 0, \pm 1, \pm 2, \ldots),$$

i.e., the interior of the angle of $n(\varphi_1 - \varphi_0)$ radians with its vertex at the origin of the w-plane (see Figure 4.1). Not only is the function $w = (z - a)^n$ conformal on G, as already noted, but it is also one-to-one on G. In fact, since $w = (z - a)^n$ is single-valued, we need only verify that every point w has only one inverse image in G. Since the n inverse images of the point w lie at the vertices of a regular n-gon in the z-plane with center at a, two inverse images can belong to the interior of the same angle with vertex at a only if the angle exceeds $2\pi/n$. But $0 < \varphi_1 - \varphi_0 \leqslant 2\pi/n$ by hypothesis, and hence every point of $\mathscr{G}$ has only one inverse image in G, as asserted. Thus, *the function* $w = (z - a)^n$ *is a one-to-one conformal mapping of the interior of one angle onto the interior of another angle which is n times larger.*

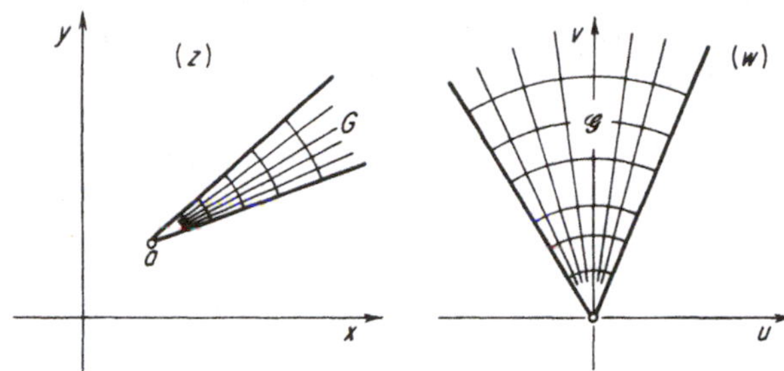

FIGURE 4.1

Of course, it would be quite incorrect to conclude that the function $w = (z - a)^n$ maps every straight line into a straight line, and every circle into a circle. For example, suppose $a = 0$, $n = 2$, so that $w = (z - a)^n$ reduces to $w = z^2$, and consider the effect of this mapping on straight lines parallel to the coordinate axes. Every line parallel to the imaginary axis (except the axis itself) has an equation of the form

$$z = b + it \qquad (-\infty < t < \infty),$$

where $b \neq 0$ is a real number, and the image of this line under the mapping $w = z^2$ has the equation

$$w = (b + it)^2. \tag{4.9}$$

Separating real and imaginary parts of (4.9), we obtain the corresponding parametric equations

$$u = b^2 - t^2, \qquad v = 2bt,$$

where u and v are the rectangular coordinates of the point w. Elimination of the parameter t leads to

$$v^2 = 4b^2(b^2 - u), \tag{4.10}$$

which is the equation of a parabola opening to the left, with axis lying along the real axis and focus at the origin. Similarly, every line parallel to the real axis (except the axis itself) has an equation of the form

$$z = t + ic \qquad (-\infty < t < \infty),$$

where $c \neq 0$ is a real number, and it is easily verified that the image of this line under the mapping $w = z^2$ is the parabola

$$v^2 = 4c^2(c^2 + u) \tag{4.11}$$

opening to the right, with axis lying along the real axis and focus at the origin. Thus, *the mapping* $w = z^2$ *transforms the two one-parameter families of straight lines parallel to (but distinct from) the coordinate axes into two one-parameter families of parabolas* (4.10) *and* (4.11), *with axes lying along the real axis and a common focus at the origin* (see Figure 4.2). Moreover, since

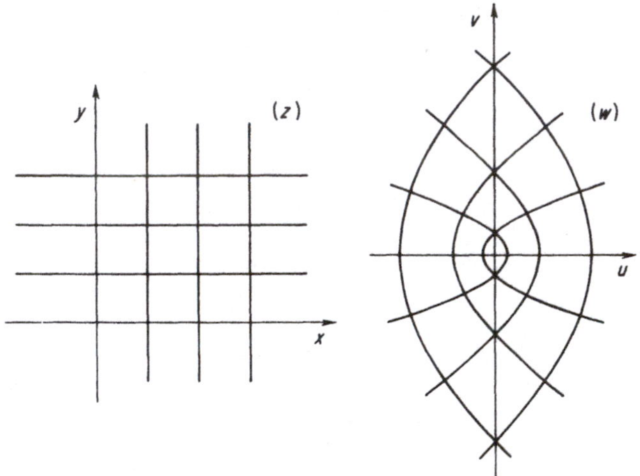

FIGURE 4.2

the two families of straight lines in the z-plane form an *orthogonal system* (i.e., every curve in one family is orthogonal to every curve in the other family, and vice versa), and since the mapping $w = z^2$ is conformal (except at $z = 0$), it follows that the two families of parabolas in the w-plane also form an orthogonal system.[3]

[3] The x and y-axes themselves are orthogonal, but their images, i.e., the line segments $u \geqslant 0$, $v = 0$ and $u \leqslant 0$, $v = 0$ (each traversed twice) are not orthogonal, since the derivative of $w = z^2$ vanishes at $z = 0$ (cf. Theorem 4.2).

Remark. It should be kept in mind that $w = z^2$ is not a one-to-one mapping, and in fact, every point in the w-plane except $w = 0$ and $w = \infty$ has two inverse images. In particular, the inverse image of the parabola (4.10) consists of the two straight lines $z = b + it$ and $z = -b + it$ which are symmetric with respect to the imaginary axis, while the inverse image of the parabola (4.11) consists of the two straight lines $z = t + ic$ and $z = t - ic$ which are symmetric with respect to the real axis. However, if we confine ourselves to a half-plane G whose boundary consists of a line passing through the origin (i.e., the interior of an angle of π radians with vertex at the origin), then the correspondence between G and its image $\mathcal{G}$ under $w = z^2$ is one-to-one. In fact, $\mathcal{G}$ is the interior of an angle of 2π radians with vertex at the origin, both of whose sides coalesce to form a single ray emanating from the origin.

Problem 1. What are the inverse images under the mapping $w = z^2$ of the lines $\operatorname{Re} w = \text{const}$, $\operatorname{Im} w = \text{const}$?

Problem 2. Verify *directly* that any parabola of the family (4.10) is orthogonal to any parabola of the family (4.11).

20. THE MAPPING $w = \sqrt[n]{z}$

The time has come to say more about multiple-valued functions (defined in Sec. 5). Let $w = f(z)$ be a multiple-valued function, with inverse $z = f^{-1}(w)$, and suppose $z = f^{-1}(w)$ is one-to-one on a domain $\mathcal{G}$ in the w-plane. Then $\mathcal{G}$ is called a *domain of univalence* for $z = f^{-1}(w)$. Let $\varphi(z)$ be the function assigning z the unique point $f(\{z\}) \cap \mathcal{G}$, i.e., the unique value of $f(z)$ which belongs to $\mathcal{G}$. Then $w = \varphi(z)$ is called a *single-valued branch* of the multiple-valued function $w = f(z)$.[4]

These new concepts will be illustrated by making a detailed study of the particularly simple multiple-valued function

$$w = \sqrt[n]{z},$$

whose inverse $z = w^n$ has already been discussed in the preceding section (with the roles of z and w reversed). Suppose we draw any n rays from the point $w = 0$, such that the angles between adjacent rays all equal $2\pi/n$. Then the interiors $\mathcal{G}_1, \ldots, \mathcal{G}_n$ of the n angles of $2\pi/n$ radians formed by these rays are all domains of univalence for the function $z = w^n$. In fact, this is the same result as on p. 71, expressed in our new terminology. The image (under $z = w^n$) of each of these domains $\mathcal{G}_k$ is the same domain G in the z-plane, whose boundary is some ray drawn from the point $z = 0$. In

[4] Preliminary contact with this notion has already been made on p. 54.

fact, if the boundary of $\mathscr{G}_k$ consists of the rays with inclinations

$$\varphi_0 + \frac{2k\pi}{n} \quad \text{and} \quad \varphi_0 + \frac{2(k+1)\pi}{n},$$

the boundary of G consists of the single ray L with inclination $n\varphi_0$. In this way, we obtain n single-valued branches

$$(\sqrt[n]{z})_1, \ldots, (\sqrt[n]{z})_n \tag{4.12}$$

of the function $\sqrt[n]{z}$, all defined on the same domain G, where $(\sqrt[n]{z})_k$ denotes the branch which maps G onto $\mathscr{G}_k$. Moreover, since $w = (\sqrt[n]{z})_k$ is a one-to-one continuous mapping of G onto $\mathscr{G}_k$, and since $z = w^n$ has a nonzero derivative nw^{n-1} on $\mathscr{G}_k$, the branches $(\sqrt[n]{z})_k$ all have nonzero derivatives on G, i.e.,

$$\frac{d}{dz}(\sqrt[n]{z})_k = \frac{1}{nw^{n-1}} = \frac{1}{n(\sqrt[n]{z})_k^{n-1}} \qquad (k = 1, \ldots, n)$$

(cf. Rule 5, p. 48).

Now suppose we rotate our family of n rays through an angle α about the origin, where $0 < \alpha < 2\pi/n$, thereby obtaining a new family of rays, which divides the w-plane into a new family of domains $\mathscr{D}_1, \ldots, \mathscr{D}_n$. Each domain $\mathscr{D}_k$ intersects two domains $\mathscr{G}_k$ and $\mathscr{G}_{k+1}$, with $\mathscr{G}_{n+1} = \mathscr{G}_1$ by definition (see Figure 4.3 illustrating the case $n = 6$, where boundaries of the domains $\mathscr{G}_k$

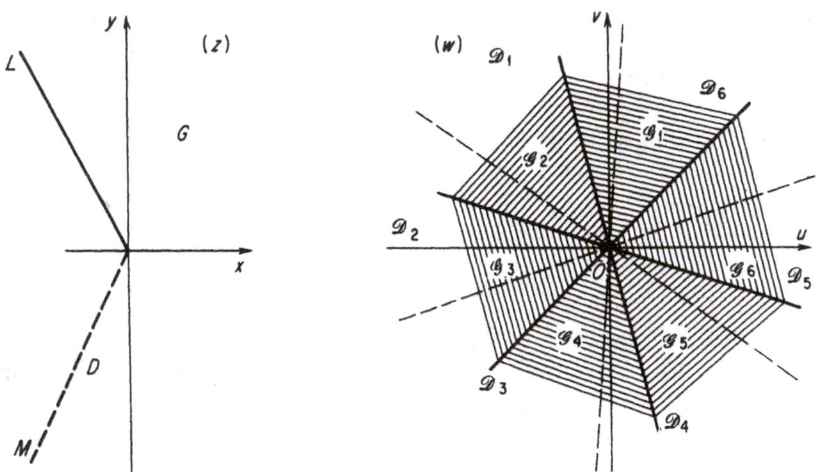

FIGURE 4.3

are indicated by solid lines, and boundaries of the domains $\mathscr{D}_k$ by dashed lines). The inverse image in the z-plane of each of the domains $\mathscr{D}_k$ is the same domain D, whose boundary is the single ray M drawn from the origin with inclination $n\varphi_0 + n\alpha$. As before, we can define n single-valued branches[5]

$$\{\sqrt[n]{z}\}_1, \ldots, \{\sqrt[n]{z}\}_n \tag{4.13}$$

of the function $w = \sqrt[n]{z}$, where now $\{\sqrt[n]{z}\}_k$ is the branch mapping D onto $\mathscr{D}_k$. Again, each of the branches (4.13) is differentiable on D, and

$$\frac{d}{dz}\{\sqrt[n]{z}\}_k = \frac{1}{n\{\sqrt[n]{z}\}_k^{n-1}} \qquad (k = 1, \ldots, n).$$

Moreover, it is clear that $\{\sqrt[n]{z}\}_k$ coincides with $(\sqrt[n]{z})_k$ on the set $\mathscr{D}_k \cap \mathscr{G}_k$ and with $(\sqrt[n]{z})_{k+1}$ on the set $\mathscr{D}_k \cap \mathscr{G}_{k+1}$. Thus, when we go from one family of domains of univalence to another such family, each new single-valued branch is obtained by combining two of the old single-valued branches, where on the part of $\mathscr{D}_k$ belonging to the common boundary Γ of $\mathscr{G}_k$ and $\mathscr{G}_{k+1}$, $\{\sqrt[n]{z}\}_k$ is the appropriate limit of either $(\sqrt[n]{z})_k$ or $(\sqrt[n]{z})_{k+1}$. More precisely, we have

$$\begin{aligned} \{\sqrt[n]{z}\}_k &= (\sqrt[n]{z})_k \qquad \text{if} \quad z \in \mathscr{D}_k \cap \mathscr{G}_k, \\ \{\sqrt[n]{z}\}_k &= (\sqrt[n]{z})_{k+1} \qquad \text{if} \quad z \in \mathscr{D}_k \cap \mathscr{G}_{k+1}, \\ \{\sqrt[n]{z}\}_k &= \lim_{\zeta \to z} (\sqrt[n]{\zeta})_k = \lim_{\zeta \to z} (\sqrt[n]{\zeta})_{k+1} \qquad \text{if} \quad z \in \mathscr{D}_k \cap \Gamma, \end{aligned}$$

where $\Gamma = \overline{\mathscr{G}}_k \cap \overline{\mathscr{G}}_{k+1}$.

Remark. If the angle of rotation α is zero, then $D = G$ and $\mathscr{D}_k = \mathscr{G}_k$ $(k = 1, \ldots, n)$, while if $\alpha = 2\pi/n$, $D = G$ again but $\mathscr{D}_k = \mathscr{G}_{k+1}$ $(k = 1, \ldots, n)$, where $\mathscr{G}_{n+1} = \mathscr{G}_1$. As α increases continuously from 0 to $2\pi/n$, the domain $\mathscr{D}_k$ overlaps the domain $\mathscr{G}_{k+1}$ more and more, until it finally coincides with $\mathscr{G}_{k+1}$, and the ray M representing the boundary of D undergoes a counterclockwise rotation of 2π radians, where its initial and final positions coincide with the ray L representing the boundary of G. At the same time, the branch $\{\sqrt[n]{z}\}_k$, which originally coincides with $(\sqrt[n]{z})_k$, shares more and more of its domain of definition with the branch $(\sqrt[n]{z})_{k+1}$, until it finally coincides with $(\sqrt[n]{z})_{k+1}$. In this sense, we can say that as α increases continuously from 0 to $2\pi/n$, the branch $(\sqrt[n]{z})_k$ changes continuously into the branch $(\sqrt[n]{z})_{k+1}$.

We can also keep track of the way one branch $(\sqrt[n]{z})_k$ changes into another branch $(\sqrt[n]{z})_{k+1}$ by making the point z describe a complete circle with center at the point $z = 0$. Suppose that at the point $z_0 \in G$ we choose a value of $\sqrt[n]{z}$ belonging to the branch $(\sqrt[n]{z})_k$ and represented by the point

$$w_0 = \sqrt[n]{|z_0|}\left(\cos\frac{\theta_0}{n} + i \sin\frac{\theta_0}{n}\right),$$

[5] Note the vital distinction between the parentheses in (4.12) and the braces in (4.13).

belonging to the domain $\mathscr{G}_k$. Then, as the point z moves continuously around the circle $|z| = |z_0|$ in the counterclockwise direction, starting from the point z_0, the value of

$$w = \sqrt[n]{|z_0|}\left(\cos\frac{\theta}{n} + i\sin\frac{\theta}{n}\right) \tag{4.14}$$

varies continuously with θ, and when z returns to its original value z_0, (4.14) goes into the value

$$w_1 = \sqrt[n]{|z_0|}\left(\cos\frac{\theta_0 + 2\pi}{n} + i\sin\frac{\theta_0 + 2\pi}{n}\right),$$

obtained by rotating w_0 through the angle $2\pi/n$ about the point $w = 0$.[6] Therefore w_1 belongs to the domain $\mathscr{G}_{k+1}$ adjacent to $\mathscr{G}_k$, and w_1 is the value of the branch $(\sqrt[n]{z})_{k+1}$ at the point z_0. Since the point $z_0 \in G$ is arbitrary, we can say that one circuit around the origin $z = 0$ in the counterclockwise direction causes the branch $(\sqrt[n]{z})_k$ to change continuously into the branch $(\sqrt[n]{z})_{k+1}$.[7] Moreover, it is easy to see that in this sense n circuits around the origin in the counterclockwise sense cause the branch $(\sqrt[n]{z})_k$ to undergo the sequence of transformations

$$(\sqrt[n]{z})_k \to (\sqrt[n]{z})_{k+1},\ (\sqrt[n]{z})_{k+1} \to (\sqrt[n]{z})_{k+2},\ \ldots,$$
$$(\sqrt[n]{z})_n \to (\sqrt[n]{z})_1,\ \ldots,\ (\sqrt[n]{z})_{k-1} \to (\sqrt[n]{z})_k,$$

which carry it continuously into itself after "going through" all the other branches in succession. Since $(\sqrt[n]{z})_k$ is arbitrary, n circuits around the origin carry any branch into itself.

Given a multiple-valued function $w = f(z)$ with continuous single-valued branches defined on a domain G, we say that a point $\zeta \in \bar{G}$ is a *branch point* of $f(z)$ if there exists a neighborhood $\mathscr{N}(\zeta)$ such that one complete circuit around an arbitrary closed Jordan curve $\gamma \subset \mathscr{N}(\zeta)$ with $\zeta \in I(\gamma)$, carries every branch of $f(z)$ into another branch of $f(z)$. If a finite number of circuits around γ (in the same direction) carries every branch of $f(z)$ into itself, and if n is the smallest such number, we say that ζ is a branch point *of finite order*, specifically, *of order* $n - 1$. In this case, the point ζ is also called an *algebraic branch point* of $f(z)$, provided that $f(z)$ has a limit (finite or infinite) at ζ. Thus we have just shown that the point $z = 0$ is an algebraic branch point of order $n - 1$ of the function $w = \sqrt[n]{z}$.

---

[6] Of course, in making the circuit around the circle $|z| = |z_0|$, we allow z to pass through the ray L, which is excluded from the domain G.

[7] More precisely, every value of $\sqrt[n]{z}$ on the branch $(\sqrt[n]{z})_k$ changes continuously into the corresponding value of $\sqrt[n]{z}$ on the branch $(\sqrt[n]{z})_{k+1}$.

Remark 1. It is clear that the point $z = \infty$ can also be regarded as an algebraic branch point of order $n - 1$ of the function $w = \sqrt[n]{z}$, since every circuit around the point at infinity along a circle of arbitrarily large radius with center at the origin is simultaneously a circuit around the origin. Therefore the multiple-valued function $w = \sqrt[n]{z}$ has two branch points in the z-plane, i.e., $z = 0$ and $z = \infty$, both of order $n - 1$.

Remark 2. The single-valued branches described above were constructed for a domain like G or D, whose boundary is a rectilinear ray joining the two branch points 0 and ∞. More generally, let γ be any unbounded Jordan curve (see Sec. 12, Prob. 2) in the extended z-plane joining the points 0 and ∞, and this time let G be the domain with boundary γ. As the point z traces out the curve γ from its initial point 0 to its final point ∞, the n points corresponding to the n values of $w = \sqrt[n]{z}$ trace out n Jordan curves $\Gamma_1, \ldots, \Gamma_n$ joining 0 and ∞. These curves have no points in common other than 0 and ∞, and each set $\Gamma_k \cup \Gamma_{k+1}$ (where $\Gamma_{k+1} = \Gamma_1$) represents a closed Jordan curve in the extended w-plane. Of the two domains with boundary $\Gamma_k \cup \Gamma_{k+1}$, let $\mathscr{G}_k$ be the domain which does not contain the other curves $\Gamma_1, \ldots, \Gamma_{k-1}$, $\Gamma_{k+2}, \ldots, \Gamma_n$. By construction, when the w-plane is rotated through the angle $2\pi/n$ about the origin, Γ_k goes into Γ_{k+1} and Γ_{k+1} goes into Γ_{k+2}, and hence the domain $\mathscr{G}_k$ goes into the domain $\mathscr{G}_{k+1}$ ($\mathscr{G}_{n+1} = \mathscr{G}_1$). Since

$$\mathscr{G}_k \cap \mathscr{G}_{k+1} = 0 \qquad (k = 1, \ldots, n),$$

the rotation cannot carry any point of $\mathscr{G}_k$ into another point of $\mathscr{G}_k$. Therefore the domains $\mathscr{G}_1, \ldots, \mathscr{G}_n$ are all domains of univalence for the function $z = w^n$, and we obtain n single-valued branches of the function $w = \sqrt[n]{z}$, all defined on the domain G, by requiring that the kth branch take its values in the domain $\mathscr{G}_k$ ($k = 1, \ldots, n$). To specify a branch, it is sufficient to indicate the value of $\sqrt[n]{z}$ at some point $z_0 \in G$; if this value is w_0, there is a unique domain $\mathscr{G}_k$ containing w_0, and a unique branch of $\sqrt[n]{z}$ taking the value w_0 at the point z_0.

Now let $[\sqrt[n]{z}]_k$ and $[\sqrt[n]{z}]_l$ be two single-valued branches of the function $\sqrt[n]{z}$, which are defined on the domain G and take values w_0' and w_0'', respectively, at a point $z_0 \in G$. Since

$$w_0' = [\sqrt[n]{z_0}]_k = \sqrt[n]{|z_0|}\left(\cos\frac{\theta_0 + 2m'\pi}{n} + i\sin\frac{\theta_0 + 2m'\pi}{n}\right),$$

$$w_0'' = [\sqrt[n]{z_0}]_l = \sqrt[n]{|z_0|}\left(\cos\frac{\theta_0 + 2m''\pi}{n} + i\sin\frac{\theta_0 + 2m''\pi}{n}\right),$$

where $\theta_0 = \arg z$, and m', m'' are integers, it follows that w_0'' equals w_0'

multiplied by

$$\eta = \cos\frac{2(m'' - m')\pi}{n} + i\sin\frac{2(m'' - m')\pi}{n},$$

i.e., by a value of $\sqrt[n]{1}$. But $\eta[\sqrt[n]{z}]_k$ is obviously a single-valued continuous function on G such that $(\eta[\sqrt[n]{z}]_k)^n = z$, i.e., $\eta[\sqrt[n]{z}]_k$ is one of the single-valued branches of $\sqrt[n]{z}$ defined on G, in fact, the branch $[\sqrt[n]{z}]_l$ containing the point $\eta[\sqrt[n]{z_0}]_k = [\sqrt[n]{z_0}]_l = w_0''$. In other words, any single-valued branch of $\sqrt[n]{z}$ defined on G can be obtained by multiplying any other single-valued branch defined on G by an appropriate nth root of unity.

Remark 3. The conclusions of this section apply (with certain obvious modifications) to the somewhat more general functions

$$w = \sqrt[n]{z - a} \quad \text{and} \quad w = \sqrt[n]{\frac{z - a}{z - b}}, \tag{4.15}$$

which are the inverses of the functions

$$z = w^n + a \quad \text{and} \quad z = \frac{bw^n - a}{w^n - 1},$$

respectively. The first of the functions (4.15) has branch points a and ∞

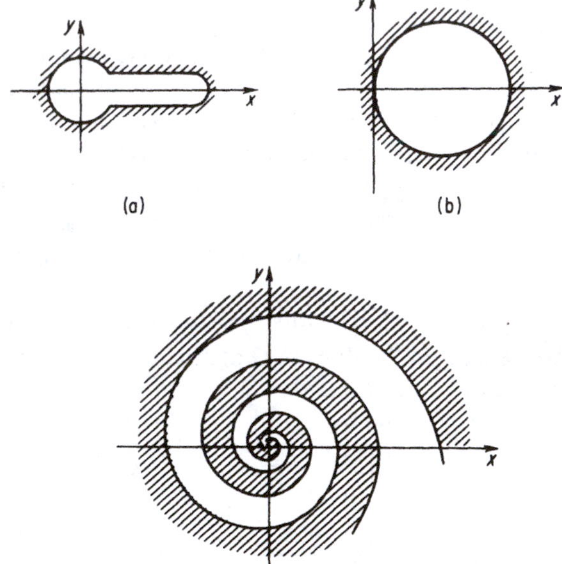

FIGURE 4.4

while the second has branch points a and b. Moreover, single-valued branches of each of the functions (4.15) can be defined on any domain whose boundary is a Jordan curve joining the appropriate branch points.

Problem 1. On which of the (unshaded) domains shown in Figure 4.4 can single-valued branches of the function $w = \sqrt{z}$ be defined?

Problem 2. Verify in detail Remark 3 above.

21. RATIONAL FUNCTIONS

In Sec. 18 we introduced the class of entire functions. A more general class of functions is the class of meromorphic functions. By a *meromorphic function* $f(z)$ we mean a function which can be written as a quotient

$$f(z) = \frac{g(z)}{h(z)}$$

of two entire functions $g(z)$ and $h(z)$, where $h(z) \not\equiv 0$. Obviously, every entire function $f(z)$ is meromorphic, since it can be written in the form $f(z)/1$, but the converse is not true, as shown by the function $1/z$ which is meromorphic but not entire, since it becomes infinite for $z = 0$.

The simplest members of the class of meromorphic functions in the full sense of the term (i.e., meromorphic functions which do not reduce to entire functions) are the rational functions. By a *rational function* we mean a function which can be written as a ratio

$$f(z) = \frac{P(z)}{Q(z)} = \frac{a_0 + a_1 z + \cdots + a_m z^m}{b_0 + b_1 z + \cdots + b_n z^n} \qquad (a_m \neq 0,\ b_n \neq 0) \quad (4.16)$$

of two polynomials $P(z)$ and $Q(z)$, where it is assumed that the fraction $P(z)/Q(z)$ is in *lowest terms*, i.e., that the equations $P(z) = 0$ and $Q(z) = 0$ have no common roots. Let $\alpha_1, \ldots, \alpha_p$ denote the distinct zeros of the polynomial $P(z)$, and let α_s be of order k_s $(s = 1, \ldots, p)$. Similarly, let $\beta_1, \ldots, \beta_q$ denote the distinct zeros of the polynomial $Q(z)$, and let β_t be of order l_t $(t = 1, \ldots, q)$. Then (4.16) can be written in the form

$$f(z) = \frac{P(z)}{Q(z)} = \frac{a_m(z - \alpha_1)^{k_1} \cdots (z - \alpha_p)^{k_p}}{b_n(z - \beta_1)^{l_1} \cdots (z - \beta_q)^{l_q}}.$$

Obviously

$$k_1 + \cdots + k_p = m, \qquad l_1 + \cdots + l_q = n,$$

and none of the numbers $\alpha_1, \ldots, \alpha_p$ equals any of the numbers $\beta_1, \ldots, \beta_q$, since $f(z)$ is assumed to be in lowest terms. The function $f(z)$ vanishes at each point α_s $(s = 1, \ldots, p)$, called a *zero* (or *zero-point*) *of order* k_s [of $f(z)$], and becomes infinite at each point β_t ($t = 1, \ldots, q$), called a *pole of order* l_t. The zero α_s is said to be *simple* if $k_s = 1$ and *multiple* if $k_s > 1$; similarly, the pole β_t is said to be *simple* if $l_t = 1$ and *multiple* if $l_t > 1$. It follows from our definition that the zeros of $f(z)$ are poles of $1/f(z)$, while the poles of $f(z)$ are zeros of $1/f(z)$, and moreover that orders are preserved, i.e., a zero of $f(z)$ of a given order becomes a pole of $1/f(z)$ of the same order, and vice versa.

To define $f(z)$ when z is the point at infinity, we use the relation

$$f(\infty) = \lim_{z\to\infty} f(z),$$

obtaining

$$\begin{aligned} &1) \quad f(\infty) = 0 \qquad \text{if} \quad m < n; \\ &2) \quad f(\infty) = \frac{a_m}{b_m} \qquad \text{if} \quad m = n; \\ &3) \quad f(\infty) = \infty \qquad \text{if} \quad m > n. \end{aligned} \tag{4.17}$$

In the first case, we say that $f(z)$ has a *zero at* ∞, and in the third case we say that $f(z)$ has a *pole at* ∞. To assign a definite order to a zero or pole at ∞, we make the preliminary transformation $\zeta = 1/z$ (cf. p. 24), which carries the point $z = \infty$ into the point $\zeta = 0$. Then

$$f^*(\zeta) = f(1/\zeta) = \frac{a_0 + a_1\dfrac{1}{\zeta} + \cdots + a_m\dfrac{1}{\zeta^m}}{b_0 + b_1\dfrac{1}{\zeta} + \cdots + b_n\dfrac{1}{\zeta^n}} = \frac{\zeta^n}{\zeta^m}\,\frac{a_m + a_{m-1}\zeta + \cdots + a_0\zeta^m}{b_n + b_{n-1}\zeta + \cdots + b_0\zeta^n},$$

and analyzing the three cases (4.17), we obtain the following results:

1. If $m < n$, then

$$f^*(\zeta) = \frac{\zeta^{n-m}(a_m + a_{m-1}\zeta + \cdots + a_0\zeta^m)}{b_n + b_{n-1}\zeta + \cdots + b_0\zeta^n}.$$

Since this rational function has a zero of order $n - m$ at $\zeta = 0$, we say that $f(z)$ has a zero of order $n - m$ at $z = \infty$.

2. If $m = n$, then

$$f^*(\zeta) = \frac{a_m + a_{m-1}\zeta + \cdots + a_0\zeta^m}{b_m + b_{m-1}\zeta + \cdots + b_0\zeta^m}.$$

Since this rational function has neither a zero nor a pole at $\zeta = 0$ [in fact, $f^*(0) = f(\infty) = a_m/b_m$], we say that $f(z)$ has neither a pole nor a zero at $z = \infty$.

3. If $m > n$, then

$$f^*(\zeta) = \frac{a_m + a_{m-1}\zeta + \cdots + a_0\zeta^m}{\zeta^{m-n}(b_n + b_{n-1}\zeta + \cdots + b_0\zeta^n)}.$$

Since this rational function has a pole of order $m - n$ at $\zeta = 0$, we say that $f(z)$ has a pole of order $m - n$ at $z = \infty$.

Remark. It is easy to see that $f(z)$ is analytic at $z = \infty$ (see Remark 1, p. 63) if $m < n$ or $m = n$, but not if $m > n$.

By the *order* N of a rational function $f(z)$, we mean the total number of zeros (or poles) of the function in the extended plane, counting each zero (or pole) a number of times equal to its order. It is easy to see that

$$N = \max(m, n),$$

i.e., N is the larger of the two integers m and n (or their common value if $m = n$). In fact, consider the three possibilities in turn:

1. If $m < n$, $f(z)$ has $k_1 + \cdots + k_p = m$ zeros at finite points and a zero of order $n - m$ at infinity, so that $f(z)$ has a total of

$$N = m + (n - m) = n = \max(m, n)$$

zeros. Moreover, $f(z)$ has $l_1 + \cdots + l_q = n$ poles at finite points and no pole at infinity, so that $f(z)$ also has a total of $N = n$ poles.

2. If $m = n$, $f(z)$ has neither a zero nor a pole at infinity, and precisely

$$N = k_1 + \cdots + k_p = l_1 + \cdots + l_q = m$$

zeros and poles at finite points.

3. If $m > n$, $f(z)$ has $l_1 + \cdots + l_q = n$ poles at finite points and a pole of order $m - n$ at infinity, so that $f(z)$ has a total of

$$N = n + (m - n) = m = \max(m, n)$$

poles. Moreover, $f(z)$ has $k_1 + \cdots + k_p = m$ zeros at finite points and no zero at infinity, so that $f(z)$ also has a total of $N = m$ zeros.

Now let A be an arbitrary complex number, and consider the roots of the equation

$$f(z) = \frac{P(z)}{Q(z)} = A, \tag{4.18}$$

or equivalently, the roots of the equation

$$F(z) = \frac{P(z) - AQ(z)}{Q(z)} = 0. \tag{4.19}$$

By the multiplicity of a root z_0 of equation (4.18), we mean the multiplicity of the same root z_0 of equation (4.19).

THEOREM 4.3. *Given a rational function $f(z)$ of order N and an arbitrary complex number A, the total number of roots of equation* (4.18) *is N* (*with due regard for multiplicity*).

Proof. If $A = 0$ or $A = \infty$, the result follows directly from the definition of the order of $f(z)$. Let r be the degree of the polynomial $P(z) - AQ(z)$ and s the order of the rational function $F(z)$. If $A \neq 0$, $A \neq \infty$, $m \neq n$, then $r = \max(m, n) = N$, and hence $s = \max(r, n) = N$. If $A \neq 0$, $A \neq \infty$, $m = n$, then $r \leqslant \max(m, n) = N$, and hence $s = \max(r, n) = N$.

Remark. As we know from Sec. 18, if $P_n(z)$ is a polynomial of degree n, the equation $P_n(z) = 0$ has exactly n roots (with due regard for multiplicity). Thus we see that this same property holds for rational functions, with the concept of *order* replacing that of *degree.*

Example. The rational function

$$f(z) = \frac{z^2 + 1}{z^2 - 1}$$

has two simple zeros $\pm i$ and two simple poles ± 1. If $A = 1$, the equation $f(z) = A$ becomes

$$\frac{z^2 + 1}{z^2 - 1} = 1$$

or

$$\frac{2}{z^2 - 1} = 0, \tag{4.20}$$

which obviously has no finite roots. However, (4.20) has a double root at ∞, since the degree of the denominator is two higher than that of the numerator.

It follows from Theorem 4.3 that a *rational function* $w = f(z)$ *of order* $N > 0$ *maps the extended plane onto itself in such a way that every point* w *has at most* N *distinct inverse images* (recall the analogous property of polynomials, proved on p. 66). Moreover, only certain exceptional values of w can have fewer than N distinct inverse images [e.g., the point $w = 0$ if $f(z)$ has multiple zeros, or the point $w = \infty$ if $f(z)$ has multiple poles]. The following result is the natural generalization of Theorem 4.1:

THEOREM 4.4. *Let* $f(z)$ *be a rational function whose numerator* $P(z)$ *is of degree* m *and whose denominator* $Q(z)$ *is of degree* n. *Then there are at most* $m + n$ *points in the extended* w*-plane with fewer than* $N = \max(m, n)$ *distinct inverse images* ($N > 1$).

Proof. First suppose $A \neq 0$, $A \neq \infty$, and suppose that A has fewer than N distinct inverse images. Then the equation

$$f(z) = \frac{P(z)}{Q(z)} = A,$$

or equivalently

$$\frac{P(z) - AQ(z)}{Q(z)} = 0, \tag{4.21}$$

must have a multiple root. Any finite multiple root of (4.21) is a multiple root of the equation

$$P(z) - AQ(z) = 0, \tag{4.22}$$

and conversely. Moreover, any multiple root of (4.22) satisfies the equation

$$P'(z) - AQ'(z) = 0,$$

and hence also satisfies the equation

$$P'(z)Q(z) - P(z)Q'(z) = 0, \tag{4.23}$$

of degree no higher than $m + n - 1$. Equation (4.23) has no more than $m + n - 1$ distinct roots $\gamma_1, \ldots, \gamma_r$ $(1 \leqslant r \leqslant m + n - 1)$. Since any finite multiple zero of $f(z)$ satisfies the equations

$$P(z) = 0, \qquad P'(z) = 0,$$

and since any finite multiple pole of $f(z)$ satisfies the equations

$$Q(z) = 0, \qquad Q'(z) = 0,$$

all the finite multiple roots of (4.21) satisfy (4.23), and are therefore already included among the numbers $\gamma_1, \ldots, \gamma_r$, even if we let $A = 0$ or $A = \infty$. In other words, the numbers

$$f(\gamma_1), \ldots, f(\gamma_r), f(\infty) \qquad (1 \leqslant r \leqslant m + n - 1)$$

are the only values of A for which the equation $f(z) = A$ can have a multiple root. Since at most $m + n$ of these numbers are distinct, the theorem is proved.[8]

Remark. The mapping $w = f(z)$ is conformal at all but a finite number of points, since the condition that z should not equal ∞, any of the numbers $\gamma_1, \ldots, \gamma_r$ or any of the zeros of $Q(z)$ is certainly sufficient to guarantee that the derivative

$$f'(z) = \frac{P'(z)Q(z) - P(z)Q'(z)}{[Q(z)]^2}$$

is finite and nonzero.[9]

---

[8] If $N = 0$ or $N = 1$, the theorem is meaningless.

[9] However, the condition is not necessary (for further details, see Probs. 1 and 2).

Problem 1. Given a rational function

$$f(z) = \frac{P(z)}{Q(z)}, \tag{4.24}$$

prove that the mapping $w = f(z)$ is conformal at any simple zero of $Q(z)$, and also at $z = \infty$ if the equation $f(z) = f(\infty)$ has no multiple roots.

Problem 2. Prove that the mapping (4.24) actually fails to be conformal at each of the points $\gamma_1, \ldots, \gamma_r$ which are the roots of the equation

$$P'(z)Q(z) - P(z)Q'(z) = 0$$

[see (4.23)], and also at the point $\gamma_0 = \infty$ if the equation $f(z) = f(\infty)$ has multiple roots. In particular, prove that an angle with vertex at γ_j is enlarged a number of times equal to the multiplicity of the root γ_j of the equation $f(z) = f(\gamma_j)$, $j = 0, 1, \ldots, r$.

22. THE MAPPING $w = \frac{1}{2}\left(z + \frac{1}{z}\right)$

A mapping by a rational function of order higher than 1 was studied in Sec. 19, in connection with the function

$$w = (z - a)^n \qquad (n > 1).$$

However, this function is not meromorphic in the full sense of the word, since it is actually entire. We now study the rational function

$$w = \lambda(z) = \frac{1}{2}\left(z + \frac{1}{z}\right) = \frac{z^2 + 1}{2z}, \tag{4.25}$$

which comes up in the course of solving a variety of problems. In fact, because of the use which the Russian scientist Joukowski made of this function in aerodynamics, it is often referred to as the *Joukowski function.* Obviously, $w = \lambda(z)$ is a rational function of order 2, which does not reduce to an entire function and which satisfies the condition

$$\lambda(z) = \lambda\left(\frac{1}{z}\right). \tag{4.26}$$

It follows from (4.26) that under the mapping $w = \lambda(z)$ every point of the w-plane except $w = \pm 1$ has two (and only two) distinct inverse images z_1 and z_2 satisfying the relation

$$z_1 z_2 = 1. \tag{4.27}$$

Let γ be the unit circle $|z| = 1$, with interior $I(\gamma)$ and $E(\gamma)$. Then, according to (4.27), $z_1 \in I(\gamma)$ if and only if $z_2 \in E(\gamma)$. Moreover $\lambda[I(\gamma)] = \lambda[E(\gamma)]$, i.e., (4.25) maps both the interior and the exterior of the unit circle into the same set in the w-plane.

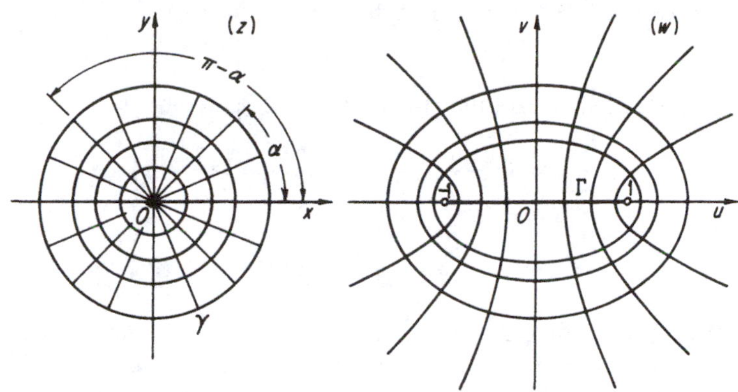

FIGURE 4.5

To study the mapping $w = \lambda(z)$ in more detail, we find the images of the circles $|z| = r$ and the rays $\operatorname{Arg} z = \alpha + 2k\pi$ (see Figure 4.5), confining ourselves to the domain $G = I(\gamma)$. Setting

$$z = re^{it} \qquad (0 \leqslant t \leqslant 2\pi),$$

where $0 < r < 1$, we find that

$$w = u + iv = \frac{1}{2}\left(re^{it} + \frac{1}{r}e^{-it}\right) = \frac{1}{2}\left(\frac{1}{r} + r\right)\cos t - i\,\frac{1}{2}\left(\frac{1}{r} - r\right)\sin t,$$

or

$$u = \frac{1}{2}\left(\frac{1}{r} + r\right)\cos t, \qquad v = -\frac{1}{2}\left(\frac{1}{r} - r\right)\sin t, \tag{4.28}$$

where $0 \leqslant t \leqslant 2\pi$. Eliminating t from (4.28), we obtain

$$\frac{u^2}{\left[\dfrac{1}{2}\left(\dfrac{1}{r} + r\right)\right]^2} + \frac{v^2}{\left[\dfrac{1}{2}\left(\dfrac{1}{r} - r\right)\right]^2} = 1, \tag{4.29}$$

which is the equation of an ellipse with semiaxes $a = \frac{1}{2}(r^{-1} + r)$, $b = \frac{1}{2}(r^{-1} - r)$ and foci at ± 1. It follows from (4.28) that as t increases continuously from 0 to 2π, i.e., as the point z describes the whole circle $|z| = r$ once in the counterclockwise direction, the image point w describes the whole ellipse once in the clockwise direction. By varying the radius of the circle $|z| = r$ from 0 to 1, we cause a to decrease from ∞ to 1 and b to decrease from ∞ to 0; as a result, the ellipses (4.29) range over the whole set of ellipses with foci ± 1. It follows that $w = \lambda(z)$ is a one-to-one mapping of the unit disk $G = I(\gamma)$ onto the domain $\mathscr{G} = \Gamma^c$, where Γ denotes the segment $-1 \leqslant u \leqslant 1$, $v = 0$ of the real axis and Γ^c its complement.

Next we consider the images of the ray

$$z = te^{i\alpha} \qquad (0 \leqslant t < 1) \tag{4.30}$$

with inclination α. Substituting (4.30) into (4.25), we obtain

$$w = u + iv = \frac{1}{2}\left(\frac{1}{t} + t\right) \cos \alpha - i \frac{1}{2}\left(\frac{1}{t} - t\right) \sin \alpha$$

or

$$u = \frac{1}{2}\left(\frac{1}{t} + t\right) \cos \alpha, \quad v = -\frac{1}{2}\left(\frac{1}{t} - t\right) \sin \alpha \qquad (0 \leqslant t < 1). \tag{4.31}$$

It follows that the images of two radii symmetric with respect to the real axis (i.e., such that if one radius has inclination α, the other has inclination $-\alpha$) are themselves symmetric with respect to the real axis, while the images of two radii symmetric with respect to the imaginary axis (i.e., such that if one radius has inclination α, the other has inclination $\pi - \alpha$) are themselves symmetric with respect to the imaginary axis. Thus it is sufficient to consider the images of radii lying in the first quadrant $0 \leqslant \alpha \leqslant \pi/2$.

For $\alpha = 0$ we have

$$u = \frac{1}{2}\left(\frac{1}{t} + t\right), \quad v = 0 \qquad (0 \leqslant t < 1),$$

which represents the infinite interval $1 < u \leqslant +\infty$. The infinite interval $-\infty \leqslant u < -1$ is the image of the radius with inclination π. For $\alpha = \pi/2$ we have

$$u = 0, \quad v = -\frac{1}{2}\left(\frac{1}{t} - t\right) \qquad (0 \leqslant t < 1),$$

which represents the negative imaginary axis $-\infty \leqslant v < 0$. The positive imaginary axis $0 < v \leqslant +\infty$ is the image of the radius with inclination $\alpha = -\pi/2$. Thus the image of the "horizontal" diameter of the unit disk $G: |z| < 1$ is the infinite interval of the real axis going from -1 to $+1$ through the point at infinity and excluding the points ± 1, while the image of the "vertical" diameter of G is the whole imaginary axis including the point at infinity but excluding the origin.

Suppose now that $0 < \alpha < \pi/2$. Then, eliminating the parameter t from (4.31), we obtain

$$\frac{u^2}{\cos^2 \alpha} - \frac{v^2}{\sin^2 \alpha} = 1,$$

which is the equation of a hyperbola, which we denote by H, with semiaxes $a = \cos \alpha$, $b = \sin \alpha$ and foci at ± 1. Let H_1, H_2, H_3 and H_4 denote the intersections of H with the first, second, third and fourth quadrants, respectively, excluding the two points $(\pm a, 0)$ of H belonging to the real axis.

Moreover, let R_α, $R_{\pi-\alpha}$, $R_{\pi+\alpha}$ and $R_{-\alpha}$ denote the sets of points belonging to the radii (4.30), with inclinations α, $\pi - \alpha$, $\pi + \alpha$, and $-\alpha$, respectively. Then it is easy to see that

$$H_1 = \lambda(R_{-\alpha}), \quad H_2 = \lambda(R_{\pi+\alpha}), \quad H_3 = \lambda(R_{\pi-\alpha}), \quad H_4 = \lambda(R_\alpha).$$

In particular, the image of each of the diameters $R_\alpha \cup R_{\pi+\alpha}$ and $R_{\pi-\alpha} \cup R_{-\alpha}$ of G is a set consisting of two "quarter-branches" of the hyperbola H, joined at infinity and minus the points $(\pm a, 0)$.

To summarize, *the function*

$$w = \lambda(z) = \frac{1}{2}\left(z + \frac{1}{z}\right)$$

is a one-to-one continuous mapping of both the interior and the exterior of the unit circle γ *onto the complement of the segment* $-1 \leqslant u \leqslant +1$ *of the real axis. Under this mapping, the one-parameter family of circles* $|z| = r$ $(0 < r < 1)$ *is transformed into a one-parameter family of confocal ellipses, with semiaxes* $\frac{1}{2}(r^{-1} \pm r)$ *and foci at* ± 1, *and the one-parameter family of pairs of diameters of* γ *symmetric with respect to the real axis, formed of the radii*

$$z = \pm te^{\pm i\alpha} \qquad (0 \leqslant t < 1),$$

where $0 < \alpha < \pi/2$,[10] *is transformed into a one-parameter family of confocal hyperbolas (minus their vertices), with semiaxes* $\cos\alpha$, $\sin\alpha$ *and foci at* ± 1.

Remark. Since the derivative

$$\lambda'(z) = \frac{1}{2}\left(1 - \frac{1}{z^2}\right)$$

is nonzero for $z \neq \pm 1$, the mapping is conformal at all points of the domains $I(\gamma)$ and $E(\gamma)$. It follows that the hyperbolas intersect the ellipses at the same angles as the radii intersect the circles, i.e., at right angles.

Problem. Let G be the exterior of the unit circle $|z| = 1$. What are the inverse images under the Joukowski function of the families of ellipses and hyperbolas in the w-plane just described?

---

[10] The cases $\alpha = 0$ and $\alpha = \pi/2$ warrant special discussion (see above).

CHAPTER 5

MÖBIUS TRANSFORMATIONS

23. THE GROUP PROPERTY OF MÖBIUS TRANSFORMATIONS

It follows from Theorems 4.3 and 4.4 that the rational function of order 1, i.e., the *fractional linear transformation* or *Möbius transformation*

$$w = L(z) = \frac{az + b}{cz + d} \qquad (ad - bc \neq 0)$$

is the only rational function which maps the extended plane onto itself in a one-to-one fashion. Here, we assume that the quantity $ad - bc$, called the *determinant* of the function $L(z)$, is nonzero, since otherwise $L(z)$ is identically equal to a constant mapping the extended plane into a single point. As we know from Sec. 17, the mapping $w = L(z)$ is conformal at all points of the extended plane. This mapping (or its various special cases) plays an important role in a variety of problems encountered in the theory of functions of a complex variable, and it therefore merits a detailed study (to which we devote the present chapter).

Let $\mathscr{M}$ denote the set of all Möbius transformations. Two such transformations

$$L_1(z) = \frac{a_1 z + b_1}{c_1 z + d_1}, \qquad L_2(z) = \frac{a_2 z + b_2}{c_2 z + d_2} \tag{5.1}$$

are regarded as *identical* if and only if $L_1(z) = L_2(z)$ for all z.

THEOREM 5.1 *A necessary and sufficient condition for the two Möbius transformations* (5.1) *to be identical is that*

$$a_2 = \lambda a_1, \quad b_2 = \lambda b_1, \quad c_2 = \lambda c_1, \quad d_2 = \lambda d_1 \qquad (\lambda \neq 0). \tag{5.2}$$

Proof. The sufficiency of the condition is obvious. To prove the necessity, suppose $L_1(z) \equiv L_2(z)$. Then, in particular,

$$L_1(0) = L_2(0), \quad L_1(1) = L_2(1), \quad L_1(\infty) = L_2(\infty),$$

which means that[1]

$$\frac{b_1}{d_1} = \frac{b_2}{d_2} = p, \quad \frac{a_1 + b_1}{c_1 + d_1} = \frac{a_2 + b_2}{c_2 + d_2}, \quad \frac{a_1}{c_1} = \frac{a_2}{c_2} = q. \tag{5.3}$$

Substituting

$$b_1 = pd_1, \quad b_2 = pd_2, \quad a_1 = qc_1, \quad a_2 = qc_2$$

into the second of the equations (5.3), we obtain

$$\frac{qc_1 + pd_1}{c_1 + d_1} = \frac{qc_2 + pd_2}{c_2 + d_2}$$

or

$$(c_1 d_2 - c_2 d_1)(q - p) = 0.$$

But $q \neq p$, since otherwise

$$\frac{a_1}{c_1} = \frac{b_1}{d_1}, \qquad \text{i.e.,} \quad a_1 d_1 - b_1 c_1 = 0,$$

contrary to hypothesis, and therefore

$$\frac{c_1}{d_1} = \frac{c_2}{d_2}.$$

Together with (5.3), this implies (5.2), as required.

Remark. A Möbius transformation is not characterized by the value of its determinant, since the determinant is multiplied by λ^2 when the coefficients change as described by (5.2). It can only be asserted that the determinant remains nonzero under any substitution (5.2).

---

[1] Here we allow p or q to take the improper value ∞.

The transformation

$$U(z) = z,$$

which obviously belongs to the set $\mathscr{M}$, is called the *unit transformation* (or the *identity transformation*). By the inverse of a given transformation

$$w = \frac{az + b}{cz + d}, \tag{5.4}$$

we mean the transformation which assigns to each w its inverse image z under the transformation (5.4). Thus the transformation

$$z = \frac{dw - b}{-cw + a}$$

(whose coefficients are unique only to within a multiplicative constant, as in Theorem 5.1) is the inverse of the transformation (5.4), We denote the inverse of the transformation L by L^{-1}.

Given two arbitrary Möbius transformations

$$L_1(z) = \frac{a_1 z + b_1}{c_1 z + d_1}, \qquad L_2(z) = \frac{a_2 z + b_2}{c_2 z + d_2},$$

we define the *product* of L_1 and L_2 as the result of first carrying out one transformation and then carrying out the other. There are two possible products corresponding to the two orders in which the transformations can be carried out. One product, written $L_1L_2(z)$, equals

$$\begin{aligned} L_1L_2(z) &= \frac{a_1[(a_2 z + b_2)/(c_2 z + d_2)] + b_1}{c_1[(a_2 z + b_2)/(c_2 z + d_2)] + d_1} \\ &= \frac{a_1(a_2 z + b_2) + b_1(c_2 z + d_2)}{c_1(a_2 z + b_2) + d_1(c_2 z + d_2)} \\ &= \frac{(a_1a_2 + b_1c_2)z + (a_1b_2 + b_1d_2)}{(c_1a_2 + d_1c_2)z + (c_1b_2 + d_1d_2)}, \end{aligned} \tag{5.5}$$

and the other, written $L_2L_1(z)$, is obtained from (5.5) by permuting the indices 1 and 2. Moreover, since

$$(a_1a_2 + b_1c_2)(c_1b_2 + d_1d_2) - (a_1b_2 + b_1d_2)(c_1a_2 + d_1c_2) \\ = (a_1d_1 - b_1c_1)(a_2d_2 - b_2c_2) \neq 0,$$

each of the transformations $L_1L_2(z)$ and $L_2L_1(z)$ belongs to the set $\mathscr{M}$. In general $L_1L_2(z) \neq L_2L_1(z)$, but obviously

$$LL^{-1}(z) = L^{-1}L(z) = U(z),$$

and

$$LU(z) = UL(z) = L(z)$$

for any $L(z) \in \mathcal{M}$.

Example. If

$$L_1(z) = \frac{z}{z+1}, \qquad L_2(z) = \frac{z+1}{z-1},$$

then

$$L_1L_2(z) = \frac{z+1}{2z}, \qquad L_2L_1(z) = -2z - 1.$$

Multiplication of transformations, as just defined, is an *associative* operation, i.e.,

$$(L_1L_2)L_3(z) = L_1(L_2L_3)(z). \tag{5.6}$$

To see this, we merely write $z_3 = L_3(z)$, and then both sides of (5.6) reduce at once to $L_1L_2(z_3)$. This associative property generalizes immediately to the product of an arbitrary number of transformations, and makes it unnecessary to use parentheses when writing products. For example, we have

$$L_1[L_2(L_3L_4)](z) = L_1L_2(L_3L_4)(z) = L_1(L_2L_3)L_4(z) = \cdots = L_1L_2L_3L_4(z).$$

Thus we have shown that the set $\mathcal{M}$ of Möbius transformations has the following properties:[2]

1. $\mathcal{M}$ is *closed under multiplication*, i.e., if $L_1 \in \mathcal{M}$, $L_2 \in \mathcal{M}$, then $L_1L_2 \in \mathcal{M}$, $L_2L_1 \in \mathcal{M}$.
2. Multiplication is associative.
3. There is an element $U \in \mathcal{M}$ such that $LU = UL = L$ for any $L \in \mathcal{M}$.
4. For each $L \in \mathcal{M}$, there is an element $L^{-1} \in \mathcal{M}$ such that $LL^{-1} = L^{-1}L = U$.

In algebraic language, these four properties are summarized by saying that $\mathcal{M}$ is a group of transformations.[3]

Problem. Prove that the Möbius transformations of the special form

$$L_1 = z, \qquad L_2 = \frac{1}{z}, \qquad L_3 = 1 - z, \tag{5.7}$$

$$L_4 = \frac{1}{1-z}, \qquad L_5 = \frac{z-1}{z}, \qquad L_6 = \frac{z}{z-1}$$

form a group.

---

[2] Henceforth, for simplicity, we shall often omit the argument z, writing L_1 instead of $L_1(z)$, U instead of $U(z)$, etc.

[3] See e.g., G. Birkhoff and S. MacLane, *op. cit.*, Chap. 6, Sec. 2.

Comment. This fact is summarized by saying that the transformations (5.7) are a *subgroup* of $\mathscr{M}$.[4]

24. THE CIRCLE-PRESERVING PROPERTY OF MÖBIUS TRANSFORMATIONS

We now prove that any Möbius transformation carries a straight line or a circle into another straight line or circle. We call this the *circle-preserving property*, since a straight line can be regarded as a limiting case of a circle (corresponding to infinite radius). The entire linear transformation $L(z) = \alpha z + \beta$ $(\alpha \neq 0)$ is obviously circle-preserving, since the mapping $w = L(z)$ is just a shift (if $\alpha = 1$), or a shift combined with a rotation and a uniform magnification (if $\alpha \neq 1$).

LEMMA. *The transformation*

$$w = \Lambda(z) = \frac{1}{z} \tag{5.8}$$

is circle-preserving.

Proof. The equation of any straight line or circle in the z-plane can be written in the form

$$A(x^2 + y^2) + 2Bx + 2Cy + D = 0, \tag{5.9}$$

where we have a straight line if $A = 0$ and at least one of the numbers B, C is nonzero, and a circle if $A \neq 0$ and $B^2 + C^2 - AD > 0$. Since

$$x^2 + y^2 = z\bar{z}, \quad 2x = z + \bar{z}, \quad 2y = -i(z - \bar{z}),$$

where $\bar{z} = x - iy$ is the complex conjugate of $z = x + iy$, we can rewrite (5.9) as

$$Az\bar{z} + \bar{E}z + E\bar{z} + D = 0, \tag{5.10}$$

where $E = B + iC$. It is easy to see that equation (5.10), where A and D are real and E is complex, is the equation of a straight line if and only if $A = 0$, $E \neq 0$, and the equation of a circle if and only if $A \neq 0$, $E\bar{E} - AD > 0$.

We now find the image of the curve with equation (5.10) under the transformation (5.8). Replacing z by $1/w$ in (5.10), we obtain

$$A\frac{1}{w\bar{w}} + \bar{E}\frac{1}{w} + E\frac{1}{\bar{w}} + D = 0$$

or

$$Dw\bar{w} + Ew + \bar{E}\bar{w} + A = 0. \tag{5.11}$$

[4] See e.g., G. Birkhoff and S. MacLane, *op. cit.*, Chap. 6, Sec. 7.

Equation (5.11) has the same form as equation (5.10), with D, $\bar{E}$ and A substituted for A, E and D, respectively. It follows that (5.11) is the equation of a straight line if $D = 0$, since then either $A = 0$ and $E \neq 0$ if (5.10) is the equation of a straight line, or else $A \neq 0$ and $E\bar{E} - AD = E\bar{E} > 0$ (so that $E \neq 0$ again) if (5.10) is the equation of a circle. Moreover, (5.11) is the equation of a circle if $D \neq 0$, since then either $A \neq 0$ and $E\bar{E} - AD > 0$ if (5.10) is the equation of a circle, or else $A = 0$ and $E \neq 0$ (so that $E\bar{E} - AD = E\bar{E} > 0$ again) if (5.10) is the equation of a straight line.

THEOREM 5.2. *Every Möbius transformation*

$$w = L(z) = \frac{az + b}{cz + d} \qquad (ad - bc \neq 0) \tag{5.12}$$

is circle-preserving.

Proof. If $c = 0$, (5.12) reduces to an entire linear transformation and hence is circle-preserving. If $c \neq 0$, (5.12) can be written in the form

$$w = \frac{a}{c} + \frac{bc - ad}{c(cz + d)}.$$

Setting

$$z_1 = L_1(z) = cz + d, \quad z_2 = \Lambda(z_1) = \frac{1}{z_1},$$

$$w = L_2(z_2) = \frac{a}{c} + \frac{bc - ad}{c} z_2,$$

we can write $L(z)$ as a product

$$L = L_2 \Lambda L_1$$

of three transformations which are all circle-preserving (use the lemma). It follows that L itself is circle-preserving.

COROLLARY. *Let $\delta = -d/c$ be the pole of the (rational) function (5.12). Then (5.12) transforms every straight line or circle which passes through δ into a straight line, and every other straight line or circle into a circle.*

Proof. If the circle or straight line passes through δ, its image under (5.12) contains the point at infinity, and hence must be a straight line, since it cannot be a circle (no circle contains ∞). Similarly, if the circle or straight line does not pass through δ, its image does not contain the point at infinity, and hence must be a circle, since it cannot be a straight line (every straight line contains ∞).

Remark. Let $w = L(z)$ be any Möbius transformation, let γ be a straight line or circle in the z-plane, and let $\Gamma = L(\gamma)$ be the image of γ in the w-plane

(Γ is itself a straight line or a circle). The two domains G_1 and G_2 with boundary γ are either two half-planes or the interior and exterior of a circle. Let $L(G_1)$ and $L(G_2)$ be the images of these two domains under the mapping $w = L(z)$. We now show that $L(G_1)$ and $L(G_2)$ are the two domains whose common boundary is the curve Γ.

First suppose $z_1 \in G_1$, $z_2 \in G_2$, and let $w_1 = L(z_1)$, $w_2 = L(z_2)$. Then $w_1 \notin \Gamma$, $w_2 \notin \Gamma$, since $z_1 \notin \gamma$, $z_2 \notin \gamma$, and hence w_1 and w_2 must belong to the union of the two (disjoint) domains into which Γ divides the extended w-plane. If w_1 and w_2 both belong to one of these two domains, we can join w_1 to w_2 by a line segment or circular arc Δ which does not intersect Γ (see Figure 5.1). The inverse image of Δ in the z-plane must be a line segment or circular

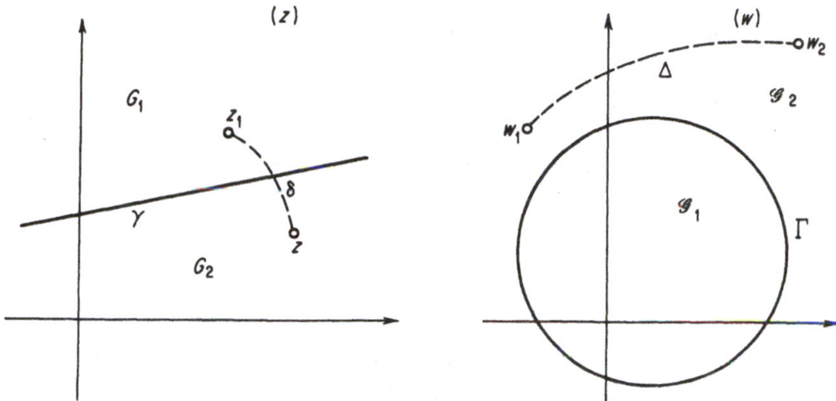

FIGURE 5.1

arc δ, which joins z_1 to z_2 and does not intersect γ. But the existence of δ contradicts the assumption that z_1 and z_2 belong to different domains G_1 and G_2. Therefore, if z_1 and z_2 belong to different domains with boundary γ, their images w_1 and w_2 must belong to different domains with boundary Γ.

We now denote the domains containing w_1 and w_2 by $\mathscr{G}_1$ and $\mathscr{G}_2$, respectively. If z is an arbitrary point in G_1, then, since z and z_2 belong to different domains G_1 and G_2, their images w and w_2 belong to different domains $\mathscr{G}_1$ and $\mathscr{G}_2$. But $w_2 \in \mathscr{G}_2$, and hence $w \in \mathscr{G}_1$, i.e., $L(z) \in \mathscr{G}_1$ if $z \in G_1$. Similarly, $L(z) \in \mathscr{G}_2$ if $z \in G_2$, and hence

$$\mathscr{G}_1 \supset L(G_1), \qquad \mathscr{G}_2 \supset L(G_2). \tag{5.13}$$

Conversely, let w be an arbitrary point in $\mathscr{G}_1$. Then w must be the image of a point z in G_1 or G_2. But $z \in G_2$ implies $w \in \mathscr{G}_2$, contrary to hypothesis, and hence $z \in G_1$, i.e., $\mathscr{G}_1 \subset L(G_1)$. Similarly, we find that $\mathscr{G}_2 \subset L(G_2)$. It follows by comparison with (5.13) that

$$\mathscr{G}_1 = L(G_1), \qquad \mathscr{G}_2 = L(G_2),$$

i.e., the two domains with boundary Γ are just the images of the two domains G_1 and G_2, as asserted. Moreover, to determine which of the two domains with boundary Γ is actually the image of a given domain G_1 with boundary γ, it is sufficient to locate the image w_1 of any point $z_1 \in G_1$, for then the domain $\mathscr{G}_1$ containing w_1 is the image of G_1.

Problem. Find the images of the following domains under the indicated Möbius transformations:

a) The quadrant $x > 0$, $y > 0$ if $w = \dfrac{z - i}{z + i}$;

b) The half-disk $|z| < 1$, $\operatorname{Im} z > 0$ if $w = \dfrac{2z - i}{2 + iz}$;

c) The sector $0 < \arg z < \dfrac{\pi}{4}$ if $w = \dfrac{z}{z - 1}$;

d) The strip $0 < x < 1$ if $w = \dfrac{z - 1}{z}$ or if $w = \dfrac{z - 1}{z - 2}$.

25. FIXED POINTS OF A MÖBIUS TRANSFORMATION. INVARIANCE OF THE CROSS RATIO

By a *fixed point* of a transformation or mapping $w = f(z)$, we mean a point which is carried into itself by the transformation. Obviously, every such point is a solution of the equation

$$z = f(z).$$

Moreover, every point of the z-plane is trivially a fixed point of the unit transformation $U(z) = z$.

THEOREM 5.3. *Every Möbius transformation different from the unit transformation has two fixed points, which in certain cases coalesce into a single fixed point.*

Proof. First let $c = 0$ $(d \neq 0)$, so that $L(z)$ reduces to the entire linear transformation

$$L(z) = \alpha z + \beta \qquad \left(\alpha = \frac{a}{d}, \quad \beta = \frac{b}{d}\right).$$

Then, since $L(\infty) = \infty$, one fixed point is the point at infinity. If $\alpha \neq 1$, there is another fixed point determined from the equation

$$z = \alpha z + \beta,$$

i.e., the point $\beta/(1-\alpha)$, but if $\alpha = 1$, $\beta \neq 0$, there is no finite fixed point. Moreover, if $\alpha \neq 1$, $\beta \neq 0$, the finite fixed point $\beta/(1-\alpha)$ approaches ∞ as $\alpha \to 1$. Therefore, in the case of the transformation

$$L(z) = z + \beta \qquad (\beta \neq 0),$$

the point at infinity can be regarded as two fixed points which have coalesced.

Now let $c \neq 0$, so that

$$L(\infty) = \frac{a}{c} \neq \infty,$$

i.e., the point at infinity is not a fixed point.[5] Similarly, the pole $\delta = -d/c$ of the transformation is not a fixed point, since

$$L(\delta) = \infty \neq \delta.$$

Assuming that $z \neq \infty$ and $z \neq \delta$, we solve the equation

$$z = \frac{az + b}{cz + d}$$

or

$$cz^2 - (a - d)z - b = 0,$$

obtaining

$$z = \frac{a - d \pm \sqrt{(a - d)^2 + 4bc}}{2c}.$$

If $(a - d)^2 + 4bc \neq 0$, we obtain two different finite fixed points; if $(a - d)^2 + 4bc = 0$, these two points coalesce to form a single finite fixed point $(a - d)/2c$.

Corollary. *The only Möbius transformation with more than two fixed points is the unit transformation $U(z) = z$, for which all points are fixed points.*

Theorem 5.4. *A sufficient condition for two Möbius transformations $L(z)$ and $\Lambda(z)$ to be identical is that the equation*

$$L(z) = \Lambda(z)$$

hold for three distinct points z_1, z_2 and *z_3. In particular, there cannot exist two distinct Möbius transformations taking three given values w_1, w_2, w_3 at three given distinct points z_1, z_2, z_3.*

---

[5] Therefore, in the class of Möbius transformations, an entire linear transformation is characterized by the fact that at least one of its fixed points is the point at infinity.

Proof. It follows from

$$L(z_k) = \Lambda(z_k) = w_k \qquad (k = 1, 2, 3)$$

that

$$\Lambda^{-1}(w_k) = z_k \qquad (k = 1, 2, 3),$$

and hence

$$\Lambda^{-1}L(z_k) = z_k \qquad (k = 1, 2, 3).$$

Therefore the transformation $\Lambda^{-1}L$ has three distinct fixed points, which, according to the corollary, implies

$$\Lambda^{-1}L = U, \tag{5.14}$$

where U is the unit transformation. Multiplying both sides of (5.14) by Λ from the left, we obtain

$$\Lambda(\Lambda^{-1}L) = \Lambda U,$$

which implies

$$L = \Lambda,$$

as asserted.

We now set about determining the (unique) Möbius transformation $w = L(z)$ carrying the points z_1, z_2, z_3 into the points w_1, w_2, w_3. First we consider the problem of finding the special Möbius transformation $w = \Lambda(z)$ carrying three finite points z_1, z_2, z_3 into the points $0, \infty, 1$. Since the function

$$w = \Lambda(z) = \frac{az + b}{cz + d}$$

vanishes for $z = z_1$ and becomes infinite for $z = z_2$ if and only if z_1 is a zero of the numerator and z_2 is a zero of the denominator, it follows that

$$w = \Lambda(z) = \frac{a}{c}\frac{z - z_1}{z - z_2}.$$

But w must equal 1 for $z = z_3$, i.e.,

$$1 = \frac{a}{c}\frac{z_3 - z_1}{z_3 - z_2},$$

which implies[6]

$$\frac{a}{c} = 1 : \frac{z_3 - z_1}{z_3 - z_2}.$$

Therefore the Möbius transformation carrying z_1, z_2, z_3 into $0, \infty, 1$ is

$$w = \Lambda(z) = \frac{z - z_1}{z - z_2} : \frac{z_3 - z_1}{z_3 - z_2}. \tag{5.15}$$

Next we determine the more general Möbius transformation $w = L(z)$ satisfying

$$L(z_1) = w_1, \quad L(z_2) = w_2, \quad L(z_3) = w_3,$$

[6] By $x:y$ is meant the *ratio of x to y*, i.e., the quantity x/y.

where w_1, w_2, w_3 are three arbitrary (but distinct) finite points. As we have just seen, the transformation

$$\Lambda_1(z) = \frac{z - w_1}{z - w_2} : \frac{w_3 - w_1}{w_3 - w_2} \tag{5.16}$$

carries the points w_1, w_2, w_3 into the points $0, \infty, 1$. Therefore the transformation $\Lambda_1 L$ carries the points z_1, z_2, z_3 into the points $0, \infty, 1$, so that

$$\Lambda_1 L(z) = \Lambda(z) = \frac{z - z_1}{z - z_2} : \frac{z_3 - z_1}{z_3 - z_2}. \tag{5.17}$$

Multiplying both sides of the equation

$$\Lambda_1 L(z) = \Lambda(z) \tag{5.18}$$

by Λ_1^{-1} from the left, we obtain

$$L(z) = \Lambda_1^{-1}\Lambda(z).$$

This solves the problem, since the functions $\Lambda(z)$ and $\Lambda_1(z)$, and hence $\Lambda_1^{-1}(z)$, are known [cf. (5.17) and (5.16)]. However, it is more convenient to use (5.18) directly, after writing $w = L(z)$. The result is

$$\Lambda_1(w) = \Lambda(z)$$

or

$$\frac{w - w_1}{w - w_2} : \frac{w_3 - w_1}{w_3 - w_2} = \frac{z - z_1}{z - z_2} : \frac{z_3 - z_1}{z_3 - z_2}, \tag{5.19}$$

which expresses the Möbius transformation $w = L(z)$ in implicit form.

Remark. In finding the Möbius transformation carrying the points z_1, z_2, z_3 into the points w_1, w_2, w_3, it was assumed that all six points are finite. However, the case of infinite points is easily handled. For example, the transformation carrying the points ∞, z_2, z_3 into the points $0, \infty, 1$ has the form

$$w = \Lambda(z) = \frac{1}{z - z_2} : \frac{1}{z_3 - z_2},$$

which can be found by inspection or by writing (5.15) in the form

$$w = \frac{\dfrac{z}{z_1} - 1}{z - z_2} : \frac{\dfrac{z_3}{z_1} - 1}{z_3 - z_2}$$

and then taking the limit as $z_1 \to \infty$. Therefore (5.19) is replaced by

$$\frac{w - w_1}{w - w_2} : \frac{w_3 - w_1}{w_3 - w_2} = \frac{1}{z - z_2} : \frac{1}{z_3 - z_2},$$

where it is assumed that the points w_1, w_2, w_3 are finite. Similarly, the transformation carrying the points z_1, ∞, z_3 into the points 0, ∞, 1 has the form

$$\Lambda(z) = (z - z_1):(z_3 - z_1),$$

and hence we have

$$\frac{w - w_1}{w - w_2} : \frac{w_3 - w_1}{w_3 - w_2} = (z - z_1) : (z_3 - z_1),$$

instead of (5.19). Finally, the transformation carrying the points z_1, z_2, ∞ into the points 0, ∞, 1 has the form

$$\Lambda(z) = \frac{z - z_1}{z - z_2},$$

and hence we have

$$\frac{w - w_1}{w - w_2} : \frac{w_3 - w_1}{w_3 - w_2} = \frac{z - z_1}{z - z_2},$$

instead of (5.19).

In just the same way, we have to replace the left-hand side of (5.19) by

$$\frac{1}{w - w_2} : \frac{1}{w_3 - w_2}, \quad (w - w_1) : (w_3 - w_1) \quad \text{or} \quad \frac{w - w_1}{w - w_2},$$

depending on whether $w_1 = \infty$, $w_2 = \infty$ or $w_3 = \infty$. As a result, we arrive at the following mnemonic rule: If $z_k = \infty$ or $w_l = \infty$ $(k, l = 1, 2, 3)$, the differences involving z_k or w_l have to be replaced by 1. The reader can easily verify this rule by taking the appropriate limits (as $z_k \to \infty$ or $w_l \to \infty$) in equation (5.19).

Equation (5.19) implies an important general property of Möbius transformations. If a, b, c and d are arbitrary distinct finite complex numbers, then the ratio

$$\frac{c - a}{c - b} : \frac{d - a}{d - b},$$

denoted by (a, b, c, d), is called the *cross ratio* (or *anharmonic ratio*) of the four numbers (or points) a, b, c and d. If one of the four points a, b, c, d is the point at infinity, we define the cross ratio as the limit of the cross ratio of four finite points, three of which coincide with the three given finite points, as the fourth point approaches infinity. Thus, according to this definition, we have

$$(\infty, b, c, d) = \frac{1}{c - b} : \frac{1}{d - b},$$

$$(a, \infty, c, d) = (c - a):(d - a),$$

$$(a, b, \infty, d) = 1 : \frac{d - a}{d - b},$$

$$(a, b, c, \infty) = \frac{c - a}{c - b}.$$

Now let $w = L(z)$ be an arbitrary Möbius transformation, and let A, B, C, D be the points into which $L(z)$ maps four arbitrary (but distinct) points a, b, c, d. Since the points A, B and D are the images of the points a, b and d, it follows from (5.19) that the relation between z and $w = L(z)$ is given by

$$\frac{w - A}{w - B} : \frac{D - A}{D - B} = \frac{z - a}{z - b} : \frac{d - a}{d - b},$$

where differences involving the point at infinity have to be replaced by 1. Moreover, since the point C is the image of the point c, we have

$$\frac{C - A}{C - B} : \frac{D - A}{D - B} = \frac{c - a}{c - b} : \frac{d - a}{d - b},$$

(where again differences involving the point at infinity have to be replaced by 1), or equivalently

$$(A, B, C, D) = (a, b, c, d).$$

In other words, *the cross ratio of any four distinct points is invariant under a Möbius transformation.*

Problem 1. Prove that every Möbius transformation $w = L(z)$ with a single finite fixed point z_0 satisfies a relation of the canonical form

$$\frac{1}{w - z_0} = \frac{1}{z - z_0} + h \qquad (h \neq 0).$$

Problem 2. Prove that every Möbius transformation $w = L(z)$ with two distinct finite fixed points z_1 and z_2 satisfies a relation of the canonical form

$$\frac{w - z_1}{w - z_2} = k\,\frac{z - z_1}{z - z_2}.$$

Problem 3. Suppose the Möbius transformation $L(z)$ has two distinct finite fixed points z_1 and z_2. Prove that

$$L'(z_1)L'(z_2) = 1.$$

Hint. Use the preceding problem.

Problem 4. Find the Möbius transformation which carries the points -1, i, $1 + i$ into the points

a) $0, 2i, 1 - i$; b) $i, \infty, 1$.

Ans. a) $w = -\dfrac{2i(z + 1)}{4z - 1 - 5i}$.

Problem 5. Find the Möbius transformation which carries the points -1, ∞, i into the points

a) $i, 1, 1 + i$; b) $\infty, i, 1$; c) $0, \infty, 1$.

Ans. b) $w = \dfrac{iz + 2 + i}{z + 1}$.

Problem 6. Find the Möbius transformation such that
a) The points 1 and i are fixed, but the point 0 goes into the point -1;
b) The points $\frac{1}{2}$ and 2 are fixed, but the point $\frac{5}{4} + \frac{3}{4}i$ goes into ∞;
c) The point i is the only fixed point and the point 1 goes into ∞.

Ans. b) $w = \dfrac{(1 - 4i)z - 2(1 - i)}{2(1 - i)z - (4 - i)}$.

26. MAPPING OF A CIRCLE ONTO A CIRCLE[7]

Using the circle-preserving property of Möbius transformations and the possibility of mapping any given triple of distinct points z_1, z_2, z_3 into any other given triple of distinct points w_1, w_2, w_3, we obtain the following basic

THEOREM 5.5. *Let γ and Γ be any two straight lines or circles, and let z_1, z_2, z_3 and w_1, w_2, w_3 be any two triples of distinct points belonging to γ and Γ, respectively. Then there exists a Möbius transformation $w = L(z)$ mapping γ onto Γ in such a way that*

$$L(z_k) = w_k \qquad (k = 1, 2, 3). \tag{5.20}$$

Proof. Construct the Möbius transformation $w = L(z)$ satisfying the conditions (5.20), which according to Theorem 5.4 and the subsequent construction, exists and is unique. According to Theorem 5.2, $w = L(z)$ maps the straight line or circle γ onto another straight line or circle Γ^*. But since γ goes through the points z_1, z_2 and z_3, Γ^* must go through the points w_1, w_2 and w_3. Moreover, since two different straight lines or circles cannot be drawn through the same three points, Γ^* must coincide with Γ, as asserted.

Remark. Again consider two arbitrary straight lines or circles γ and Γ (which may coincide). Let G be one of the two domains with boundary γ, and let $\mathscr{G}$ be one of the two domains with boundary Γ, so that G is either a half-plane, the interior of a circle or the exterior of a circle, and the same is true of $\mathscr{G}$. We now show how to map G onto $\mathscr{G}$. Choose an arbitrary triple of distinct points z_1, z_2, z_3 on γ, and suppose an observer moving along γ in the direction from z_1 to z_3 through z_2 finds the domain G on his *left*, say. Next choose a triple of distinct points w_1, w_2, w_3 on Γ such that an observer moving along Γ in the direction from w_1 to w_3 through w_2 finds the domain $\mathscr{G}$ on his left, but let w_1, w_2, w_3 be otherwise arbitrary. As in Theorem 5.5, we form the Möbius transformation $w = L(z)$ which satisfies the conditions (5.20) and hence maps γ onto Γ. Then $w = L(z)$ also maps G onto $\mathscr{G}$, i.e., $\mathscr{G} = L(G)$. In fact, if δ is a segment of the normal to the curve γ

[7] As usual, a straight line is regarded as a limiting case of a circle (cf. p. 93).

drawn from the point z_2 and pointing into the interior of G, so that an observer at z_2 facing in the direction established on γ finds δ on his left, then, since the mapping $w = L(z)$ is conformal, an observer at w_2 facing in the direction established on Γ will also find the image $\Delta = L(\delta)$, which is a line segment or circular arc, on his left (see Figure 5.2).[8] Therefore $\Delta \cap \mathscr{G} \neq 0$

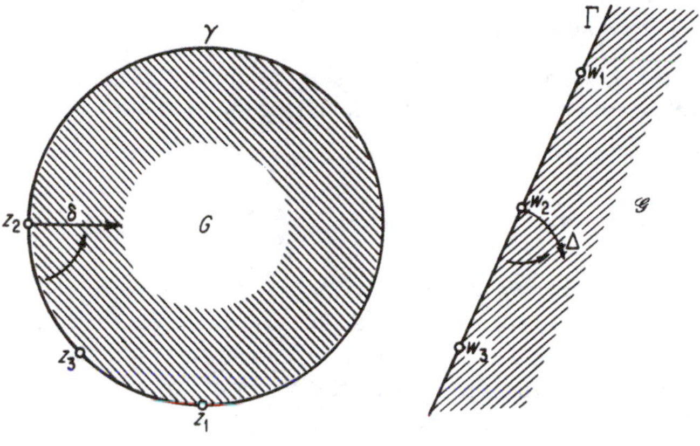

FIGURE 5.2

and hence $\mathscr{G}$ contains images of certain points belonging to G (i.e., points of the segment δ). But, according to the remark on pp. 94–96, $L(G)$ is one of the two domains with boundary $\Gamma = L(\gamma)$, in fact just the domain containing the image of any point in G. In other words $\mathscr{G} = L(G)$, as asserted.

Example. *Find a conformal mapping of the upper half-plane* Im $z > 0$ *onto the interior of the unit circle.*

To solve this problem, we choose $z_1 = -1$, $z_2 = 0$, $z_3 = 1$, say, so that the upper half-plane is on the left of an observer moving along the real axis in the direction from z_1 to z_3 through z_2. We also choose three points w_1, w_2, w_3 on the unit circle, such that the interior of the circle is on the left of an observer moving along the circle in the direction from w_1 to w_3 through w_2. For simplicity, let $w_1 = 1$, $w_2 = i$, $w_3 = -1$. Then the desired Möbius transformation satisfies the conditions $L(z_k) = w_k$, $k = 1, 2, 3$ and can be

---

[8] Note that the assertion that the observer at z_2 finds G on his left is equivalent to the assertion that in order to enter G along the segment δ at the point z_2, the observer must make a "left turn," i.e., a counterclockwise rotation through 90°. But $w = L(z)$ has a nonzero derivative at z_2, and hence is a conformal mapping of the first kind which preserves not only angles but also the directions in which they are measured (recall p. 57). Therefore, in order to enter $\mathscr{G}$ along the arc Δ at the point w_2, the observer at w_2 must also make a left turn and hence finds $\mathscr{G}$ on his left.

represented in the form

$$\frac{w-1}{w-i}:\frac{-1-1}{-1-i}=\frac{z+1}{z}:\frac{1+1}{1}$$

or

$$w=\frac{z-i}{iz-1},$$

where we have used (5.19).

Problem. Find the Möbius transformation $w = L(z)$ mapping the upper half-plane $\operatorname{Im} z > 0$ onto itself and satisfying the conditions $w(0) = 1$, $w(1) = 2$, $w(2) = \infty$.

27. SYMMETRY TRANSFORMATIONS

Let z_1 and z_2 be two points which are symmetric with respect to a given straight line γ, i.e., such that γ is the perpendicular bisector of the line segment joining z_1 and z_2. By definition, the straight line passing through z_1 and z_2 is orthogonal to γ. Moreover, the center of any circle δ passing through z_1 and z_2 lies on γ, and hence δ is also orthogonal to γ. It is easy to see that the converse is true as well, i.e., if every straight line or circle passing through a pair of points z_1 and z_2 is orthogonal to a given straight line γ, then z_1 and z_2 are symmetric with respect to γ. Generalizing the concept of symmetry with respect to a straight line, we introduce the following definition: *Two points z_1 and z_2 are symmetric with respect to a given circle γ if and only if every straight line or circle passing through z_1 and z_2 is orthogonal to γ.*

THEOREM 5.6. *Let z_1 and z_2 be any two points symmetric with respect to a given straight line or circle γ, and let $w = L(z)$ be any Möbius transformation. Then the points $w_1 = L(z_1)$ and $w_2 = L(z_2)$ are symmetric with respect to the straight line or circle $\Gamma = L(\gamma)$.*[9]

Proof. We have to show that an arbitrary straight line or circle Δ passing through w_1 and w_2 is orthogonal to Γ. Let $z = L^{-1}(w)$ be the inverse of the transformation $w = L(z)$. Clearly, L^{-1} is also a Möbius transformation, and

$$L^{-1}(w_1) = z_1, \quad L^{-1}(w_2) = z_2, \quad L^{-1}(\Gamma) = \gamma.$$

Moreover, $\delta = L^{-1}(\Delta)$ is a straight line or circle passing through z_1 and z_2. Since z_1 and z_2 are symmetric with respect to γ, by hypothesis, it follows that δ is orthogonal to γ. But then, since the mapping $w = L(z)$ is conformal, $\Delta = L(\delta)$ is orthogonal to Γ, and the proof is complete.

[9] It is in this sense that Möbius transformations are said to be *symmetry-preserving*.

COROLLARY. *There is only one point z_2 symmetric to a given point z_1 with respect to a given straight line or circle γ.*

Proof. If γ is a straight line, the statement is obvious. Thus let γ be a circle, and suppose that besides z_2, there is another point $z_2' \neq z_2$ symmetric to z_1 with respect to γ. Choosing a Möbius transformation $w = L(z)$ mapping γ onto a straight line Γ, we find that $w_2 = L(z_2)$ and $w_2' = L(z_2')$ are two distinct points symmetric with respect to Γ, which is impossible.

Remark. Suppose $w = L(z)$ maps a straight line or circle γ onto a circle Γ with center w_1, and let z_1 be the inverse image of w_1. Then the point z_2 symmetric to z_1 with respect to γ must be mapped into the point at infinity. To see this, we note that $w_2 = \infty$ is symmetric to w_1 with respect to the circle Γ, since any straight line or circle passing through 0 and ∞, i.e., any straight line passing through the center of Γ, is orthogonal to Γ. The uniqueness of w_2 follows from the corollary.

Let γ be an arbitrary straight line or circle. A transformation of the extended plane into itself, which carries each point z into the point z^* symmetric to z with respect to γ is called a *symmetry transformation* with respect to γ or a *reflection* in γ. In the case where γ is a circle, the transformation is also called an *inversion* in γ. We now derive analytical expressions for symmetry transformations.

First let γ be a straight line with an assigned direction, and consider reflection in γ. The straight line γ is completely characterized by one of its points a and by the unit vector

$$\tau = \cos\theta + i\sin\theta \qquad (\bar{\tau} = \tau^{-1})$$

pointing in the direction of γ. Suppose we carry out the entire linear transformation

$$z = L(w) = a + \tau w, \tag{5.21}$$

which obviously maps the real axis onto γ, since (5.21) corresponds to a shift by the vector a (carrying the origin of coordinates into the point a), followed by a rotation through the angle θ about the point a. Since the inverse transformation $w = L^{-1}(z)$ maps γ onto the real axis, it maps every pair of points z and z^* symmetric with respect to γ into a pair of points w and w^* symmetric with respect to the real axis. But the points w and w^* are represented by two conjugate complex numbers, i.e.,

$$w = t, \qquad w^* = \bar{t}. \tag{5.22}$$

Therefore $z - a = \tau t$, and

$$\overline{z - a} = \bar{\tau}\bar{t}, \qquad z^* - a = \tau w^* = \tau\bar{t}. \tag{5.23}$$

Eliminating $\bar{t}$ from (5.23), we obtain

$$z^* - a = \tau^2(\overline{z - a}). \tag{5.24}$$

According to (5.24), reflection in a straight line γ going through a point a at an angle θ with the real axis can be accomplished by first constructing the vector $\overline{z - a}$ which is the reflection of the vector $z - a$ in the real axis, and then rotating $\overline{z - a}$ through the angle 2θ about the point a.

Next let γ be a circle, and consider inversion in γ. Let R $(0 < R < \infty)$ be the radius and a the center of γ. We begin by finding a Möbius transformation which maps γ onto the real axis. The simplest approach is to choose the transformation

$$z = L(w) = a + R\,\frac{1 + iw}{1 - iw},$$

which maps the three points $w_1 = -1$, $w_2 = 0$, $w_3 = 1$ of the real axis into the points $z_1 = a - iR$, $z_2 = a + R$, $z_3 = a + iR$ of the circle γ. The inverse transformation $w = L^{-1}(z)$ maps γ onto the real axis, and maps every pair of points z and z^* symmetric with respect to γ into a pair of points w and w^* symmetric with respect to the real axis. As before, the points w and w^* are represented by two conjugate complex numbers (5.22). Therefore

$$z - a = R\,\frac{1 + it}{1 - it}.$$

and

$$\overline{z - a} = R\,\frac{1 - i\bar{t}}{1 + i\bar{t}}, \qquad z^* - a = R\,\frac{1 + i\bar{t}}{1 - i\bar{t}}. \tag{5.25}$$

Multiplying the two equations (5.25), we obtain

$$(\overline{z - a})(z^* - a) = R^2$$

or

$$z^* - a = \frac{R^2}{\overline{z - a}}. \tag{5.26}$$

In particular, it follows from (5.26) that

$$\operatorname{Arg}(z^* - a) = -\operatorname{Arg}(\overline{z - a}) = \operatorname{Arg}(z - a)$$

and

$$|z - a|\,|z^* - a| = R^2.$$

Therefore the points z and z^* lie on the same ray emanating from the center of γ, and the product of their distances from the center of γ equals the square of the radius of γ. These two conditions, equivalent to formula (5.26), determine the position of one of the points z, z^* with respect to the other, and completely characterize the operation of inversion in the circle γ with equation $|z - a| = R$.

Remark. We note that any symmetry transformation reduces to consecutive application of a linear transformation (entire or fractional), followed by reflection in the real axis. In fact, according to (5.24), reflection in a straight line can be represented in the form

$$z_1 = \bar{a} + \bar{\tau}^2(z - a), \qquad z^* = \bar{z}_1,$$

while, according to (5.26), reflection in a circle can be represented in the form

$$z_1 = \bar{a} + \frac{R^2}{z - a}, \qquad z^* = \bar{z}_1.$$

Since any Möbius transformation is conformal and circle-preserving, and since reflection in the real axis has the same properties, except that while preserving the magnitudes of angles it reverses the directions in which they are measured, we see that the most general symmetry transformation is a conformal mapping of the second kind (see p. 57) which is also circle-preserving.

Problem 1. Find the point symmetric to the point $2 + i$ with respect to a) The circle $|z| = 1$; b) The circle $|z - i| = 3$.

Problem 2. Invert each of the following curves in the unit circle:

a) $|z| = \frac{1}{2}$; b) $|z - 1| = 1$; c) $y = 2$; d) $|z - z_0| = |z_0|$;

e) $|z - z_0| = \sqrt{|z_0|^2 - 1}$ $(|z_0| > 1)$; f) $x^2 - y^2 = 1$;

g) The boundary of the triangle with nonzero vertices z_1, z_2, z_3.

28. EXAMPLES

We now give two examples illustrating the use of the symmetry-preserving property of Möbius transformations.

Example 1. *Find a conformal mapping of the upper half-plane* $\Pi_U : \operatorname{Im} z > 0$ *onto the disk* $K : |w| < R$, *such that a given point* $\alpha \in \Pi_U$ *is mapped into the center of* K.[10]

Any such mapping

$$w = L(z) = \frac{az + b}{cz + d},$$

provided it exists, vanishes for $z = \alpha$, and hence α is the zero of $L(z)$. But the point $\bar{\alpha}$ symmetric to α with respect to the real axis must be mapped into

[10] In expressions like $\Pi_U : \operatorname{Im} z > 0$ and $K : |w| < R$, the colon means "defined by the condition."

the point symmetric to the center $w = 0$ of K with respect to the boundary of K, i.e., the circle $C : |w| = R$. Therefore $\bar{\alpha}$ must be mapped into the point at infinity (see the remark on p. 105), so that $L(\bar{\alpha}) = \infty$ and $\bar{\alpha}$ is the pole of $L(z)$. It follows that $L(z)$ has the form

$$w = L(z) = \frac{a(z - \alpha)}{c(z - \bar{\alpha})} = \lambda \frac{z - \alpha}{z - \bar{\alpha}}, \tag{5.27}$$

where λ is a nonzero complex number.

We now show that (5.27) maps Π_U onto K, with $z = \alpha$ going into $w = 0$, if we choose $|\lambda| = R$. Since $L(\alpha) = 0$ for any λ, by construction, it is sufficient to show that (5.27) maps the real axis onto the circle C. If $z = x$ is an arbitrary real number, then $x - \alpha$ and $x - \bar{\alpha}$ are complex conjugates, and hence

$$|w| = |L(x)| = \left|\lambda \frac{x - \alpha}{x - \bar{\alpha}}\right| = |\lambda| \left|\frac{x - \alpha}{x - \bar{\alpha}}\right| = |\lambda| = R.$$

Therefore (5.27) maps the real axis *into* C. But since any three distinct points of the real axis are mapped into three distinct points of C, it follows from Theorem 5.5 that the real axis is mapped *onto* C.

In (5.27) the argument of λ is left unspecified. The geometric reason for this indeterminacy is clear: Going from one value of λ to another in (5.27), while keeping $|\lambda| = R$ fixed, is equivalent to changing the arguments of all points w by the same quantity, i.e., to rotating the disk K about its center $w = 0$. Such a rotation transforms K into itself while leaving its center fixed, and hence does not violate the conditions of the problem. Thus, if the problem is to have a unique solution, we must impose an extra condition on $L(z)$. For example, we might require either that

1. *A given point $x = x_0$ of the real axis should go into the point $w = R$ of the circle C*, or that

2. *The derivative $L'(\alpha)$ should be a positive real number.* (Geometrically, this means that the mapping does not change the slopes of tangents to curves passing through the point α.)

Imposing condition 1, we find from (5.27) that

$$R = L(x_0) = \lambda \frac{x_0 - \alpha}{x_0 - \bar{\alpha}},$$

so that

$$\lambda = R \frac{x_0 - \bar{\alpha}}{x_0 - \alpha},$$

and hence

$$L(z) = R \frac{x_0 - \bar{\alpha}}{x_0 - \alpha} \frac{z - \alpha}{z - \bar{\alpha}}.$$

Moreover, we still have

$$|\lambda| = R\left|\frac{x_0 - \bar{\alpha}}{x_0 - \alpha}\right| = R,$$

as required. Imposing condition 2, with $\alpha = \xi + i\eta$ $(\eta > 0)$, we have

$$L'(\alpha) = \frac{\lambda}{\alpha - \bar{\alpha}} = \frac{\lambda}{2i\eta},$$

so that λ/i is a positive real number. But, on the other hand, $|\lambda|$ must equal R. Therefore $\lambda = iR$, and

$$L(z) = iR\,\frac{z - \alpha}{z - \bar{\alpha}}.$$

Example 2. *Find a conformal mapping of the disk $K : |z| < R$ onto itself such that a given point $\alpha \in K$ is mapped into the center of K.*

Any such mapping

$$w = L(z) = \frac{az + b}{cz + d},$$

provided it exists, vanishes for $z = \alpha$, and hence α is the zero of $L(z)$. But the point α^* symmetric to α with respect to the boundary of K, i.e., the circle $C : |z| = R$, must be mapped into a point symmetric to the center $w = 0$ of K with respect to C. Therefore α^* must be mapped into the point at infinity, so that $L(\alpha^*) = \infty$ and α^* is the pole of $L(z)$. It follows that $L(z)$ has the form

$$w = L(z) = \lambda\,\frac{z - \alpha}{z - \alpha^*} \qquad (5.28)$$

[cf. (5.27)], where λ is a nonzero complex number. According to (5.26), the point α^* symmetric to α with respect to C is

$$\alpha^* = \frac{R^2}{\bar{\alpha}},$$

and hence (5.28) becomes

$$w = L(z) = -\lambda\bar{\alpha}\,\frac{z - \alpha}{R^2 - \bar{\alpha}z} = \mu\,\frac{z - \alpha}{R^2 - \bar{\alpha}z}. \qquad (5.29)$$

We now show that (5.29) maps K onto K, with $z = \alpha$ going into $w = 0$, if we choose $|\mu| = R^2$. Since $L(\alpha) = 0$ for any μ, by construction, it is sufficient to show that (5.29) maps the circle $C : |z| = R$ onto itself. If

$$z = R\tau$$

is an arbitrary point of C, where

$$\tau = \cos\theta + i\sin\theta \qquad (0 \leqslant \theta < 2\pi),$$

then

$$w = L(R\tau) = \mu\frac{R\tau - \alpha}{R^2 - \bar{\alpha}R\tau} = \frac{\mu}{R\tau}\frac{R\tau - \alpha}{R\bar{\tau} - \bar{\alpha}},$$

and hence

$$|w| = |L(R\tau)| = \left|\frac{\mu}{R\tau}\frac{R\tau - \alpha}{R\bar{\tau} - \bar{\alpha}}\right| = \frac{|\mu|}{R} = \frac{R^2}{R} = R.$$

Therefore (5.29) maps C *into* C, and hence *onto* C, by the same argument as before. This example is considered further in Probs. 4–6.

Problem 1. Map the upper half-plane $\operatorname{Im} z > 0$ onto the unit disk $|w| < 1$ in such a way that

a) $w(i) = 0$, $\arg w'(i) = -\pi/2$;
b) $w(2i) = 0$, $\arg w'(2i) = 0$;
c) $w(\alpha) = 0$, $\arg w'(\alpha) = \theta$ ($\operatorname{Im}\alpha > 0$).

Ans. b) $w = i\dfrac{z - 2i}{z + 2i}$.

Problem 2. Map the disk $|z| < 2$ onto the half-plane $\operatorname{Re} w > 0$ in such a way that $w(0) = 1$, $\arg w'(0) = \pi/2$.

Problem 3. Map the disk $|z - 4i| < 2$ onto the half-plane $v > u$ in such a way that the center of the disk goes into the point -4, while the point $2i$ on the circumference of the disk goes into the origin.

Problem 4. In formula (5.29), the argument of μ is left unspecified. If the problem is to have a unique solution, we must impose an extra condition on $L(z)$. Find the form of the mapping (5.29) if

a) A given point a on the circle $C: |z| = R$ goes into the point $w = R$ on C;
b) The derivative $L'(\alpha)$ is a positive real number.

Comment. Cf. the corresponding conditions on p. 108 for the mapping (5.27).

Problem 5. Find the Möbius transformation $w = L(z)$ mapping the unit disk $|z| < 1$ onto itself in such a way that

a) $w(\frac{1}{2}) = 0$, $\arg w'(\frac{1}{2}) = 0$; b) $w(\frac{1}{2}i) = 0$, $\arg w'(\frac{1}{2}i) = \pi/2$;
c) $w(0) = 0$, $\arg w'(0) = -\pi/2$; d) $w(\alpha) = \alpha$, $\arg w'(\alpha) = \theta$ ($|\alpha| < 1$).

Problem 6. Find the Möbius transformation mapping the unit disk $|z| < 1$ onto itself in such a way that two given points z_1 and z_2 of the disk go into the points $\pm\alpha$ ($|\alpha| < 1$). Find the expression for α in terms of z_1 and z_2.

CHAPTER 6

EXPONENTIALS AND LOGARITHMS

29. THE EXPONENTIAL

An entire function which is not a polynomial is called an *entire transcendental function.* The simplest example of such a function is the *exponential (function)* e^z or $\exp z$, obtained by suitably extending the familiar function e^x, defined for the real variable x, to the case where x takes arbitrary complex values, i.e., is replaced by the complex variable $z = x + iy$. It is not hard to show that the real exponential $f(x) = e^x$ is the unique function with the following properties:

1. $f(x)$ is defined and single-valued for all real x, takes only real values, and in particular takes the value e when $x = 1$;
2. $f(x)$ satisfies the *addition theorem*

$$f(x_1 + x_2) = f(x_1)f(x_2);$$

3. $f(x)$ is continuous for all x.[1]

The complex function $f(z) = e^z$ can be characterized in much the same way:

[1] This result follows at once by a specialization of the proof of Theorem 6.1.

THEOREM 6.1. *There is a unique function $f(z) = e^z$ with the following properties:*

1′. *$f(z)$ is defined and single-valued for all (finite) complex z, takes real values when z is real, and in particular takes the value e when $z = 1$;*

2′. *$f(z)$ satisfies the addition theorem*

$$f(z_1 + z_2) = f(z_1)f(z_2); \tag{6.1}$$

3′. *$f(z)$ is differentiable for all z, i.e., $f(z)$ is entire.*[2]

Proof. Since

$$f(z)f(1 - z) = f(1) = e \neq 0,$$

the function $f(z)$ cannot vanish for any z. Therefore $|f(z)| > 0$ and $\operatorname{Arg} f(z)$ are defined for all z (the latter function is defined only to within an arbitrary integral multiple of 2π). It follows from (6.1) that

$$|f(z_1 + z_2)| = |f(z_1)|\,|f(z_2)|$$

and

$$\operatorname{Arg} f(z_1 + z_2) = \operatorname{Arg} f(z_1) + \operatorname{Arg} f(z_2),$$

and hence the real functions of a complex variable $\ln |f(z)|$ and $\operatorname{Arg} f(z)$ both satisfy the functional equation

$$F(z_1 + z_2) = F(z_1) + F(z_2). \tag{6.2}$$

When $F(z)$ is the multiple-valued function $\operatorname{Arg} f(z)$, equation (6.2) means that given any value of $F(z_1 + z_2)$, values of $F(z_1)$ and $F(z_2)$ can be found whose sum is $F(z_1 + z_2)$, and conversely, given any values of $F(z_1)$ and $F(z_2)$, their sum is one of the values of $F(z_1 + z_2)$.

It follows from property 3′ that $f(z)$ is continuous, and therefore $\ln |f(z)|$ and $\operatorname{Arg} f(z)$ are also continuous. As applied to $\operatorname{Arg} f(z)$, this means that given any value of $\operatorname{Arg} f(z)$ and any sequence $\{z_n\}$ converging to z, there is a sequence of values of $\operatorname{Arg} f(z_n)$ such that $\operatorname{Arg} f(z_n) \to \operatorname{Arg} f(z)$ as $n \to \infty$ (cf. footnote 4, p. 55). Thus $\ln |f(z)|$ and $\operatorname{Arg} f(z)$ are to be found among the continuous real solutions of (6.2). To find these solutions, we need only solve (6.2) for the case where the independent variable is restricted to real values. In fact,

$$F(x + iy) = F(x) + F(iy), \tag{6.3}$$

where $F(x)$ and $F(iy)$ are functions of the real variables x and y, respectively, satisfying the equations

$$F(x_1 + x_2) = F(x_1) + F(x_2),$$
$$F[i(y_1 + y_2)] = F(iy_1) + F(iy_2).$$

[2] It will be apparent in the course of the proof that the assumption that $f(z)$ is merely *continuous* for all z is not sufficient to determine $f(z)$ completely (unlike the real case).

Therefore, once we know $F(x)$ and $F(iy)$ separately, we can immediately write down $F(x + iy)$ by forming the sum (6.3).

First suppose $F(x)$ is a single-valued continuous real function, and let $F(1) = a$. Then

$$\underbrace{F\left(\frac{1}{n}\right) + \cdots + F\left(\frac{1}{n}\right)}_{n \text{ times}} = F(1) = a,$$

which implies

$$F\left(\frac{1}{n}\right) = \frac{a}{n}$$

and

$$\underbrace{F\left(\frac{1}{n}\right) + \cdots + F\left(\frac{1}{n}\right)}_{m \text{ times}} = F\left(\frac{m}{n}\right) = \frac{m}{n}a.$$

Moreover

$$F(0) = 0,$$

since

$$F(x + 0) = F(x) + F(0) = F(x).$$

It follows that

$$F(x) + F(-x) = F(0) = 0,$$

and hence

$$F\left(-\frac{m}{n}\right) = -\frac{m}{n}a.$$

Therefore $F(x) = ax$ for all rational x. If $x = \xi$ is irrational, and if $\{x_n\}$ is a sequence of rational numbers converging to ξ, we find that

$$F(\xi) = \lim_{n\to\infty} F(x_n) = a\xi,$$

since $F(x)$ is continuous. Thus we see that $F(x) = ax$ for all real x. Similarly, we have $F(iy) = by$, so that

$$F(z) = F(x) + F(iy) = ax + by.$$

Moreover, since F is real, so are the constants a and b.

Next suppose $F(x)$ is a multiple-valued continuous real function, all of whose values are obtained from one value by adding arbitrary integral multiples of 2π, where $F(x)$ is continuous in the sense just described in connection with the function Arg $f(z)$. Then from the relation

$$F(x) + F(0) = F(x),$$

we deduce that the values of $F(0)$ are all integral multiples of 2π, and hence that there exists a value of $F(0)$ equal to 0. Therefore, since $F(x)$ is continuous, there must exist a $\delta > 0$ such that

$$|F(x)| < \frac{\pi}{2} \tag{6.4}$$

for all $|x| \leqslant \delta$. Because of the way in which $F(x)$ is multiple-valued, it follows that $F(x)$ has only one value satisfying (6.4). Let this value be denoted by $\Phi(x)$. Obviously, $\Phi(x)$ is continuous (in the ordinary sense) in the interval $[-\delta, \delta]$. If x_1 and x_2 are two numbers in $[-\delta, \delta]$ such that $x_1 + x_2$ also belongs to $[-\delta, \delta]$, then $\Phi(x_1) + \Phi(x_2)$, being one of the values of $F(x_1) + F(x_2) = F(x_1 + x_2)$, can differ from $\Phi(x_1 + x_2)$ only by a multiple of 2π. But

$$|\Phi(x_1) + \Phi(x_2)| < \pi, \qquad |\Phi(x_1 + x_2)| < \pi,$$

and therefore

$$\Phi(x_1 + x_2) = \Phi(x_1) + \Phi(x_2). \tag{6.5}$$

Writing $\Phi(\delta) = \alpha\delta$ and applying our previous argument to the single-valued continuous function $\Phi(x)$ satisfying the functional equation (6.5), we find that

$$\Phi(x) = \alpha x$$

for $|x| \leqslant \delta$. If we use this formula to *define* $\Phi(x)$ for arbitrary real x, then, as before, one of the values of $F(x)$ equals $\Phi(x)$. In fact, any $x \geqslant 0$ can be uniquely represented in the form $2n\delta + \xi$, where n is a nonnegative integer and $|\xi| \leqslant \delta$. Therefore, if $x \geqslant 0$, one value of the sum

$$\underbrace{F(\delta) + \cdots + F(\delta)}_{2n \text{ times}} + F(\xi) = F(x)$$

is $2n\Phi(\delta) + \Phi(\xi) = \Phi(x)$. In other words, if $x \geqslant 0$, all the values of $F(x)$ are of the form

$$\Phi(x) + 2j\pi \qquad (j = 0, \pm 1, \pm 2, \ldots).$$

On the other hand, if $x \leqslant 0$, we deduce from the relation

$$F(x) + F(-x) = F(0) = 2k\pi \qquad (k = 0, \pm 1, \pm 2, \ldots)$$

that

$$\begin{aligned} F(x) = 2k\pi - F(-x) &= 2k\pi - \Phi(-x) - 2j\pi = \Phi(x) + 2(k - j)\pi \\ &= \Phi(x) + 2l\pi \qquad (k, j, l = 0, \pm 1, \pm 2, \ldots). \end{aligned}$$

Thus, if $F(x)$ is a multiple-valued function of the type under discussion, we have

$$F(x) = \alpha x + 2j\pi \qquad (j = 0, \pm 1, \pm 2, \ldots)$$

for all real x. Similarly, we have

$$F(iy) = \beta y + 2k\pi \qquad (k = 0, \pm 1, \pm 2, \ldots),$$

so that

$$F(z) = F(x) + F(iy) = \alpha x + \beta y + 2l\pi \qquad (l = 0, \pm 1, \pm 2, \ldots).$$

Moreover, since F is real, so are the constants α and β.

It follows from these considerations that any single-valued continous real function satisfying (6.2) must have the form $ax + by$, whereas any multiple-valued continous real function satisfying (6.2), with the "same kind of multiple-valuedness" as $\operatorname{Arg} f(z)$, must have the form $\alpha x + \beta y + 2l\pi$ ($l = 0, \pm 1, \pm 2, \ldots$), where a, b, α and β are all real constants. Therefore, any continuous complex function which satisfies property 2′ (the addition theorem), and is not identically zero or unity, must be of the form

$$\begin{aligned} f(z) &= e^{\ln|f(z)|}\, e^{i \operatorname{Arg} f(z)} \\ &= e^{ax+by}[\cos(\alpha x + \beta y) + i \sin(\alpha x + \beta y)]. \end{aligned} \tag{6.6}$$

We now apply property 1′, which states that $f(z)$ takes real values when z is real. Setting $z = x$, $y = 0$ in (6.6), we see that $\cos \alpha x + i \sin \alpha x$ must be real for all x, which implies $\alpha = 0$. Moreover, property 1′ also requires that $f(1) = e$, so that $e^a = e$ or $a = 1$. Therefore any continuous function with properties 1′ and 2′ must have the form

$$f(z) = e^{x+by}(\cos \beta y + i \sin \beta y). \tag{6.7}$$

Finally we impose the full force of property 3′ upon $f(z)$, i.e., we require that $f(z)$ be *differentiable*. Applying the Cauchy-Riemann equations to the function (6.7), we find that

$$e^{x+by} \cos \beta y = be^{x+by} \sin \beta y + \beta e^{x+by} \cos \beta y,$$

$$be^{x+by} \cos \beta y - \beta e^{x+by} \sin \beta y = -e^{x+by} \sin \beta y,$$

or

$$b \sin \beta y + (\beta - 1) \cos \beta y = 0, \tag{6.8}$$

$$b \cos \beta y - (\beta - 1) \sin \beta y = 0. \tag{6.9}$$

Each of these relations by itself is sufficient to determine b and β. In fact, setting first $y = 0$ and then $y = \pi/2$ in (6.8), we find that $\beta = 1$ and $b = 0$, and these values also satisfy (6.9). Thus, finally, the unique function $f(z) = e^z$ with all the properties listed in the statement of the theorem is

$$f(z) \equiv e^z = e^x(\cos y + i \sin y). \tag{6.10}$$

Moreover, the fact that (6.10) has all the desired properties is obvious by inspection.

Remark. If instead of requiring $f(z)$ to be differentiable (property 3′), we had only required $f(z)$ to be continuous, then we would have obtained an infinite set of functions of the form (6.7), where b and β are arbitrary real numbers. However, only the function (6.10) is entire, i.e., analytic everywhere in the finite plane.

Problem 1. Prove that property 3, p. 111 can be replaced by the condition that $f(x)$ be continuous at $x = 0$, while property 3′, p. 112 can be replaced by the condition that $f(z)$ be differentiable at $z = 0$.

Problem 2. Give another proof of Theorem 6.1.

Hint. First prove that $f(0) = f'(0) = 1$ and that $f'(z) = f(z)$. Integration of the differential equation then gives (6.10).

30. THE MAPPING $w = e^z$

It follows from (6.10) that e^z is nonzero for all z and

$$|e^z| = e^x, \qquad \operatorname{Arg} e^z = y + 2k\pi.$$

For $z = iy$ ($x = 0$) we obtain *Euler's formula*

$$e^{iy} = \cos y + i \sin y. \tag{6.11}$$

Using (6.11), we can replace the trigonometric form of a complex number

$$z = r(\cos \Phi + i \sin \Phi)$$

by the more concise *polar form*

$$z = re^{i\Phi}.$$

It is apparent from (6.10) that the exponential is *periodic* in z with *period* $2\pi i$. In other words, if z is changed by $2\pi i$, so that y is changed by 2π, the value of e^z does not change:

$$e^{z+2\pi i} = e^z.$$

We now show that $2\pi i$ is the *fundamental* (or *primitive*) *period* of the function e^z, i.e., that any other period ω of e^z must be of the form $2k\pi i$, where k is an integer. To see this, let $\omega = \alpha + i\beta$. Then

$$e^{z+\omega} = e^z$$

for any z, and in particular,

$$e^{\omega} = e^{\alpha+i\beta} = e^{\alpha}(\cos \beta + i \sin \beta) = 1$$

for $z = 0$. But this means that $|e^{\omega}| = e^{\alpha} = 1$ which implies $\alpha = 0$, and hence $\cos \beta + i \sin \beta = 1$ which implies $\beta = 2k\pi i$, so that

$$\omega = \alpha + i\beta = 2k\pi i,$$

as asserted.

The expression e^∞ will be regarded as meaningless, since

$$\lim_{z\to\infty} e^z$$

does not exist. This can be seen from the fact that $e^x \to \infty$ as $x \to +\infty$, whereas $e^x \to 0$ as $x \to -\infty$. In particular, it follows that e^z cannot coincide with any polynomial, i.e., e^z is actually an entire *transcendental* function, since any polynomial (excluding the trivial case of a constant) approaches infinity as $z \to \infty$.

Next we study the geometric behavior of the mapping $w = e^z$. As already noted, e^z is nonzero for all z. This means that the origin of coordinates in the w-plane does not belong to the image of the finite z-plane under the mapping $w = e^z$. However, as we now show, any other finite point of the w-plane does belong to this image. In fact, from the equation $w = e^z$, where $w \neq 0$ is given and $z = x + iy$ is unknown, we obtain

$$|w| = e^x \quad \text{or} \quad x = \ln |w|$$

and

$$\operatorname{Arg} w = y + 2k\pi \quad \text{or} \quad y = \operatorname{Arg} w.$$

Therefore the inverse images of the point w can only be points of the form

$$z = \ln |w| + i \operatorname{Arg} w. \tag{6.12}$$

Obviously there are infinitely many points (6.12), since Arg w takes infinitely many values, all differing by integral multiples of 2π. Moreover, each of these points is actually an inverse image of w, since

$$\begin{aligned} \exp [\ln |w| + i \operatorname{Arg} w] &= e^{\ln|w|}(\cos \operatorname{Arg} w + i \sin \operatorname{Arg} w) \\ &= |w| (\cos \operatorname{Arg} w + i \sin \operatorname{Arg} w) = w. \end{aligned}$$

Therefore the set of all roots of the equation $e^z = w$ $(w \neq 0)$ is given by the formula

$$z = \ln |w| + i \operatorname{Arg} w = \ln |w| + i(\arg w + 2k\pi), \tag{6.13}$$

where $k = 0, \pm 1, \pm 2, \ldots$. These points all lie on the same straight line parallel to the imaginary axis, and the distance between any two consecutive points along the line is 2π. Thus the function $w = e^z$ maps the finite z-plane onto the domain obtained from the finite w-plane by deleting the single point $w = 0$, but the mapping is not one-to-one, since every point $w \neq 0$ has an infinite number of inverse images (6.13). On the other hand, the mapping is conformal at every point of the finite z-plane, since the derivative

$$(e^z)' = \frac{\partial(e^x \cos y)}{\partial x} + i\,\frac{\partial(e^x \sin y)}{\partial x} = e^x(\cos y + i \sin y) = e^z$$

does not vanish for any value of z.

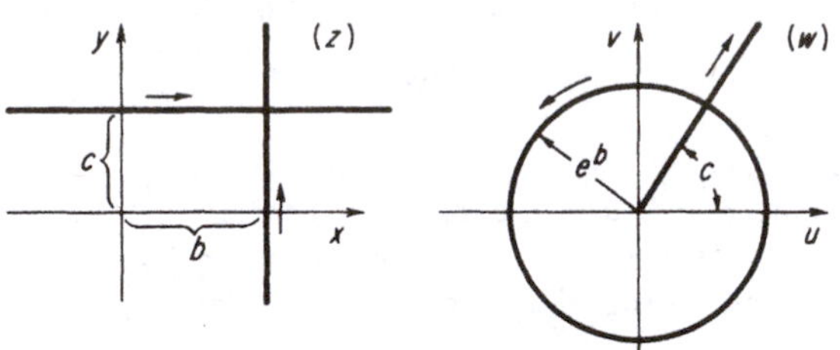

FIGURE 6.1

Now suppose z traces out a straight line parallel to one of the coordinate axes (see Figure 6.1). For example, consider the line

$$z = b + it, \tag{6.14}$$

parallel to the imaginary axis. Then the image of (6.14) under the mapping $w = e^z$ is the curve

$$w = e^b(\cos t + i \sin t), \tag{6.15}$$

i.e., w traces out a circle of radius e^b with its center at the origin. Moreover, as z describes the line (6.14) once in such a way that t, the ordinate of z, increases continuously from $-\infty$ to $+\infty$, w describes the circle (6.15) an infinite number of times in the positive (counterclockwise) direction.

Next consider the line

$$z = t + ic, \tag{6.16}$$

parallel to the real axis. Then the image of (6.16) under the mapping $w = e^z$ is the curve

$$w = e^t(\cos c + i \sin c), \tag{6.17}$$

i.e., w traces out a ray of slope $\tan c$ emanating from the origin. Moreover, as z describes the line (6.16) once in such a way that t, the abscissa of z, increases continuously from $-\infty$ to $+\infty$, w describes the ray (6.17) once in such a way that the distance of w from the origin increases continuously from 0 to ∞ (of course, the limits 0 and ∞ are excluded, since $|w| = e^t$). Thus, *under the mapping $w = e^z$, a family of lines parallel to the imaginary axis is transformed into a family of concentric circles with the origin as center, and a family of lines parallel to the real axis is transformed into a family of rays emanating from the origin.*

Now consider the domain G consisting of all points z such that

$$\varphi_0 < \operatorname{Im} z < \varphi_1,$$

where $\varphi_1 - \varphi_0 = h$; such a domain will be called an (*open*) *strip* of width h. Suppose $0 < h < 2\pi$, and let $\mathscr{G}$ be the image of G under the mapping $w = e^z$.

It follows from the considerations just given that $\mathscr{G}$ is the interior of the angle of h radians with vertex at the origin, formed by the rays

$$\text{Arg } w = \varphi_0 + 2k\pi, \quad \text{Arg } w = \varphi_1 + 2k\pi \qquad (k = 0, \pm 1, \pm 2, \ldots)$$

(see Figure 6.2). Moreover, the correspondence between the domains G and $\mathscr{G}$ under the mapping $w = e^z$ is *one-to-one*. To see this, we recall that the inverse images of a point $w \in \mathscr{G}$ are all of the form (6.13), and hence differ

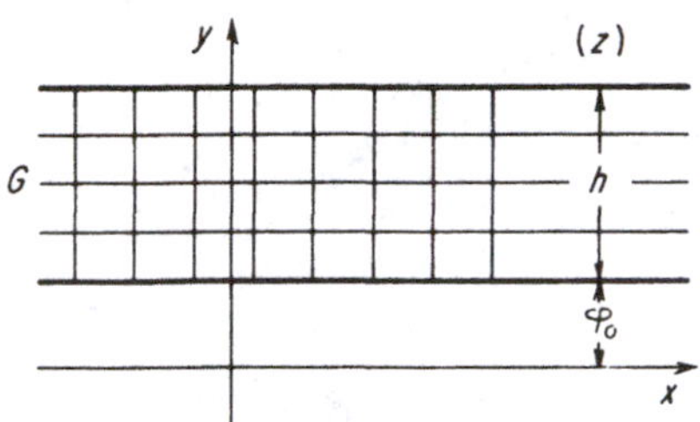

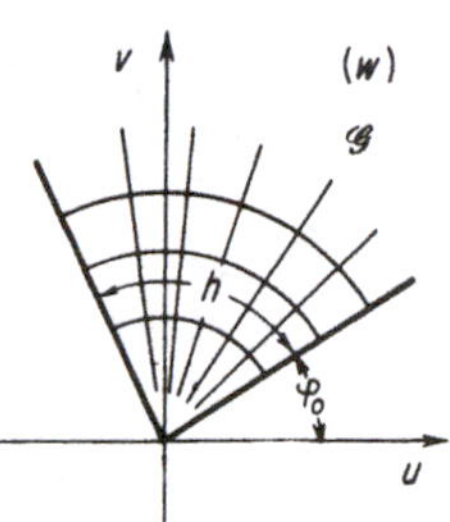

FIGURE 6.2

only in the values of their imaginary parts. In fact, any two points (6.13) lie on a line parallel to the imaginary axis, and the distance between them is an integral multiple of 2π. However, by assumption, the width h of our strip does not exceed 2π, and G can contain only one inverse image of the point w, i.e., not only is $w = f(z)$ a single-valued function on G, but its inverse $z = f^{-1}(w)$ is a single-valued function on $\mathscr{G} = f(G)$. Thus, *the exponential function $w = e^z$ is a one-to-one conformal mapping of an open strip of width $h \leqslant 2\pi$ with sides parallel to the real axis onto the interior of an angle of h radians with vertex at the origin.*

Next consider a straight line with equation

$$z = (1 + i\alpha)t + ib \qquad (-\infty < t < \infty), \tag{6.18}$$

which is not parallel to one of the coordinate axes. Here $\alpha \neq 0$ is the slope of the line (6.18), and b is its y-intercept. The image of (6.18) under the mapping $w = e^z$ is the curve

$$w = \exp\,[t + i(\alpha t + b)] = e^t[\cos\,(\alpha t + b) + i \sin\,(\alpha t + b)].$$

Therefore

$$|w| = r = e^t, \qquad \varphi = \text{Arg } w = \alpha t + b + 2k\pi,$$

and eliminating the parameter t, we obtain

$$r = \exp\,[(\varphi - b - 2k\pi)/\alpha]. \tag{6.19}$$

If we set $\theta = \varphi - 2k\pi$, (6.19) becomes

$$r = ce^{\theta/\alpha}, \tag{6.20}$$

where $c = e^{-b/\alpha}$. This is the equation (in polar form) of a *logarithmic spiral.* Since the mapping $w = e^z$ is conformal, and since (6.20) is the image of the line (6.18) intersecting all lines parallel to the real axis at the same angle arc tan α, it follows that the logarithmic spiral intersects the images of all these lines, i.e., all rays emanating from the origin, at the same angle arc tan α, a property which characterizes the logarithmic spiral (see Figure 6.3).

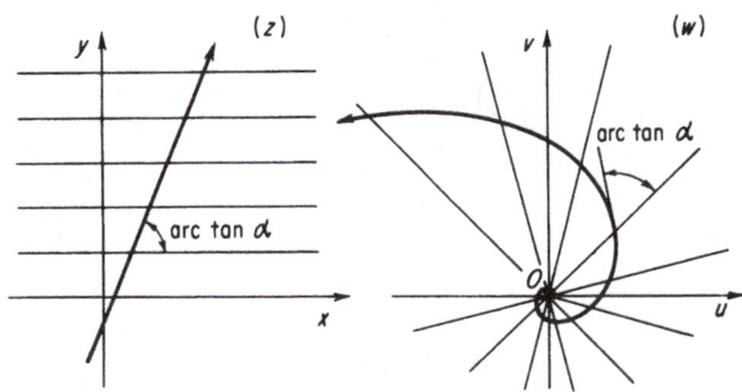

FIGURE 6.3

Problem 1. Express the complex numbers ± 1, $\pm i$, $1 \pm i$, $-1 \pm i$ in polar form.

Problem 2. Find the moduli and principal values of the arguments of the following complex numbers:

a) e^{2+i}; b) e^{2-3i}; c) e^{3+4i}; d) e^{-3-4i}; e) $-ae^{i\varphi}$ $(a > 0, |\varphi| \leqslant \pi)$.

Problem 3. What is the image under the mapping $w = e^z$ of the strip in the z-plane whose sides consist of the parallel inclined lines

$$y = k(x - a_1), \qquad y = k(x - a_2) \qquad (k \neq 0, \infty)?$$

What is the condition for the mapping to be one-to-one?

Problem 4. What is the image under the mapping $w = z + e^z$ of the closed strip $-\pi \leqslant y \leqslant \pi$ in the z-plane?

31. SOME FUNCTIONS RELATED TO THE EXPONENTIAL

According to the formulas

$$e^{ix} = \cos x + i \sin x, \qquad e^{-ix} = \cos x - i \sin x, \tag{6.21}$$

we have

$$\cos x = \frac{e^{ix} + e^{-ix}}{2}, \qquad \sin x = \frac{e^{ix} - e^{-ix}}{2i} \tag{6.22}$$

for arbitrary real x. If z is an arbitrary finite complex number, we define two (entire) *trigonometric functions* $\cos z$ and $\sin z$, called the *cosine* and *sine*, by simply changing x to z everywhere in (6.22):

$$\cos z = \frac{e^{iz} + e^{-iz}}{2}, \qquad \sin z = \frac{e^{iz} - e^{-iz}}{2i}. \tag{6.23}$$

This seems quite natural, since the functions $\cos z$ and $\sin z$ are obviously analytic for all z, and reduce to the familiar functions $\cos x$ and $\sin x$ when $z = x$ is real. It follows from the definitions (6.23) that $\cos z$ is even and $\sin z$ is odd, i.e., that

$$\cos(-z) = \cos z, \qquad \sin(-z) = -\sin z.$$

Moreover, (6.23) implies the formulas

$$e^{iz} = \cos z + i \sin z, \qquad e^{-iz} = \cos z - i \sin z, \tag{6.24}$$

which generalize (6.21).

The functions $\cos z$ and $\sin z$ are both periodic with period 2π, since changing z to $z + 2\pi$ in (6.23) amounts to multiplying the exponentials by $e^{\pm 2\pi i} = 1$. Actually, 2π is the fundamental period of $\cos z$ and $\sin z$, i.e., any other period is an integral multiple of 2π, as we now verify for $\cos z$. If ω is any period of $\cos z$, then

$$\cos(z + \omega) = \cos z,$$

and hence, setting $z = \pi/2$, we obtain

$$\cos\left(\omega + \frac{\pi}{2}\right) = 0.$$

But this implies

$$\exp\left[i\left(\omega + \frac{\pi}{2}\right)\right] + \exp\left[-i\left(\omega + \frac{\pi}{2}\right)\right] = 0,$$

or

$$\exp[i(2\omega + \pi)] = -1.$$

Therefore, according to (6.13),

$$i(2\omega + \pi) = \ln|-1| + i \operatorname{Arg}(-1) = i(\pi + 2k\pi),$$

so that $\omega = k\pi$. But $\cos \omega = \cos 0 = 1$, and hence k is even, i.e.,

$$\omega = 2m\pi,$$

as asserted. Similarly, it can easily be verified that 2π is the fundamental period of $\sin z$.

Next we derive *addition theorems* for the functions $\cos z$ and $\sin z$, i.e., formulas relating the quantities $\cos(z_1 + z_2)$ and $\sin(z_1 + z_2)$ to the quantities $\cos z_1$, $\sin z_1$, $\cos z_2$ and $\sin z_2$, where z_1 and z_2 are arbitrary complex numbers. As might be expected, the required relations are immediate consequences of the addition theorem

$$\exp(z_1 + z_2) = \exp z_1 \exp z_2$$

for the exponential. In fact, replacing z by $z_1 + z_2$ in the formulas (6.24), we find that

$$\begin{aligned}\cos(z_1 + z_2) + i\sin(z_1 + z_2) &= \exp[i(z_1 + z_2)] = \exp(iz_1)\exp(iz_2)\\ &= (\cos z_1 + i\sin z_1)(\cos z_2 + i\sin z_2)\\ &= (\cos z_1\cos z_2 - \sin z_1\sin z_2) + i(\sin z_1\cos z_2 + \cos z_1\sin z_2)\end{aligned} \tag{6.25}$$

and

$$\begin{aligned}\cos(z_1 + z_2) - i\sin(z_1 + z_2) &= \exp[-i(z_1 + z_2)] = \exp(-iz_1)\exp(-iz_2)\\ &= (\cos z_1 - i\sin z_1)(\cos z_2 - i\sin z_2)\\ &= (\cos z_1\cos z_2 - \sin z_1\sin z_2) - i(\sin z_1\cos z_2 + \cos z_1\sin z_2).\end{aligned} \tag{6.26}$$

First adding (6.26) to (6.25), and then subtracting (6.26) from (6.25), we obtain the addition theorems

$$\begin{aligned}\cos(z_1 + z_2) &= \cos z_1\cos z_2 - \sin z_1\sin z_2,\\ \sin(z_1 + z_2) &= \sin z_1\cos z_2 + \cos z_1\sin z_2,\end{aligned} \tag{6.27}$$

which are basic in the theory of trigonometric functions. In particular, the so-called *reduction formulas* are implicit in (6.27). For example, setting $z_1 = z$, $z_2 = \pi/2$ in (6.27) gives

$$\cos\left(z + \frac{\pi}{2}\right) = \cos z\cos\frac{\pi}{2} - \sin z\sin\frac{\pi}{2} = -\sin z,$$

$$\sin\left(z + \frac{\pi}{2}\right) = \sin z\cos\frac{\pi}{2} + \cos z\sin\frac{\pi}{2} = \cos z,$$

setting $z_1 = z$, $z_2 = \pi$, gives

$$\cos(z + \pi) = -\cos z,$$
$$\sin(z + \pi) = -\sin z,$$

and so on. Moreover, substituting $z_1 = z$, $z_2 = -z$ into the first of the formulas (6.27), we obtain the following basic relation between $\cos z$ and $\sin z$:

$$\cos^2 z + \sin^2 z = 1. \tag{6.28}$$

Thus we see that all the familiar trigonometric formulas involving cosines and sines of real arguments remain valid in the complex domain. However, we cannot infer from (6.28) that $|\cos z| \leqslant 1$ and $|\sin z| \leqslant 1$, since $\cos^2 z$ and $\sin^2 z$ are generally not nonnegative real numbers.

Closely related to the trigonometric functions $\cos z$ and $\sin z$ are the *hyperbolic functions* $\cosh z$ and $\sinh z$, called the *hyperbolic cosine* and the *hyperbolic sine*, respectively, and defined by the formulas

$$\cosh z = \frac{e^z + e^{-z}}{2}, \qquad \sinh z = \frac{e^z - e^{-z}}{2}. \tag{6.29}$$

Clearly, the functions (6.29) are analytic for all finite z, and reduce to the familiar hyperbolic functions $\cosh x$ and $\sinh x$ when $z = x$ is real. It will be recalled that the function $\cosh x$ is even and decreases monotonically from $+\infty$ to 1 as x increases from $-\infty$ to 0, and then increases monotonically from 1 to $+\infty$ again as x increases from 0 to $+\infty$. On the other hand, the function $\sinh x$ is odd, and increases monotonically from $-\infty$ to $+\infty$ as x increases from $-\infty$ to $+\infty$. Comparing (6.29) and (6.23), we see that the trigonometric and hyperbolic functions are connected by the relations

$$\cosh z = \cos(iz), \qquad \sinh z = -i\sin(iz). \tag{6.30}$$

In particular, it follows that

$$\cosh^2 z - \sinh^2 z = \cos^2(iz) + \sin^2(iz) = 1. \tag{6.31}$$

Next we derive expressions for the real and imaginary parts, and also the moduli, of the functions $\cos z$ and $\sin z$. Setting $z = x + iy$, and using formulas (6.27) and (6.30), we find that

$$\begin{aligned} \cos(x + iy) &= \cos x \cos(iy) - \sin x \sin(iy) = \cos x \cosh y - i \sin x \sinh y, \\ \sin(x + iy) &= \sin x \cos(iy) + \cos x \sin(iy) = \sin x \cosh y + i \cos x \sinh y, \end{aligned} \tag{6.32}$$

which imply

$$\operatorname{Re}\cos(x + iy) = \cos x \cosh y, \qquad \operatorname{Im}\cos(x + iy) = -\sin x \sinh y,$$
$$\operatorname{Re}\sin(x + iy) = \sin x \cosh y, \qquad \operatorname{Im}\sin(x + iy) = \cos x \sinh y.$$

Hence for the moduli of $\cos z$ and $\sin z$ we obtain the expressions

$$\begin{aligned} |\cos z| &= \sqrt{(\cos x \cosh y)^2 + (\sin x \sinh y)^2} \\ &= \sqrt{\cosh^2 y(1 - \sin^2 x) + \sin^2 x \sinh^2 y} = \sqrt{\cosh^2 y - \sin^2 x}, \\ |\sin z| &= \sqrt{(\sin x \cosh y)^2 + (\cos x \sinh y)^2} \\ &= \sqrt{\sin^2 x(1 + \sinh^2 y) + \cos^2 x \sinh^2 y} = \sqrt{\sinh^2 y + \sin^2 x}, \end{aligned}$$

which in turn imply the inequalities

$$\begin{aligned} |\sinh y| = \sqrt{\cosh^2 y - 1} \leqslant |\cos z| \leqslant \cosh y, \\ |\sinh y| \leqslant |\sin z| \leqslant \sqrt{\sinh^2 y + 1} = \cosh y. \end{aligned} \tag{6.33}$$

Thus we see that the moduli of the functions $\cos z$ and $\sin z$ increase without limit as $|y|$ increases, i.e., as the distance from z to the real axis increases, but this increase is no faster than that of $\cosh y$ and no slower than that of $\sinh y$.

Since $\sinh y \neq 0$ for $y \neq 0$, it follows from the inequalities (6.33) that $\cos z$ and $\sin z$ cannot vanish except on the real axis, i.e., the equations $\cos z = 0$ and $\sin z = 0$ have no imaginary roots. Therefore the roots of the equation $\cos z = 0$ reduce to

$$z = (k + \tfrac{1}{2})\pi \qquad (k = 0, \pm 1, \pm 2, \ldots),$$

while those of the equation $\sin z = 0$ reduce to

$$z = k\pi \qquad (k = 0, \pm 1, \pm 2, \ldots),$$

which are just the roots familiar from elementary trigonometry.

Finally we introduce the functions

$$\begin{gathered} \tan z = \frac{\sin z}{\cos z}, \quad \cot z = \frac{\cos z}{\sin z}, \quad \sec z = \frac{1}{\cos z}, \quad \csc z = \frac{1}{\sin z}, \\ \tanh z = \frac{\sinh z}{\cosh z}, \quad \coth z = \frac{\cosh z}{\sinh z}, \quad \operatorname{sech} z = \frac{1}{\cosh z}, \quad \operatorname{csch} z = \frac{1}{\sinh z}, \end{gathered} \tag{6.34}$$

which for real $z = x$ reduce to the familiar trigonometric functions $\tan x$, $\cot x$, $\sec x$, $\csc x$ and hyperbolic functions $\tanh x$, $\coth x$, $\operatorname{sech} x$, $\operatorname{csch} x$. Each of the functions (6.34) is a *transcendental meromorphic function*, by which we mean a meromorphic function which is not a rational function. In fact, unlike the case of rational functions, each of the functions (6.34) becomes infinite at infinitely many points, i.e., at the infinitely many zeros of $\cos z$, $\sin z$, $\cosh z$ or $\sinh z$.

Problem 1. Find the real and imaginary parts of the following complex numbers:

a) $\cos(2 + i)$; b) $\sin 2i$; c) $\cosh(2 - i)$; d) $\sinh e^i$.

Problem 2. Prove the formulas

$$\frac{d}{dz}\cos z = -\sin z, \quad \frac{d}{dz}\sin z = \cos z, \quad \frac{d}{dz}\cosh z = \sinh z, \quad \frac{d}{dz}\sinh z = \cosh z.$$

Find formulas for the derivatives of $\tan z$, $\cot z$, $\sec z$, $\csc z$ and the corresponding hyperbolic functions.

Problem 3. (M1, Secs. 41 and 42). Make a detailed study of the mapping $w = \cos z$. In particular, find the images of straight lines parallel to the coordinate axes, and find domains of univalence for $w = \cos z$.

Problem 4. Deduce the estimates (6.33) directly from (6.23).

Problem 5. (M1, p. 205). Make a detailed study of the mapping $w = \tan z$. In particular, find the image of the strip $x_0 < x < x_0 + h$ $(0 < h \leqslant \pi)$.

Hint. Regard $w = \tan z$ as the result of making the following four mappings in succession:

$$\zeta = iz, \qquad t = e^{\zeta}, \qquad \tau = t^2, \qquad w = \frac{1}{i}\frac{\tau - 1}{\tau + 1}.$$

Ans. A circular lune with angles of $2h$ radians and vertices $\pm i$ (see M1, p. 206).

32. THE LOGARITHM

The inverse of the function

$$z = e^w = e^u(\cos v + i \sin v)$$

is defined for any value of z different from 0 and ∞, and is represented by the formula

$$w = \ln |z| + i \operatorname{Arg} z$$

[cf. (6.13)]. This function, which is obviously multiple-valued (in fact, *infinite-valued*), is called the *logarithm* and is denoted by $\operatorname{Ln} z$, i.e.,

$$\operatorname{Ln} z = \ln |z| + i \operatorname{Arg} z, \tag{6.35}$$

by definition. The value

$$\ln |z| + i \arg z$$

of the logarithm is called the *principal value*, and is denoted by $\ln z$. Then (6.35) can be written in the form

$$\operatorname{Ln} z = \ln z + 2k\pi i \qquad (k = 0, \pm 1, \pm 2, \ldots). \tag{6.36}$$

It follows that *every complex number different from* 0 *and* ∞ *has infinitely many logarithms (i.e., values of the function* $\operatorname{Ln} z$*), and any two of these logarithms differ by an integral multiple of* $2\pi i$.

If z is a positive real number, the principal value of the logarithm is just $\ln |z|$, which is exactly what is meant by the logarithm in elementary mathematics; for example,

$$\ln 1 = 0, \quad \ln e = 1, \quad \ln 2 = 0.69315\ldots$$

For negative real numbers and for imaginary numbers, the principal value of the logarithm is an imaginary number

$$\ln |z| + i \arg z \qquad (\arg z \neq 0, -\pi < \arg z \leqslant \pi)$$

and all the other values of the logarithm are also imaginary numbers, calculated by using (6.36); for example,

$$\operatorname{Ln}(-1) = (2k+1)\pi i,$$

$$\operatorname{Ln}(-2) = 0.69315\ldots + (2k+1)\pi i,$$

$$\operatorname{Ln}(1-i) = \ln\sqrt{2} - \frac{\pi i}{4} + 2k\pi i = 0.34657\ldots + (8k-1)\frac{\pi i}{4}.$$

The familiar rules for finding logarithms of products and quotients remain valid for the multiple-valued logarithms of complex numbers, since

$$\begin{aligned}\operatorname{Ln}(z_1z_2) &= \ln|z_1z_2| + i\operatorname{Arg}(z_1z_2)\\ &= \ln|z_1| + \ln|z_2| + i(\operatorname{Arg} z_1 + \operatorname{Arg} z_2)\\ &= \operatorname{Ln} z_1 + \operatorname{Ln} z_2\end{aligned} \tag{6.37}$$

and

$$\begin{aligned}\operatorname{Ln}\frac{z_1}{z_2} &= \ln\frac{z_1}{z_2} + i\operatorname{Arg}\frac{z_1}{z_2}\\ &= \ln|z_1| - \ln|z_2| + i(\operatorname{Arg} z_1 - \operatorname{Arg} z_2)\\ &= \operatorname{Ln} z_1 - \operatorname{Ln} z_2,\end{aligned} \tag{6.38}$$

where z_1 and z_2 are arbitrary nonzero complex numbers. In (6.37) and (6.38), both the left and right-hand sides (for fixed z_1 and z_2) represent infinite sets of complex numbers, and the equalities have to be understood in the sense that these two sets are equal, i.e., have the same members. Failure to remember this fact can lead to paradoxical results. For example, in a sophism constructed by John Bernoulli, it is claimed that $\operatorname{Ln}(-z) = \operatorname{Ln} z$ for arbitrary $z \neq 0$, and the following chain of equalities is adduced as "proof":

1. $\operatorname{Ln}[(-z)^2] = \operatorname{Ln}(z^2)$;
2. $\operatorname{Ln}(-z) + \operatorname{Ln}(-z) = \operatorname{Ln} z + \operatorname{Ln} z$;
3. $2\operatorname{Ln}(-z) = 2\operatorname{Ln} z$;
4. $\operatorname{Ln}(-z) = \operatorname{Ln} z$.

However, the conclusion that $\operatorname{Ln}(-z) = \operatorname{Ln} z$ is false, since

$$\operatorname{Ln} z = \ln|z| + i\operatorname{Arg} z = \ln|z| + i\arg z + 2k\pi i,$$

$$\operatorname{Ln}(-z) = \ln|-z| + i\operatorname{Arg}(-z) = \ln|z| + i\arg z + (2k+1)\pi i,$$

and obviously none of the numbers representing the values of $\operatorname{Ln} z$ is the same as any of the numbers representing the values of $\operatorname{Ln}(-z)$. The fallacy in the "proof" occurs in going from equality 2 to equality 3. The first of these relations is based on formula (6.37), and is of course true. However, the sum $\operatorname{Ln}(-z) + \operatorname{Ln}(-z)$ cannot be replaced by $2\operatorname{Ln}(-z)$, since the sum in question is obtained from the set of numbers $\operatorname{Ln}(-z)$ by adding each of these numbers to itself and to *all the other* numbers of the set $\operatorname{Ln}(-z)$, whereas the set $2\operatorname{Ln}(-z)$ is obtained by simply doubling all the numbers $\operatorname{Ln}(-z)$, i.e., by adding each such number *to itself only.*[3] Therefore

$$\operatorname{Ln}(-z) + \operatorname{Ln}(-z) \neq 2\operatorname{Ln}(-z),$$

and by the same token,

$$\operatorname{Ln} z + \operatorname{Ln} z \neq 2\operatorname{Ln} z.$$

Setting $z_1 = z_2 = z \neq 0$ in (6.38), we obtain the relation

$$\operatorname{Ln} 1 = \operatorname{Ln} z - \operatorname{Ln} z, \qquad (6.39)$$

which is a correct formula. However, the right-hand side of (6.39) cannot be replaced by 0, since here we are talking about the set of all possible differences between values of the logarithm of the same number. This set consists of all possible multiples of $2\pi i$, so that to be perfectly explicit, we should write (6.39) as

$$\operatorname{Ln} 1 = 2k\pi i \qquad (k = 0, \pm 1, \pm 2, \ldots).$$

We now study the single-valued branches of the logarithm. We begin by finding domains of univalence for the exponential function $z = e^w$, which is the inverse of the logarithm $w = \operatorname{Ln} z$. All the numbers w for which e^w takes any given value z are given by the formula

$$w = \ln |z| + i \operatorname{Arg} z,$$

i.e., all the numbers w can be obtained by shifting any one of them by $2k\pi i$ where $k = 0, \pm 1, \pm 2, \ldots$. Therefore a domain of univalence for $z = e^w$ cannot contain any pair of points such that one point can be obtained from the other by a shift of this kind. The simplest way to satisfy this requirement is to start with an open rectilinear strip

$$\mathscr{G}_0 : v_0 < v < v_0 + 2\pi$$

of width 2π parallel to the real axis in the w-plane. Then, subjecting $\mathscr{G}_0$ to all possible shifts of the form $2k\pi i$, we obtain an infinite family of domains of univalence:

$$\mathscr{G}_k : v_0 + 2k\pi < v < v_0 + 2(k+1)\pi \qquad (k = 0, \pm 1, \pm 2, \ldots).$$

---

[3] The following simple example may help clarify the situation: Let A be the set consisting of the two numbers 0 and 1. Then $A + A$ is the set consisting of the three numbers $0 + 0 = 0$, $0 + 1 = 1$ and $1 + 1 = 2$, whereas the set $2A$ consists only of two numbers $2 \cdot 0 = 0$ and $2 \cdot 1 = 2$.

Obviously, every point of the w-plane is either an interior point of a domain $\mathscr{G}_k$ or a boundary point of two domains $\mathscr{G}_k$ and $\mathscr{G}_{k+1}$ [see Figure 6.4(b)].

The image (under $z = e^w$) of each of the strips $\mathscr{G}_k$ is the same domain G in the z-plane, i.e., the interior of an angle of 2π radians with its vertex at the

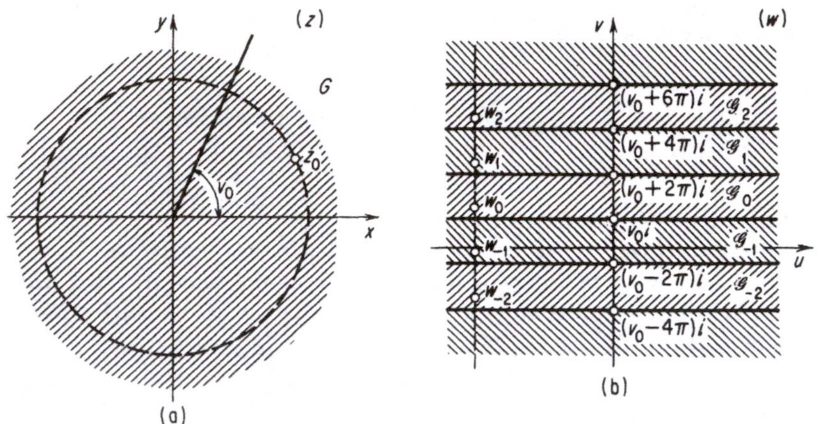

FIGURE 6.4

origin. The boundary of G is a single ray of inclination v_0 emanating from the origin [see Figure 6.4(a)]. We can define infinitely many different single-valued branches of Ln z on the domain G be specifying that the kth branch $\text{Ln}_k\, z$ ($k = 0, \pm 1, \pm 2, \ldots$) have $\mathscr{G}_k$ as its range. Each of these branches is uniquely characterized by the value w_0 which it assigns to any given point $z_0 \in G$, since one and only one of the domains $\mathscr{G}_k$ contains the point w_0. Expressed somewhat differently,

$$\text{Ln}_k\, z = \ln |z| + i \,\text{Arg}_k\, z,$$

where $\text{Arg}_k\, z$ is the value of the argument satisfying the inequality

$$v_0 + 2k\pi < \text{Arg}_k\, z < v_0 + 2(k + 1)\pi \qquad (k = 0, \pm 1, \pm 2, \ldots).$$

Moreover, since $w = \text{Ln}_k\, z$ is a one-to-one continuous mapping of G onto $\mathscr{G}_k$, and since $z = e^w$ has a nonzero derivative e^w on $\mathscr{G}_k$, the branches $\text{Ln}_k\, z$ all have nonzero derivatives on G, i.e.,

$$\frac{d}{dz} \text{Ln}_k\, z = \frac{1}{e^w} = \frac{1}{z} \qquad (k = 0, \pm 1, \pm 2, \ldots)$$

(cf. Rule 5, p. 48).

The points 0 and ∞ are both branch points of the function $\operatorname{Ln} z$, as defined on p. 76. To see this, suppose that at a point $z_0 \in G$ we choose a value of $\operatorname{Ln} z$ corresponding to the branch $\operatorname{Ln}_k z$ and represented by the point

$$w_k = \operatorname{Ln}_k z_0 = \ln |z_0| + i \operatorname{Arg}_k z_0,$$

belonging to the strip $\mathscr{G}_k$. Then, as z moves continuously around the circle $|z| = |z_0|$ in the counterclockwise direction, starting from the point z_0, the value of

$$w = \ln |z| + i \operatorname{Arg} z \tag{6.40}$$

changes continuously, and when z returns to its original value z_0, (6.40) goes into the value

$$w_{k+1} = \ln |z_0| + i \operatorname{Arg}_k z_0 + 2\pi i = \ln |z_0| + i \operatorname{Arg}_{k+1} z_0 = \operatorname{Ln}_{k+1} z_0.$$

Thus, since the point $z_0 \in G$ is arbitrary, one circuit around the origin $z = 0$ (or about the point at infinity $z = \infty$) causes the branch $\operatorname{Ln}_k z$ to change continuously into the branch $\operatorname{Ln}_{k+1} z$.[4] Obviously, as we make additional circuits around the origin in the counterclockwise direction, the branch $\operatorname{Ln}_k z$ undergoes the infinite sequence of transformations

$$\operatorname{Ln}_k z \to \operatorname{Ln}_{k+1} z, \quad \operatorname{Ln}_{k+1} z \to \operatorname{Ln}_{k+2} z, \quad \operatorname{Ln}_{k+2} z \to \operatorname{Ln}_{k+3} z, \ldots,$$

so that $\operatorname{Ln}_k z$ is never carried back into itself. For this reason, the points 0 and ∞ are called *branch points of infinite order* or *logarithmic branch points.*

Remark. Single-valued branches of the function $\operatorname{Ln} z$ can be defined on domains more general than G. Let γ be any Jordan curve joining the points $z = 0$ and $z = \infty$. Then the curve γ has infinitely many images Γ_k ($k = 0, \pm 1, \pm 2, \ldots$) under the mapping $w = \operatorname{Ln} z$. The curves Γ_k, which are all Jordan curves, divide the w-plane into infinitely many open curvilinear strips $\mathscr{D}_k$, where the boundary of $\mathscr{D}_k$ consists of the pair of curves Γ_k and Γ_{k+1}, and $\mathscr{D}_k$ does not contain any of the other curves Γ_k. Thus we can define a countably infinite family of single-valued branches of $\operatorname{Ln} z$ on the domain D with boundary γ by specifying that the kth branch $(\operatorname{Ln} z)_k$, where $k = 0, \pm 1, \pm 2, \ldots$, have $\mathscr{D}_k$ as its range.[5] Clearly,

$$(\operatorname{Ln} z)_k = (\operatorname{Ln} z)_l + 2\pi(k - l)i,$$

and, as before, we find that

$$\frac{d}{dz} (\operatorname{Ln} z)_k = \frac{1}{z}. \tag{6.41}$$

---

[4] Similarly, one circuit around the origin in the opposite (i.e., clockwise) direction causes the branch $\operatorname{Ln}_k z$ to change continuously into the branch $\operatorname{Ln}_{k-1} z$.

[5] Note the distinction between $\operatorname{Ln}_k z$ and $(\operatorname{Ln} z)_k$.

Since (6.41) is independent of how the branches of Ln z are defined, we can simply write

$$\frac{d}{dz}\operatorname{Ln} z = \frac{1}{z},$$

where by the left-hand side we mean any single-valued branch of Ln z, defined on a domain containing the given point z.

Problem 1. Calculate the following logarithms:

a) $\operatorname{Ln} 4$; b) $\operatorname{Ln}(-1)$; c) $\operatorname{Ln} i$; d) $\operatorname{Ln}\dfrac{1 \pm i}{\sqrt{2}}$; e) $\operatorname{Ln}(2 - 3i)$.

Problem 2. Suppose we make a complete circuit in the positive direction around the circle $|z| = 2$, starting and ending at the point $z = 2$. Assuming that the value of $\operatorname{Im} f(z)$ is zero at $z = 2$ and that $f(z)$ varies continuously with the point z, find the final value of

a) $f(z) = 2 \operatorname{Ln} z$; b) $f(z) = \operatorname{Ln}\dfrac{1}{z}$; c) $f(z) = \operatorname{Ln} z \pm \operatorname{Ln}(z + 1)$.

Problem 3. The *inverse trigonometric functions* Arc cos z, Arc sin z, Arc tan z, Arc cot z and the *inverse hyperbolic functions* Arc cosh z, Arc sinh z, Arc tanh z, Arc coth z are defined as the inverses of the corresponding trigonometric and hyperbolic functions. Verify the following relations:

a) $\operatorname{Arc\,cos} z = -i \operatorname{Ln}(z + \sqrt{z^2 - 1})$;

b) $\operatorname{Arc\,sin} z = -i \operatorname{Ln} i(z + \sqrt{z^2 - 1})$;

c) $\operatorname{Arc\,tan} z = \dfrac{i}{2}\operatorname{Ln}\dfrac{i + z}{i - z} = \dfrac{1}{2i}\operatorname{Ln}\dfrac{1 + iz}{1 - iz}$; d) $\operatorname{Arc\,cot} z = \dfrac{i}{2}\operatorname{Ln}\dfrac{z - i}{z + i}$;

e) $\operatorname{Arc\,cosh} z = \operatorname{Ln}(z + \sqrt{z^2 - 1})$; f) $\operatorname{Arc\,sinh} z = \operatorname{Ln}(z + \sqrt{z^2 + 1})$;

g) $\operatorname{Arc\,tanh} z = \dfrac{1}{2}\operatorname{Ln}\dfrac{1 + z}{1 - z}$; h) $\operatorname{Arc\,coth} z = \dfrac{1}{2}\operatorname{Ln}\dfrac{z + 1}{z - 1}$.

Problem 4 (M1, Sec. 58). Prove that $w = \operatorname{Arc\,cos} z$ has algebraic branch points of order 1 at ± 1 and a logarithmic branch point at ∞, but no other branch points in the extended z-plane.

Problem 5. Find all values of the following expressions:

a) $\operatorname{Arc\,sin}\frac{1}{2}$; b) $\operatorname{Arc\,cos}\frac{1}{2}$; c) $\operatorname{Arc\,cos} 2$; d) $\operatorname{Arc\,sin} i$;

e) $\operatorname{Arc\,tan}(1 + 2i)$; f) $\operatorname{Arc\,cosh} 2i$; g) $\operatorname{Arc\,tanh}(1 - i)$.

Ans. b) $2k\pi \pm \frac{1}{3}\pi$; d) $2k\pi - i \ln(\sqrt{2} - 1)$, $(2k + 1)\pi - i \ln(\sqrt{2} + 1)$;

f) $\ln(\sqrt{5} \pm 2) + (2k \pm \frac{1}{2})\pi i$.

33. THE FUNCTION z^α. EXPONENTIALS AND LOGARITHMS TO AN ARBITRARY BASE

We begin by defining the expression a^α, where $a \neq 0$ and α are arbitrary complex numbers. We first assume that α is real, and examine in turn the cases where α is an integer, a rational number and an irrational number.

Case 1. If $\alpha = n$ is an integer, then

$$a^\alpha = a^n = |a|^n[\cos (n \operatorname{Arg} a) + i \sin (n \operatorname{Arg} a)], \tag{6.42}$$

and a^α *has just one value.*

Case 2. If $\alpha = r$ is a rational number, then $r = m/n$ where m and $n > 0$ are relatively prime integers. As we already know [cf. formula (1.19)], in this case,

$$\begin{aligned} a^\alpha = a^{m/n} &= |a|^{m/n}\left[\cos\left(\frac{m}{n} \operatorname{Arg} a\right) + i \sin\left(\frac{m}{n} \operatorname{Arg} a\right)\right] \\ &= |a|^r[\cos (r \operatorname{Arg} a) + i \sin (r \operatorname{Arg} a)], \end{aligned} \tag{6.43}$$

and a^α *has n different values.*

Case 3. If $\alpha = \rho$ is an irrational number, we define a^ρ by continuity, i.e., as the limit

$$\lim_{n\to\infty} a^{r_n} = \lim_{n\to\infty} |a|^{r_n}[\cos (r_n \operatorname{Arg} a) + i \sin (r_n \operatorname{Arg} a)], \tag{6.44}$$

where $\{r_n\}$ is an arbitrary sequence of irrational numbers converging to ρ. In taking the limit (6.44), we hold $\operatorname{Arg} a$ fixed. Then

$$\lim_{n\to\infty} r_n \operatorname{Arg} a = \rho \operatorname{Arg} a,$$

and (6.44) implies

$$a^\rho = |a|^\rho[\cos (\rho \operatorname{Arg} a) + i \sin (\rho \operatorname{Arg} a)]. \tag{6.45}$$

To obtain all the values of a^ρ, we now let $\operatorname{Arg} a$ take all its values. Since two values of $\rho \operatorname{Arg} a$ differ by a number of the form $2k\rho\pi$, where k is a nonzero integer, and since $k\rho$ can never be an integer, it follows that different values of $\operatorname{Arg} a$ give rise to different values of a^ρ. Thus, in this case, a^α *has infinitely many different values.*

Remark. It should be noted that formulas (6.42), (6.44) and (6.45) are all special cases of the formula

$$a^\alpha = |a|^\alpha[\cos (\alpha \operatorname{Arg} a) + i \sin (\alpha \operatorname{Arg} a)],$$

which can be written in the form

$$\begin{aligned} a^\alpha &= \exp(\alpha \ln |a|)[\cos(\alpha \operatorname{Arg} a) + i \sin(\alpha \operatorname{Arg} a)] \\ &= \exp(\alpha \ln |a| + i\alpha \operatorname{Arg} a) = \exp(\alpha \operatorname{Ln} a). \end{aligned} \tag{6.46}$$

For the time being, we make a distinction between $\exp z$ and e^z, with only the former being used to denote the exponential function defined in Sec. 29 (see Example 2 below).

Now let α be an arbitrary complex number. Observing that in this case the right-hand side of (6.46) still has meaning, we write

$$a^\alpha = \exp(\alpha \operatorname{Ln} a),$$

by definition. If α is imaginary, all the values of a^α corresponding to different values of $\operatorname{Ln} a$ (or equivalently, to different values of $\operatorname{Arg} a$) are also different, since two distinct values of $\alpha \operatorname{Ln} a$ differ by a number of the form $2\pi i\alpha$, which cannot be an integral multiple of $2\pi i$ if α is imaginary. It follows from (6.46) and the addition theorem for the exponential that

$$\begin{aligned} a^\alpha a^\beta &= \exp(\alpha \operatorname{Ln} a) \exp(\beta \operatorname{Ln} a) = \exp(\alpha \operatorname{Ln} a + \beta \operatorname{Ln} a) \\ &= \exp[(\alpha + \beta) \ln a + 2\pi i(k\alpha + l\beta)], \\ a^{\alpha+\beta} &= \exp[(\alpha + \beta) \operatorname{Ln} a] = \exp[(\alpha + \beta) \ln a + 2\pi i(m\alpha + m\beta)], \end{aligned}$$

where k, l and m are all integers. Thus every value of $a^{\alpha+\beta}$ is a value of $a^\alpha a^\beta$, but the converse is true only for special values of α and β.

Similarly

$$\begin{aligned} (a^\alpha)^\beta &= [\exp(\alpha \operatorname{Ln} a)]^\beta = \exp[\beta(\alpha \operatorname{Ln} a + 2l\pi i)] \\ &= \exp[\alpha\beta \ln a + 2\pi i\beta(k\alpha + l)], \\ a^{\alpha\beta} &= \exp(\alpha\beta \operatorname{Ln} a) = \exp(\alpha\beta \ln a + 2\pi i\beta m\alpha), \end{aligned}$$

where k, l and m are all integers. Thus every value of $a^{\alpha\beta}$ is a value of $(a^\alpha)^\beta$, but again the converse is true only for special values of α and β.

Example 1.

$$\begin{aligned} 1^{\sqrt{2}} &= \exp(\sqrt{2} \ln 1) = \exp(2k\pi\sqrt{2}i) \\ &= \cos(2k\pi\sqrt{2}) + i \sin(2k\pi\sqrt{2}) \qquad (k = 0, \pm 1, \pm 2, \ldots). \end{aligned}$$

Example 2.

$$\begin{aligned} e^z &= \exp(z \operatorname{Ln} e) = \exp[z(1 + 2k\pi i)] \\ &= \exp z \exp(2k\pi iz) \qquad (k = 0, \pm 1, \pm 2, \ldots). \end{aligned} \tag{6.47}$$

It follows from (6.47) that only one of the values of the power e^z coincides with $\exp z$. In fact, the other values are

$$\exp z \exp(2\pi iz), \qquad \exp z \exp(-2\pi iz), \ldots$$

In particular, only one of the values of e^x (where x is a real number) coincides with the positive real number $\exp x$, and in this case, the other values are

$$\exp x \exp (2\pi i x), \qquad \exp x \exp (-2\pi i x), \ldots \tag{6.48}$$

[If x is rational, there are only a finite number of different values (6.48), but if x is irrational, there are infinitely many different values.] Despite this, *we shall continue to denote the exponential function by* e^z, *as well as by* $\exp z$. This use of the "multiple-valued symbol" e^z to denote a single number is analogous to the conventional use of the symbol $\sqrt[n]{a}$ (where a is a positive real number) to denote the unique positive value of the nth root of a (cf. footnote 7, p. 10).

We are now in a position to study the functions z^α and a^z, where α and $a \neq 0$ are arbitrary complex numbers. First we consider the function

$$f(z) = z^\alpha$$

(which in general is defined only for $z \neq 0$), and examine in turn the cases where α is an integer, a rational number and an irrational or imaginary number.

Case 1. If $\alpha = n$ is an integer, then

$$f(z) = z^\alpha = z^n$$

is a particularly simple rational function. In this case, $f(z)$ is defined for $z = 0$, where it has a zero (if $n > 0$) or a pole (if $n < 0$).

Case 2. If $\alpha = r$ is a rational number, then $r = m/n$ where m and $n > 0$ are relatively prime integers, and hence

$$f(z) = z^{m/n} = \sqrt[n]{z^m}.$$

In this case, $f(z)$ is an n-valued function, with branch points $z = 0$ and $z = \infty$ of order $n - 1$. Let γ be any Jordan curve joining 0 and ∞, and let G be the domain with boundary γ. Then we can define n single-valued differentiable branches of $f(z)$ on G, which change continuously into each other as we go around any closed Jordan curve whose interior contains the origin.

Case 3. If α is not a rational number, i.e., if α is an irrational real number or an imaginary number, then $f(z) = z^\alpha$ is an infinite-valued function, all of whose values are given by the formula

$$z^\alpha = \exp (\alpha \operatorname{Ln} z).$$

In this case, $z = 0$ and $z = \infty$ are branch points (as in Case 2), but now they are of infinite order. In fact, if we go around the point $z = 0$ (say, in the positive direction), $\operatorname{Arg} z$ varies continuously and increases by 2π. Therefore the value of $\alpha \operatorname{Ln} z$ increases by $2\pi i\alpha$, i.e., $f(z)$ is multiplied by factor $e^{2\pi i\alpha} \neq 1$.

Next we consider the *exponential to the base a*, i.e., the function

$$a^z = \exp (z \operatorname{Ln} a),$$

defined for any finite z and any $a \neq 0$. To obtain a definite single-valued branch of $f(z)$, it is sufficient to fix one of the values of Ln a, say

$$\text{Ln}\, a = b = \ln a + 2k_0\pi i. \tag{6.49}$$

After this has been done, we obtain a single-valued, everywhere differentiable function $\exp(bz)$. Taking all possible values of Ln a, we obtain all possible single-valued branches of the function a^z. Since two values of Ln a differ by a term of the form $2k\pi i$, where k is an integer, two branches of the function a^z differ by a factor of the form $\exp(2k\pi iz)$. This factor is also a single-valued, everywhere differentiable function, which takes the value 1 only when z is a rational number of the form m/k (where m is an arbitrary integer).

Remark. The branches of the multiple-valued function a^z differ in an essential way from those of the multiple-valued functions considered previously. In all the cases considered so far, we can find points of the extended plane, called *branch points*, with the property that by going around each branch point along suitable closed Jordan curves we can carry any single-valued branch into any other. However, in the case of the function a^z, this is impossible, since every branch is a single-valued continuous function defined in the whole finite plane (and not on some domain whose boundary consists of certain curves joining the branch points). In fact, after making a circuit around any (finite) closed Jordan curve whatsoever, we return to the original complex number z (perhaps with a different value of the argument) and hence to the same value of the function $\exp(bz)$, where b is a fixed value of Ln a. Thus the multiple-valued function a^z has no branch points at all, and its single-valued branches cannot be carried continuously into each other by making circuits around closed Jordan curves. In other words, the different branches can be regarded as self-contained, independent, single-valued, everywhere differentiable functions,[6] i.e., as nothing more or less than the infinite set of entire functions

$$\exp(z \ln a), \quad \exp[z(\ln a + 2\pi i)], \quad \exp[z(\ln a - 2\pi i)], \ldots$$

The fact that all these different entire functions can be represented as branches of a single infinite-valued function a^z is no more surprising than the fact that $\sin z$ and $-\sin z$ can be regarded as branches of the double-valued function

$$\sqrt{1 - \cos^2 z}, \tag{6.50}$$

[6] The same situation has already been encountered in Example 2 above, in connection with the function e^z (where $a = e$ and $\ln a = 1$). However, in the case of e^z, we agreed to interpret e^z as the particular single-valued branch $\exp z$, i.e., as the branch which takes real values when z is real (see p. 112). We shall make the same choice whenever a is a positive real number.

or that cosh z and sinh z can be regarded as branches of the double-valued function

$$\tfrac{1}{2}[\exp z + \sqrt{\exp(-2z)}]. \tag{6.51}$$

[It should be noted that just like the function a^z, the functions (6.50) and (6.51) have no branch points.]

Finally we consider the *logarithm to the base a*, denoted by $\mathrm{Log}_a\, z$ and defined as the inverse of the function

$$z = a^w = \exp(w \,\mathrm{Ln}\, a). \tag{6.52}$$

If we again write (6.49), thereby choosing a branch of a^w, (6.52) becomes

$$z = \exp(bw),$$

and hence

$$w = \mathrm{Log}_a\, z = \frac{1}{b}\,\mathrm{Ln}\, z, \tag{6.53}$$

which differs from Ln z only by the factor $1/b$. Since b is a value of Ln a, it follows from (6.53) that

$$\mathrm{Log}_a\, z = \frac{\mathrm{Ln}\, z}{\mathrm{Ln}\, a}, \tag{6.54}$$

where in the denominator we fix one of the infinitely many values of Ln a (which is then kept the same for all z). Thus, to define $\mathrm{Log}_a\, z$, we must specify not only the base a, but also a particular value of Ln a.

Example. Let $a = e$, and choose the value of Ln e equal to 1. Then

$$\mathrm{Log}_e\, z = \mathrm{Ln}\, z,$$

which is just the ordinary definition of the *natural logarithm*. However, we can also choose another value of Ln e, say $1 + 2\pi i$. In this case, (6.54) gives

$$\mathrm{Log}_e\, z = \frac{\mathrm{Ln}\, z}{1 + 2\pi i}.$$

It is easy to see that with this second definition of the natural logarithm, $\mathrm{Log}_e\, z$ will have a real value (in fact, exactly one) only if $z = e^k$, where k is an integer.

Problem 1. Prove that

$$i^i = e^{(4k-1)\pi/2} \qquad (k = 0, \pm 1, \pm 2, \ldots).$$

Problem 2. Find all values of the following expressions:

a) $(-2)^{\sqrt{2}}$; b) 2^i; c) 1^{-i}; d) $\left(\frac{1-i}{\sqrt{2}}\right)^{1+i}$; e) $(3-4i)^{1+i}$.

Problem 3. Calculate Im $\mathrm{Log}_{1+i}(1-i)$, choosing as Ln $(1+i)$ the principal value of the logarithm.

Problem 4. Find imaginary values of α and β such that the multiple-valued functions $a^\alpha a^\beta$ and $a^{\alpha+\beta}$ coincide. Do the same for the functions $(a^\alpha)^\beta$ and $a^{\alpha\beta}$.

CHAPTER 7

COMPLEX INTEGRALS. CAUCHY'S INTEGRAL THEOREM

34. RECTIFIABLE CURVES. COMPLEX INTEGRALS

Let L be a (continuous) curve with equation

$$z = \lambda(t) = \mu(t) + i\nu(t) \qquad (a \leqslant t \leqslant b), \tag{7.1}$$

and suppose we divide the interval $[a, b]$ into n subintervals

$$[t_{k-1}, t_k] \qquad (k = 1, \ldots, n) \tag{7.2}$$

by introducing $n - 1$ intermediate points $t_1, \ldots, t_{n-1}$ satisfying the inequalities

$$a = t_0 < t_1 < \cdots < t_{n-1} < t_n = b.$$

The set

$$\mathscr{P} = \{t_0, t_1, \ldots, t_n\} \tag{7.3}$$

is called a *partition* of the interval $[a, b]$, and the largest of the numbers

$$t_1 - t_0, \qquad t_2 - t_1, \ldots, t_n - t_{n-1},$$

i.e., the maximum length of the subintervals (7.2) is called the *norm* of the partition $\mathscr{P}$, denoted by $|\mathscr{P}|$. Corresponding to the partition (7.3), the curve

L is divided into n arcs[1]

$$\sigma_k = \widehat{z_{k-1}z_k} \qquad (k = 1, \ldots, n), \tag{7.4}$$

where

$$z_k = \lambda(t_k) \qquad (k = 0, 1, \ldots, n), \tag{7.5}$$

and the final point of each arc (except the last one) coincides with the initial point of the next arc. Joining each of the points $z_0, z_1, \ldots, z_n$ to the next by a straight line segment (as illustrated by Figure 7.1 for the case $n = 5$), we obtain a polygonal curve Λ *inscribed* in L. The segments of Λ are the chords joining the end points of the arcs σ_k, and the length of Λ is obviously

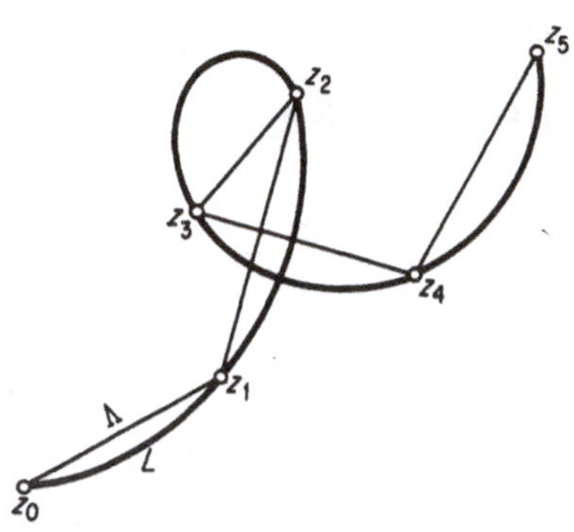

FIGURE 7.1

$$\sum_{k=1}^{n} |z_k - z_{k-1}| \tag{7.6}$$

The curve L is said to be *rectifiable* if

$$\sup_{\mathscr{P}} \sum_{k=1}^{n} |z_k - z_{k-1}| = l < \infty,$$

where the least upper bound (denoted by the symbol sup) is taken over all possible partitions (7.3). The nonnegative real number l is called the *length* of the curve L. However, if the sums (7.6) become arbitrarily large for suitably chosen partitions, the curve is said to be *nonrectifiable*. In this case, L is considered to have no length at all (or, if preferred, infinite length).

Now let L be a rectifiable curve with equation (7.1), and let $P(x, y)$ and $Q(x, y)$ be two real functions, defined and continuous at every point $z = x + iy$ of L. Given an arbitrary partition $\mathscr{P} = \{t_0, t_1, \ldots, t_n\}$ of the interval $[a, b]$, let σ_k and z_k be the same as in (7.4) and (7.5), and let

$$\zeta_k = \xi_k + i\eta_k = \lambda(\tau_k) \qquad (t_{k-1} \leqslant \tau_k \leqslant t_k)$$

be an arbitrary point of σ_k $(k = 1, \ldots, n)$. Then it can be shown (see Prob. 7) that the sum

$$\sum_{k=1}^{n} [P(\xi_k, \eta_k)(x_k - x_{k-1}) + Q(\xi_k, \eta_k)(y_k - y_{k-1})] \tag{7.7}$$

[1] By $\widehat{AB}$ we mean a curve joining the points A and B (the curve in question is always apparent from the context). In the case where the curve is the line segment joining A to B, we write $\overline{AB}$ as well as $\widehat{AB}$. If γ and δ are two curves such that the final point of γ coincides with the initial point of δ, then by $\gamma + \delta$ (as opposed to $\gamma \cup \delta$) we mean the *oriented* curve obtained by first going along γ from its initial point to its final point and then along δ from its initial point to its final point. All curves figuring in line integrals and complex integrals (see below) are regarded as oriented (recall the remark on p. 38).

approaches a unique limit, denoted by

$$\int_L \{P(x, y)\,dx + Q(x, y)\,dy\}, \tag{7.8}$$

as the norm of the partition $\mathscr{P}$ approaches zero. An expression of the form (7.8) is known as a *line integral* (*along the curve L*). Similarly, let

$$f(z) = u(x, y) + iv(x, y)$$

be a complex function, defined and continuous on L, and consider the sum

$$S = \sum_{k=1}^{n} f(\zeta_k)(z_k - z_{k-1}).$$

Introducing the additional notation

$$u_k = u(\xi_k, \eta_k), \quad v_k = v(\xi_k, \eta_k), \quad \Delta x_k = x_k - x_{k-1}, \quad \Delta y_k = y_k - y_{k-1},$$

we have

$$f(\zeta_k) = u_k + iv_k, \qquad z_k - z_{k-1} = \Delta x_k + i\,\Delta y_k,$$

and hence

$$\begin{aligned} S &= \sum_{k=1}^{n} f(\zeta_k)(z_k - z_{k-1}) = \sum_{k=1}^{n}(u_k + iv_k)(\Delta x_k + i\,\Delta y_k) \\ &= \sum_{k=1}^{n}(u_k\,\Delta x_k - v_k\,\Delta y_k) + i\sum_{k=1}^{n}(v_k\,\Delta x_k + u_k\,\Delta y_k). \end{aligned} \tag{7.9}$$

Thus we see that the real and imaginary parts of S are both sums of the form (7.7), corresponding to the partition $\mathscr{P}$, where in the first case the role of the real functions P and Q is played by the functions u and $-v$, and in the second case by the functions v and u. By the (*contour*) *integral of* $f(z)$ *along the curve* (*or contour*) L,[2] denoted by

$$\int_L f(z)\,dz, \tag{7.10}$$

we mean the quantity

$$\lim_{|\mathscr{P}| \to 0} \sum_{k=1}^{n} f(\zeta_k)(z_k - z_{k-1}), \tag{7.11}$$

i.e., the limit of the sum S as the norm of the partition $\mathscr{P}$ approaches zero. Since the real functions u and v are continuous on L [being the real and imaginary parts of the continuous function $f(z)$], and since L is rectifiable, it follows from (7.9) and the definition of the corresponding line integrals that

$$\int_L f(z)\,dz = \int_L \{u(x, y)\,dx - v(x, y)\,dy\} + i\int_L \{v(x, y)\,dx + u(x, y)\,dy\}. \tag{7.12}$$

---

[2] As in the case of a line integral, L is often called the *path of integration*. The term *contour* is used as a synonym for *curve*, mainly when the curve is closed.

In other words, evaluating the integral of a complex function reduces to evaluating two line integrals of real functions.

Example 1. If $f(z) \equiv 1$, (7.10) becomes

$$\int_L dz = \lim_{|\mathscr{P}|\to 0} \sum_{k=1}^{n}(z_k - z_{k-1}) = \lim_{|\mathscr{P}|\to 0} (z_n - z_0) = Z - z_0,$$

where z_0 is the initial point and Z the final point of the curve L. In particular, if L is closed, $Z = z_0$ and

$$\int_L dz = 0.$$

Example 2. If $f(z) = z$, we first choose ζ_k to be the initial point z_{k-1} of the arc $\sigma_k = \widehat{z_{k-1}z_k}$, and then the final point z_k of σ_k. This gives

$$\begin{aligned}\int_L z\, dz &= \lim_{|\mathscr{P}|\to 0} \sum_{k=1}^{n} z_{k-1}(z_k - z_{k-1}) = \lim_{|\mathscr{P}|\to 0} \sum_{k=1}^{n} z_k(z_k - z_{k-1}) \\ &= \frac{1}{2}\lim_{|\mathscr{P}|\to 0} \sum_{k=1}^{n}(z_k + z_{k-1})(z_k - z_{k-1}) = \frac{1}{2}\lim_{|\mathscr{P}|\to 0} \sum_{k=1}^{n}(z_k^2 - z_{k-1}^2) \\ &= \frac{1}{2}\lim_{|\mathscr{P}|\to 0}(z_n^2 - z_0^2) = \frac{1}{2}(Z^2 - z_0^2).\end{aligned}$$

In particular, if L is a closed curve, so that $Z = z_0$, we again find that (7.10) vanishes.

Example 3. Suppose $f(z) = 1/(z - a)$, and let L be the curve with equation

$$z = a + re^{it} \qquad (0 \leqslant t \leqslant 2\pi),$$

i.e., the circle of radius r with center at the point a, traversed once in the positive direction. Dividing L into n equal arcs $\sigma_k = \widehat{z_{k-1}z_k}$, where

$$z_k = a + re^{2k\pi i/n} \qquad (k = 0, 1, \ldots, n)$$

($z_n = z_0$, since L is closed), we choose ζ_k to be the midpoint of the arc σ_k:

$$\zeta_k = a + re^{(2k-1)\pi i/n} \qquad (k = 1, \ldots, n).$$

As a result,

$$f(\zeta_k) = \frac{1}{\zeta_k - a} = \frac{1}{r}e^{-(2k-1)\pi i/n},$$

and

$$\begin{aligned}\sum_{k=1}^{n} f(\zeta_k)(z_k - z_{k-1}) &= \sum_{k=1}^{n} e^{-(2k-1)\pi i/n}(e^{2k\pi i/n} - e^{2(k-1)\pi i/n}) \\ &= \sum_{k=1}^{n}(e^{\pi i/n} - e^{-\pi i/n}) = 2i\sum_{k=1}^{n}\sin\frac{\pi}{n} = 2in\sin\frac{\pi}{n}.\end{aligned}$$

It follows that

$$\int_L \frac{dz}{z-a} = \lim_{n\to\infty} 2in \sin\frac{\pi}{n} = 2\pi i \lim_{n\to\infty} \frac{\sin(\pi/n)}{\pi/n} = 2\pi i.$$

Next we derive some elementary properties of complex integrals. The first three properties can be verified immediately either by using (7.12) to go over to line integrals, or by using (7.11) directly. In each case, we assume that L is a rectifiable curve, and that $f(z)$ is defined and continuous on L.

Property 1.

$$\int_L f(z)\,dz = -\int_{-L} f(z)\,dz,$$

where $-L$ denotes the curve L traversed in the opposite direction.

Property 2. Suppose that

$$L = L_1 + L_2 + \cdots + L_m,$$

where the final point of L_k coincides with the initial point of L_{k+1} $(k = 1, \ldots, m-1)$. Then we have

$$\int_L f(z)\,dz = \int_{L_1} f(z)\,dz + \int_{L_2} f(z)\,dz + \cdots + \int_{L_m} f(z)\,dz.$$

Property 3.

$$\int_L [c_1 f_1(z) + \cdots + c_m f_m(z)]\,dz = c_1 \int_L f_1(z)\,dz + \cdots + c_m \int_L f_m(z)\,dz,$$

where $f_1(z), \ldots, f_m(z)$ are arbitrary complex functions which are defined and continuous on L, and $c_1, \ldots, c_m$ are arbitrary complex constants.

Property 4. Suppose $f(z)$ satisfies the inequality

$$|f(z)| \leqslant M$$

at all points of the curve L, e.g., suppose M is the quantity

$$\max_{z\in L} |f(z)|.$$

Then we have the estimate

$$\left| \int_L f(z)\,dz \right| \leqslant Ml, \tag{7.13}$$

where l is the length of the curve L. To prove (7.13), we need only take the limit of the inequality

$$\left| \sum_{k=1}^{n} f(\zeta_k)(z_k - z_{k-1}) \right| \leqslant \sum_{k=1}^{n} |f(\zeta_k)|\,|z_k - z_{k-1}| \leqslant M \sum_{k=1}^{n} |z_k - z_{k-1}| \leqslant Ml$$

as the norm of the partition $\mathscr{P}$ goes to zero.

Problem 1. Given a continuous curve L with end points A and B, let C be any other point of L. Prove that the curve $L = \overset{\frown}{AB}$ is rectifiable if and only if the subcurves $\overset{\frown}{AC}$ and $\overset{\frown}{CB}$ are rectifiable. Assuming that these curves are rectifiable, prove that the length of $\overset{\frown}{AB}$ is the sum of the lengths of $\overset{\frown}{AC}$ and $\overset{\frown}{CB}$.

Problem 2 (M1, p. 253). Let L be a rectifiable curve of length l, and let $\Lambda(\mathscr{P})$ be the polygonal curve inscribed in L corresponding to the partition $\mathscr{P}$. Prove that $\Lambda(\mathscr{P}) \to l$ as $|\mathscr{P}| \to 0$.

Problem 3. A real function $f(t)$ defined on a closed interval $[a, b]$ is said to be of *bounded variation* on $[a, b]$ if

$$\sup_{\mathscr{P}} \sum_{k=1}^{n} |f(t_k) - f(t_{k-1})| < \infty,$$

where the least upper bound is taken over all possible partitions $\mathscr{P} = \{t_0, t_1, \ldots, t_n\}$ of the interval $[a, b]$. Prove that the curve with equation $z = \lambda(t) = \mu(t) + i\nu(t)$, $a \leqslant t \leqslant b$ is rectifiable if and only if both $\mu(t)$ and $\nu(t)$ are of bounded variation on $[a, b]$.

Problem 4. Show that the function

$$f(t) = \begin{cases} t \sin \dfrac{\pi}{2t} & \text{for} \quad t \neq 0, \\ 0 & \text{for} \quad t = 0 \end{cases}$$

is continuous but not of bounded variation on $[0, 1]$. Give an example of a continuous nonrectifiable curve.

Problem 5. Prove that $f(t)$ is of bounded variation on $[a, b]$ if and only if

$$f(t) = f_1(t) - f_2(t)$$

on $[a, b]$, where $f_1(t)$ and $f_2(t)$ are both increasing functions on $[a, b]$ (see Sec. 12, Prob. 1).

Hint. First prove that $f(t)$ is of bounded variation on $[a, b]$ if and only if $f(t)$ has a *majorant* $F(t)$ on $[a, b]$, i.e., a bounded increasing function $F(t)$ such that

$$|f(t'') - f(t')| \leqslant F(t'') - F(t')$$

whenever $a \leqslant t' < t'' \leqslant b$.

Problem 6. Let $f(t)$ and $g(t)$ be two functions defined on the same interval $[a, b]$, let $\mathscr{P} = \{t_0, t_1, \ldots, t_n\}$ be an arbitrary partition of $[a, b]$, and let τ_k be any point in the interval $[t_{k-1}, t_k]$. Then the limit

$$\lim_{|\mathscr{P}| \to 0} \sum_{k=1}^{n} f(\tau_k)[g(t_k) - g(t_{k-1})],$$

provided it exists, it called the *Stieltjes integral* of $f(t)$ with respect to $g(t)$ from a to b, and is denoted by

$$\int_a^b f(t)\,dg(t).$$

Prove that this limit exists if $f(t)$ is continuous on $[a, b]$ and $g(t)$ is of bounded variation on $[a, b]$.

Hint. First consider the case where $g(t)$ is increasing on $[a, b]$, and repeat step by step the familiar proof that the ordinary Riemann integral

$$\int_a^b f(t)\,dt$$

exists if $f(t)$ is continuous on $[a, b]$. Then use the preceding problem.

Problem 7. Let L be a continuous rectifiable curve, with equation

$$z = \lambda(t) = \mu(t) + i\nu(t) \qquad (a \leqslant t \leqslant b),$$

and let $P(x, y)$ and $Q(x, y)$ be two real functions which are defined and continuous on L. Prove the existence of the line integral

$$\int_L \{P(x, y)\,dx + Q(x, y)\,dy\},$$

as defined on p. 139.

Hint. Write the line integral as a sum of Stieltjes integrals with respect to $\mu(t)$ and $\nu(t)$, and use Probs. 3 and 6.

Problem 8. Verify that the definite integral of a real function of a real variable is a special case of the integral of a function of a complex variable along a curve.

Problem 9. Let L be a rectifiable curve with equation $z = \lambda(t)$, $a \leqslant t \leqslant b$, and let $s(t)$ be the length of the arc of L joining $\lambda(a)$ to $\lambda(t)$ [cf. Prob. 1], where $s(a) = 0$ by definition. Prove that the nonnegative function $s(t)$ is continuous and increasing on $[a, b]$, in fact, *strictly* increasing if L is a Jordan curve.

Problem 10. Let $f(z)$ be a continuous function on a rectifiable curve L of length l, with equation $z = \lambda(t)$, $a \leqslant t \leqslant b$, and let $s(t)$ be the arc length function introduced in the preceding problem. By $\int_L f(z)\,ds$ or $\int_L f(z)\,|dz|$, we mean the Stieltjes integral

$$\int_a^b f[\lambda(t)]\,ds(t),$$

whose existence is guaranteed by the fact that $s(t)$ is of bounded variation on $[a, b]$. Prove that

$$\left|\int_L f(z)\,dz\right| \leqslant \int_L |f(z)|\,ds \leqslant Ml,$$

which sharpens the estimate (7.13).

Hint. If $\mathscr{P}$, σ_k and ζ_k are the same as on p. 138, then

$$\int_L f(z)\,ds = \lim_{|\mathscr{P}|\to 0} \sum_{k=1}^{n} f(\zeta_k) l_k,$$

where l_k is the length of σ_k.

35. THE CASE OF SMOOTH CURVES

Suppose L is *smooth*, i.e., suppose there is at least one parametric representation (7.1) such that $\lambda(t)$ has a continuous nonvanishing derivative $\lambda'(t)$ at every point of the interval $[a, b]$.[3] The geometric meaning of smoothness is clear from Sec. 15, where it was proved that if $\lambda'(t_0) \neq 0$, then L has a tangent of inclination $\operatorname{Arg} \lambda'(t_0)$ at the point $z_0 = \lambda(t_0)$. Therefore a smooth curve has a tangent at every point, and moreover the inclination of the tangent varies continuously as the point of tangency moves along the curve (why?). It is not hard to show (see Probs. 3 and 4) that a smooth curve is rectifiable, with length

$$l = \int_a^b \sqrt{[\mu'(t)]^2 + [\nu'(t)]^2}\, dt. \tag{7.14}$$

Since

$$\lambda'(t) = \mu'(t) + i\nu'(t), \qquad |\lambda'(t)|^2 = [\mu'(t)]^2 + [\nu'(t)]^2,$$

(7.14) can also be written as

$$l = \int_a^b |\lambda'(t)|\, dt.$$

Similarly, (7.8) becomes

$$\int_a^b \{P[\mu(t), \nu(t)]\mu'(t) + Q[\mu(t), \nu(t)]\nu'(t)\}\, dt \tag{7.15}$$

(see Prob. 5), while (7.12) takes the form

$$\begin{aligned}\int_L f(z)\,dz &= \int_a^b \{u[\mu(t), \nu(t)]\mu'(t) - v[\mu(t), \nu(t)]\nu'(t)\}\, dt \\ &\quad + i\int_a^b \{v[\mu(t), \nu(t)]\mu'(t) + u[\mu(t), \nu(t)]\nu'(t)\}\, dt \\ &= \int_a^b \{u[\mu(t), \nu(t)] + iv[\mu(t), \nu(t)]\}\{\mu'(t) + i\nu'(t)\}\, dt \\ &= \int_a^b f[\lambda(t)]\lambda'(t)\, dt.\end{aligned} \tag{7.16}$$

[3] At $t = a$, $\lambda'(t)$ denotes the right-hand derivative and is continuous from the right; at $t = b$, $\lambda'(t)$ denotes the left-hand derivative and is continuous from the left. Thus L has a right-hand tangent at $\lambda(a)$ and a left-hand tangent at $\lambda(b)$, in the sense of Sec. 15, Prob. 1 (verify this assertion).

Example. Formula (7.16) can be used to simplify the evaluation of the integral in Example 3, p. 140. In fact, since

$$z = \lambda(t) = a + re^{it}, \qquad \lambda'(t) = ire^{it},$$

we have

$$\int_L \frac{dz}{z-a} = \int_0^{2\pi} \frac{ire^{it}\,dt}{re^{it}} = i\int_0^{2\pi} dt = 2\pi i.$$

Remark. The formulas just written for the case of a smooth curve L remain valid if L is merely *piecewise smooth*, i.e., if $[a, b]$ can be divided into a finite number of subintervals

$$a = a_0 \leqslant t \leqslant a_1, a_1 \leqslant t \leqslant a_2, \ldots, a_{m-1} \leqslant t \leqslant a_m = b$$

such that $\lambda(t)$ has a continuous nonvanishing derivative on each subinterval, but not necessarily on $[a, b]$ itself. Thus $\lambda'(t)$ is now allowed to have discontinuities at finitely many interior points of $[a, b]$, provided nonzero one-sided derivatives (and hence one-sided tangents) exist at every such point.

Problem 1. Let L and L' be two identical curves, in the sense of p. 38. Prove that L is smooth if and only if L' is smooth, provided the parameter change $\tau = \alpha(t)$ has a continuous nonvanishing derivative on $[a, b]$. How about piecewise smoothness?

Problem 2. Give an example of a curve $z = \lambda(t), a \leqslant t \leqslant b$ with continuously differentiable $\lambda(t)$ which fails to have a tangent at some point of $[a, b]$.

Ans. Let $\lambda(t) = t^3 + i\,|t^3|$ in the interval $[-1, 1]$.

Problem 3. Verify formula (7.14) for a smooth curve L.

Hint. Use the mean value theorem to write

$$\begin{aligned}\sum_{k=1}^{n} |z_k - z_{k-1}| &= \sum_{k=1}^{n} \sqrt{[\mu(t_k) - \mu(t_{k-1})]^2 + [\nu(t_k) - \nu(t_{k-1})]^2} \\ &= \sum_{k=1}^{n} \sqrt{[\mu'(\tau_k)]^2 + [\nu'(\tilde{\tau}_k)]^2}\,\Delta t_k,\end{aligned} \tag{7.17}$$

where τ_k and $\tilde{\tau}_k$ are two points in the interval $[t_{k-1}, t_k]$, and $\Delta t_k = t_k - t_{k-1}$. Except for the fact that τ_k and $\tilde{\tau}_k$ are in general different, (7.17) is an approximating sum of the Riemann integral (7.14). Use the fact that $\mu'(t)$ and $\nu'(t)$ are continuous to circumvent this difficulty.

Comment. Note that we have not used the fact that $\lambda'(t)$ is nonvanishing on $[a, b]$.

Problem 4. Prove that if $f(t)$ is continuously differentiable on $[a, b]$, then $f(t)$ is of bounded variation on $[a, b]$.

Comment. Together with Sec. 34, Prob. 3, this gives another proof of the fact that a smooth curve is rectifiable.

Problem 5. Verify the equivalence of (7.8) and (7.15) if $L: z = \lambda(t)$ is smooth.

Hint. Recall the definition of (7.8) as a Stieltjes integral (Sec. 34, Prob. 7). Then use the mean value theorem to replace $d\mu(t)$ and $d\nu(t)$ by $\mu'(t)\,dt$ and $\nu'(t)\,dt$.

Problem 6. Evaluate the integrals

$$J_1 = \int_L x\,dz, \qquad J_2 = \int_L y\,dz$$

along the following paths:

a) The line segment joining the point $z = 0$ to the point $z = 2 + i$;
b) The semicircle $|z| = 1$, $0 \leqslant \arg z \leqslant \pi$, with initial point $z = 1$;
c) The circle $|z - a| = R$.

Ans. b) $J_1 = \frac{1}{2}\pi i$, $J_2 = -\frac{1}{2}\pi$.

Problem 7. Evaluate the integral

$$\int_L |z|\,dz$$

along

a) The circle $|z| = R$;
b) The semicircle $|z| = 1$, $-\pi/2 \leqslant \arg z \leqslant \pi/2$ with initial point $z = -i$.

36. CAUCHY'S INTEGRAL THEOREM. THE KEY LEMMA

We are now in a position to formulate

THEOREM 7.1 (*Cauchy's integral theorem*). *Let G be a simply connected domain, and let $f(z)$ be a single-valued analytic function on G. Then*

$$\int_L f(z)\,dz = 0,$$

where L is any closed rectifiable curve contained in G.

The proof of Theorem 7.1 is rather complicated, and will be given in the next section. We begin by proving the following weaker version of the theorem:

THEOREM 7.2. *Let G be a simply connected domain, and let $f(z)$ be a single-valued continuously differentiable function on G. Then*

$$\int_L f(z)\,dz = 0,$$

where L is any closed piecewise smooth Jordan curve contained in G.

Proof. According to formula (7.12),

$$\int_L f(z)\,dz = \int_L (u\,dx - v\,dy) + i\int_L (v\,dx + u\,dy), \qquad (7.18)$$

where $u = \operatorname{Re} f(z)$, $v = \operatorname{Im} f(z)$. Since $f'(z)$ is continuous on G and

$$f'(z) = \frac{\partial u}{\partial x} + i\frac{\partial v}{\partial y} = \frac{\partial v}{\partial x} - i\frac{\partial u}{\partial y},$$

it follows that the first-order partial derivatives of u and v are continuous on $\bar{D}$, where $D = I(L)$ is the interior of an arbitrary closed piecewise smooth Jordan curve $L \subset G$. We now apply Green's theorem[4]

$$\int_L (P\,dx + Q\,dy) = \iint_D \left(\frac{\partial Q}{\partial x} - \frac{\partial P}{\partial y}\right) dx\,dy \tag{7.19}$$

to each of the integrals in the right-hand side of (7.18), where the line integral in (7.19) is taken in the positive (i.e., counterclockwise) direction. The result is

$$\begin{aligned}\int_L (u\,dx - v\,dy) &= \iint_D \left(-\frac{\partial v}{\partial x} - \frac{\partial u}{\partial y}\right) dx\,dy,\\ \int_L (v\,dx + u\,dy) &= \iint_D \left(\frac{\partial u}{\partial x} - \frac{\partial v}{\partial y}\right) dx\,dy.\end{aligned} \tag{7.20}$$

But the integrand of each double integral in (7.20) vanishes, because of the Cauchy-Riemann equations (see Theorem 3.2). Therefore both sides of (7.18) vanish, and the proof is complete.

Remark 1. The strengthening of Theorem 7.2 represented by Theorem 7.1 consists in dropping the requirement that $f'(z)$ be continuous and that L be a Jordan curve, and in assuming that L is only rectifiable instead of piecewise smooth.

Remark 2. Actually, it turns out (see Theorem 8.5, Corollary) that $f'(z)$ is analytic (and hence automatically continuous) on G, if $f(z)$ is analytic on G. However, Theorem 7.1 will be needed to prove this!

In proving Theorem 7.1, we shall need the following

LEMMA (*Key lemma*). *Let $f(z)$ be a continuous function defined on a domain G, and let L be an arbitrary rectifiable curve contained in G, with equation $z = \lambda(t)$, $a \leqslant t \leqslant b$. Then, given any $\varepsilon > 0$, there exists a $\delta = \delta(\varepsilon) > 0$ such that for every partition $\mathscr{P} = \{t_0, t_1, \ldots, t_n\}$ of the interval $[a, b]$ with norm less than δ, the polygonal curve*[5]

$$\Lambda = \overline{z_0 z_1} + \cdots + \overline{z_{n-1} z_n}, \tag{7.21}$$

[4] See e.g., D. V. Widder, *op. cit.*, p. 223.

[5] Recall footnote 1, p. 138.

where $z_k = \lambda(t_k)$, *is contained in* G *and moreover*

$$\left| \int_L f(z)\, dz - \int_\Lambda f(z)\, dz \right| < \varepsilon. \tag{7.22}$$

Proof. The curve L is a compact set (see Theorem 2.6). Hence there is another compact set $E \subset G$ and a number $\rho > 0$ such that E contains all the closed neighborhoods $\overline{\mathscr{N}(z; \rho)}$, $z \in L$. This is obvious if G is the whole finite plane. Otherwise, let Γ be the boundary of G, let d be the distance between L and Γ (cf. Theorem 2.4 and Sec. 9, Prob. 10a), and choose E to be the set of all points whose distance from L does not exceed $\rho = \frac{1}{2}d$.[6] Since $f(z)$ is uniformly continuous on E (see Theorem 2.8), given any $\varepsilon > 0$, there exists an $\eta > 0$ such that

$$|f(z) - f(z')| < \frac{\varepsilon}{2l} \tag{7.23}$$

for every pair of points z, z' in E such that $|z - z'| < \eta$, where l is the length of L. Moreover, the function $z = \lambda(t)$ is uniformly continuous on $[a, b]$, and hence there exists a $\delta_1 > 0$ such that

$$|z - z'| = |\lambda(t) - \lambda(t')| < \min(\eta, \rho)$$

if $|t - t'| < \delta_1$.

Now suppose

$$\left| \int_L f(z)\, dz - \sum_{k=1}^{n} f(z_k)(z_k - z_{k-1}) \right| < \frac{\varepsilon}{2} \tag{7.24}$$

if $|\mathscr{P}| < \delta_2$ (such a δ_2 exists by the very definition of the complex integral) and let $\delta = \min(\delta_1, \delta_2)$. Choose a partition $\mathscr{P} = \{t_0, t_1, \ldots, t_n\}$ with norm less than δ, and let Λ be the corresponding polygonal curve (7.21). Clearly $\Lambda \subset G$, since if Δ_k is the segment $\overline{z_{k-1}z_k}$, then $z \in \Delta_k$ implies

$$|z - z_k| \leqslant |z_k - z_{k-1}| < \min(\eta, \rho) \leqslant \rho,$$

and hence $z \in \overline{\mathscr{N}(z_k; \rho)} \subset E \subset G$. Furthermore, as we now prove,

$$\left| \int_\Lambda f(z)\, dz - \sum_{k=1}^{n} f(z_k)(z_k - z_{k-1}) \right| < \frac{\varepsilon}{2}.$$

In fact,

$$f(z_k)(z_k - z_{k-1}) = f(z_k) \int_{\Delta_k} dz = \int_{\Delta_k} f(z_k)\, dz,$$

[6] The fact that E is contained in G and closed is somewhat less than obvious (see Probs. 2 and 4).

so that

$$\left|\int_\Lambda f(z)\,dz - \sum_{k=1}^n f(z_k)(z_k - z_{k-1})\right| = \left|\sum_{k=1}^n \int_{\Delta_k} f(z)\,dz - \sum_{k=1}^n \int_{\Delta_k} f(z_k)\,dz\right|$$

$$= \left|\sum_{k=1}^n \int_{\Delta_k} [f(z) - f(z_k)]\,dz\right|$$

$$\leqslant \sum_{k=1}^n \left|\int_{\Delta_k} [f(z) - f(z_k)]\,dz\right|. \qquad (7.25)$$

But if $z \in \Delta_k$, then, as already noted, both z and z_k belong to E, and moreover

$$|z - z_k| \leqslant |z_k - z_{k-1}| < \min(\eta, \rho) \leqslant \eta,$$

which implies

$$|f(z) - f(z_k)| < \frac{\varepsilon}{2l}$$

[cf. (7.23)]. It follows from the general estimate (7.13) that

$$\left|\int_{\Delta_k} [f(z) - f(z_k)]\,dz\right| \leqslant \frac{\varepsilon}{2l}|z_k - z_{k-1}|,$$

and hence (7.25) becomes

$$\left|\int_\Lambda f(z)\,dz - \sum_{k=1}^n f(z_k)(z_k - z_{k-1})\right| \leqslant \sum_{k=1}^n \frac{\varepsilon}{2l}|z_k - z_{k-1}| < \frac{\varepsilon}{2} \qquad (7.26)$$

(recall the definition of l). Comparing (7.26) and (7.24), we obtain the inequality (7.22), and the proof is complete.

Problem 1. Let L be a closed piecewise smooth Jordan curve enclosing an area S. Prove that

a) $\int_L x\,dz = iS$; b) $\int_L y\,dz = -S$; c) $\int_L \bar{z}\,dz = 2iS$.

Hint. Use Green's theorem.

Problem 2. Let E and G be the same as in the proof of the key lemma. Prove that $E \subset G$.

Hint. Obviously E cannot contain a boundary point of G. But if E contains an exterior point of G, then E must contain a boundary point of G. This in turn follows from the fact that a line segment joining an interior point of G to an exterior point of G must contain a boundary point of G (why?).

Problem 3. As on p. 28, let $\rho(z, E)$ be the distance between a point z and a set E. Show that $\rho(z, E)$ is a continuous function of z if E is compact, or for that matter only closed.

Hint. According to Sec. 11, Prob. 4, there is a point $\zeta \in E$ such that

$$\rho(z, E) = |z - \zeta|.$$

Clearly, given any z',

$$|z' - \zeta| = |(z' - z) + (z - \zeta)| \leqslant |z' - z| + |z - \zeta| = |z' - z| + \rho(z, E),$$

and hence

$$\rho(z', E) \leqslant |z' - \zeta| \leqslant |z' - z| + \rho(z, E),$$

or

$$\rho(z', E) - \rho(z, E) \leqslant |z' - z|.$$

Interchanging z and z', we obtain

$$\rho(z, E) - \rho(z', E) \leqslant |z - z'|$$

and hence

$$|\rho(z, E) - \rho(z', E)| \leqslant |z - z'|.$$

Problem 4. Let E and G be the same as in the proof of the key lemma. Prove that E is closed.

Hint. Use the preceding problem.

37. PROOF OF CAUCHY'S INTEGRAL THEOREM

We now turn to the proof proper of Theorem 7.1 (Cauchy's integral theorem). First we prove the theorem for closed polygonal curves, and then we use the key lemma to prove the theorem for arbitrary closed rectifiable curves. The proof will be divided into six steps, of progressively increasing complexity.

STEP 1 ("*2-gons*"). If L is a straight line segment Δ, traversed twice in opposite directions, then obviously

$$\int_L f(z)\,dz = \int_\Delta f(z)\,dz + \int_{-\Delta} f(z)\,dz = \int_\Delta f(z)\,dz - \int_\Delta f(z)\,dz = 0.$$

Thus, in this simplest case, the differentiability of the function need not be invoked at all.

STEP 2 (*Triangles*). As we shall see in the course of the proof, this is the basic step, and here essential use will be made of the fact that $f(z)$ is differentiable. Let L be a triangular contour contained in G, and suppose L is traversed once in a definite direction (say counterclockwise). Moreover, let

$$\left|\int_L f(z)\,dz\right| = M,$$

where obviously $M \geqslant 0$. Our goal is to show that $M = 0$. To this end, we draw the line segments joining midpoints of the sides of the triangle, thereby obtaining four congruent subtriangles L^{I}, L^{II}, L^{III} and L^{IV}, as shown in Figure 7.2.[7]

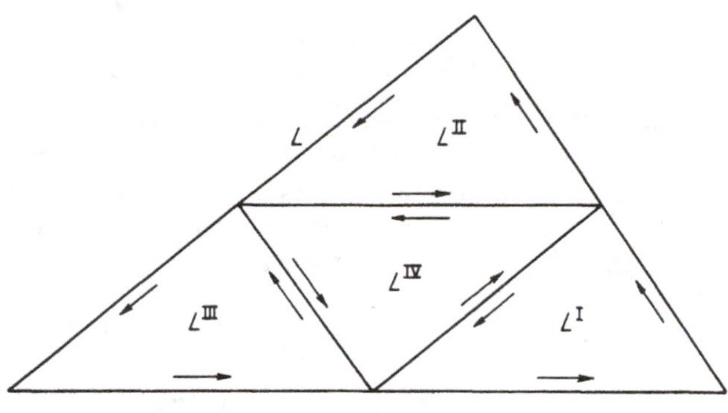

FIGURE 7.2

Forming the sum of the integrals taken along L^{I}, L^{II}, L^{III} and L^{IV} in the directions shown by the arrows in the figure (i.e., in the counterclockwise direction), we obtain

$$\int_{L^{\mathrm{I}}} f(z)\,dz + \int_{L^{\mathrm{II}}} f(z)\,dz + \int_{L^{\mathrm{III}}} f(z)\,dz + \int_{L^{\mathrm{IV}}} f(z)\,dz. \qquad (7.27)$$

Each of these four integrals can be replaced by the sum of three integrals taken along the separate sides of the subtriangles. Of the resulting integrals, six have a sum equal to $\int_L f(z)\,dz$ (i.e., the integrals along the segments whose union is L). The remaining six integrals can be grouped into three pairs, where each pair is taken along the same segment traversed twice in opposite directions. According to Step 1, the sum of each of these pairs of integrals equals zero. Therefore the whole sum (7.27) equals $\int_L f(z)\,dz$, and moreover

$$M = \left|\int_L f(z)\,dz\right|$$
$$\leqslant \left|\int_{L^{\mathrm{I}}} f(z)\,dz\right| + \left|\int_{L^{\mathrm{II}}} f(z)\,dz\right| + \left|\int_{L^{\mathrm{III}}} f(z)\,dz\right| + \left|\int_{L^{\mathrm{IV}}} f(z)\,dz\right|, \qquad (7.28)$$

[7] For simplicity, we shall use the same symbols L, L^{I}, L_1, etc., (and the same word "triangle") to denote both triangular contours and the closed domains bounded by such contours. Whether we are talking about contours or domains will always be clear from the context.

since the modulus of a sum does not exceed the sum of the moduli of the summands. It follows from (7.28) that at least one of the terms in the right-hand side must be no less than $M/4$. Denoting the corresponding triangle by L_1 (so that L_1 is one of the triangles L^{I}, L^{II}, L^{III}, L^{IV}), we have

$$\left|\int_{L_1} f(z)\,dz\right| \geqslant \frac{M}{4}.$$

Next we apply the same argument to the triangle L_1 as to the original triangle L. Thus, dividing L_1 into four congruent subtriangles L_1^{I}, L_1^{II}, L_1^{III} and L_1^{IV}, we note that the integral along L_1 equals the sum of four integrals along L_1^{I}, L_1^{II}, L_1^{III} and L_1^{IV}, taken in the same (counterclockwise) direction. It follows that the modulus of one of these integrals is no less than

$$\frac{1}{4}\cdot\frac{M}{4} = \frac{M}{4^2}.$$

If this integral is along L_2 (so that L_2 is one of the triangles L_1^{I}, L_1^{II}, L_1^{III}, L_1^{IV}), we have

$$\left|\int_{L_2} f(z)\,dz\right| \geqslant \frac{M}{4^2}.$$

Continuing this argument indefinitely, we obtain a sequence of triangles $L = L_0, L_1, L_2, \ldots, L_n, \ldots$ such that

1. L_{n+1} is a proper subset of L_n $(n = 0, 1, 2, \ldots)$, and in fact L_{n+1} is one of the four triangles obtained by drawing the line segments joining the midpoints of the sides of L_n.
2. If $l = l_0$ is the length of $L = L_0$ and l_n the length of L_n, then
$$l_n = \frac{l}{2^n}.$$
3. Each triangle is contained in G, since the curve $L \subset G$ and hence $I(L) \subset G$.
4. The integral of $f(z)$ along L_n satisfies the inequality
$$\left|\int_{L_n} f(z)\,dz\right| \geqslant \frac{M}{4^n} \qquad (n = 1, 2, \ldots). \tag{7.29}$$

It follows from the first of the properties just enumerated that there is a unique point ζ belonging to all the triangles (why?), and from the third property that $\zeta \in G$. Therefore, by hypothesis, the function $f(z)$ has a derivative $f'(\zeta)$ at the point ζ (recall the statement of Theorem 7.1), i.e., for any $\varepsilon > 0$, we can find a $\delta > 0$ such that

$$|f(z) - f(\zeta) - f'(\zeta)(z - \zeta)| < \varepsilon\,|z - \zeta| \tag{7.30}$$

whenever $|z - \zeta| < \delta$ (cf. Theorem 3.1). Since ζ belongs to all the "nested" triangles, and since the triangles become arbitrarily small, then, starting from some value of $n > N$, the triangles are all contained in the disk $|z - \zeta| < \delta$, so that the inequality (7.30) is satisfied at all their points.

We now integrate the function $f(z) - f(\zeta) - f'(\zeta)(z - \zeta)$ along the closed curve L_n, obtaining

$$\begin{aligned}\int_{L_n} [f(z) - f(\zeta) - f'(\zeta)(z - \zeta)]\, dz \\ &= \int_{L_n} f(z)\, dz - f(\zeta) \int_{L_n} dz - f'(\zeta) \int_{L_n} z\, dz + \zeta f'(\zeta) \int_{L_n} dz \\ &= \int_{L_n} f(z)\, dz,\end{aligned}$$

where we have used the fact that the integrals $\int_{L_n} dz$ and $\int_{L_n} z\, dz$ vanish (see Sec. 34, Examples 1 and 2). Therefore, according to (7.30) and (7.13), if $n > N$,

$$\left| \int_{L_n} f(z)\, dz \right| = \left| \int_{L_n} [f(z) - f(\zeta) - f'(\zeta)(z - \zeta)]\, dz \right| < \varepsilon l_n^2 = \frac{\varepsilon l^2}{4^n}, \tag{7.31}$$

where, in writing the inequality, we note that the distance between any two points of the same triangle must be less than the perimeter of the triangle. Comparing (7.29) and (7.31), we find that

$$\frac{M}{4^n} \leqslant \left| \int_{L_n} f(z)\, dz \right| < \frac{\varepsilon l^2}{4^n},$$

which implies

$$M < \varepsilon l^2,$$

or

$$M \leqslant 0$$

if we let $\varepsilon \to 0$. But $M \geqslant 0$, since M is intrinsically nonnegative. Therefore

$$M = \left| \int_L f(z)\, dz \right| = 0,$$

and the proof of Cauchy's integral theorem for triangles is complete.

STEP 3 (*Convex polygons*). Next let $L = A_0A_1 \cdots A_{n-1}A_0$ ($n \geqslant 4$) be the boundary of an arbitrary convex n-gon (see Sec. 12, Prob. 3), traversed in a definite direction.[8] If we divide the polygon into $n - 2$ triangles by drawing

---

[8] Simplifying our earlier notation somewhat, by A_iA_j we mean an arc (or line segment) joining A_i to A_j, where the arc in question is always clear from the context. By

$$A_{i_1}A_{i_2}A_{i_3} \cdots A_{i_{n-1}}A_{i_n}$$

we mean the oriented curve $A_{i_1}A_{i_2} + A_{i_2}A_{i_3} + \cdots + A_{i_{n-1}}A_{i_n}$.

diagonals from the vertex A_0 to the other vertices, the integral of $f(z)$ along L can be written in the form

$$\int_L f(z)\,dz = \int_{A_0A_1A_2A_0} f(z)\,dz + \int_{A_0A_2A_3A_0} f(z)\,dz + \cdots + \int_{A_0A_{n-2}A_{n-1}A_0} f(z)\,dz, \tag{7.32}$$

where each of the triangular contours $A_0A_kA_{k+1}A_0$ $(k = 1, \ldots, n-2)$ is traversed in the same direction (as illustrated by Figure 7.3 for the case

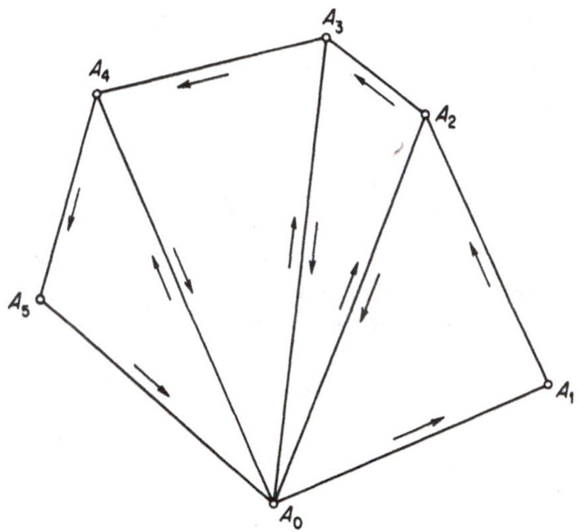

FIGURE 7.3

$n = 6$). But according to Step 2, each of the integrals in the right-hand side of (7.32) vanishes.[9] This proves Cauchy's theorem for convex polygons.

STEP 4 (*Closed polygonal Jordan curves*). Now let $L = A_0A_1 \cdots A_{n-1}A_0$ $(n \geqslant 4)$ be any closed polygonal curve (traversed once in a definite direction), which does not intersect itself. Then L is a Jordan curve, and $\overline{I(L)} \subset G$. The case where $\overline{I(L)}$ is convex has already been considered in Step 3. If $\overline{I(L)}$ is not convex, the extensions of some of the sides

$$A_0A_1,\ A_1A_2, \ldots, A_{n-1}A_0$$

[9] Note that $A_0A_kA_{k+1}A_0 \subset G$, since $L \subset G$ and $\overline{I(L)} \subset G$.

in one direction or the other (possibly both) intersect $I(L)$, as illustrated by Figure 7.4 for the case $n = 6$ (see Prob. 2). Extending each side $A_{k-1}A_k$ ($k = 1, \ldots, n$; $A_n = A_0$) which intersects $I(L)$, and terminating the extension of $A_{k-1}A_k$ (in the appropriate direction) at the point where it first intersects

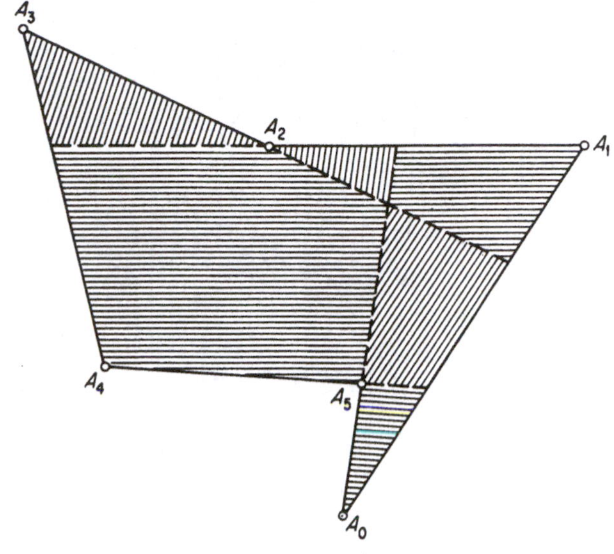

FIGURE 7.4

the boundary L, we decompose the original boundary into a finite number of subpolygons, each of which is convex (why?). Since $\int_L f(z)\, dz$ is obviously the sum of the integrals of $f(z)$ along the boundaries of these convex subpolygons, the fact that $\int_L f(z)\, dz$ vanishes follows at once from Step 3.

STEP 5 (*Arbitrary closed polygonal curves*). Next let L be an arbitrary closed polygonal curve. By definition (see p. 39), L consists of a finite number of line segments $\Delta_1, \Delta_2, \ldots, \Delta_n$ with definite directions, such that the final point of each segment Δ_k coincides with the initial point of the next segment Δ_{k+1} ($\Delta_{n+1} = \Delta_1$). Some of these segments may have points in common other than those just indicated, i.e., the curve may intersect itself. In this case, some of the segments Δ_k may be parts of other segments, or may even coincide with other segments. In other words, as we go along the polygonal curve L, some of its segments may be partially or totally traversed several times.

To keep the notation simple, we will study the polygonal curve shown in Figure 7.5, consisting of eight segments $\Delta_k = A_{k-1}A_k$ ($k = 1, \ldots, 8$; $A_8 = A_0$), where $A_6A_7 \subset A_5A_6$. Starting from A_0, we move along L until a

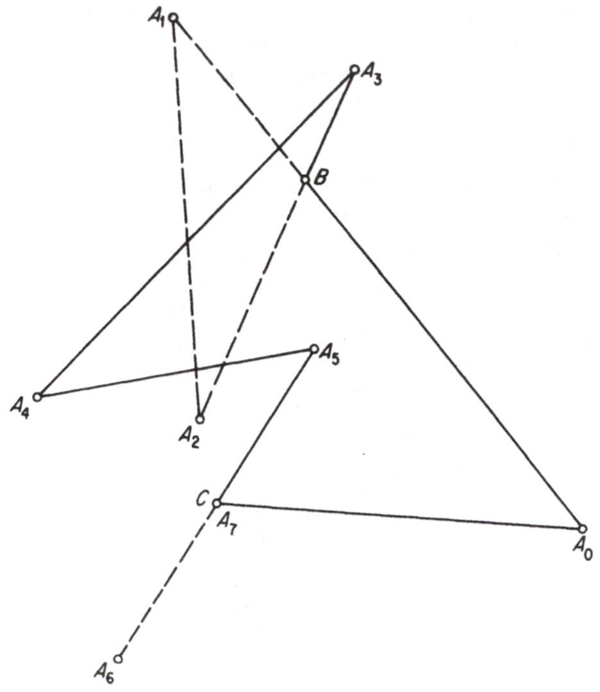

FIGURE 7.5

new segment intersects a previously traversed segment for the first time. In our case, A_2A_3 is such a segment, since it intersects A_0A_1 in the point B. Then the closed polygonal curve (here, the triangle BA_1A_2B), obtained by starting from the point B and traversing L until the *first* return to the point B, is a closed polygonal Jordan curve contained in G. Therefore

$$\int_{BA_1A_2B} f(z)\,dz = 0,$$

and we do not change the value of the integral of $f(z)$ along L if we remove BA_1A_2B from L, thereby obtaining a new polygonal curve

$$L' = A_0BA_3A_4A_5A_6A_7A_0,$$

with one vertex less than L.[10]

There is still another possibility which must be considered. Starting from A_0 and moving along the new polygonal curve L', we find that L' does not

[10] In general, the curve L' obtained by removing a closed polygonal curve like BA_1A_2B from L will have at least one vertex less than L, since we must remove at least a triangle, thereby losing at least two vertices, while creating at most one new vertex.

intersect itself before the point A_6. After A_6 we encounter previously traversed points, since A_6A_7 is a part of A_5A_6, but none of these points can be called the first, since the point A_6 is a common end point of the two consecutive segments A_5A_6 and A_6A_7, and hence is not classified as a point at which L' intersects itself. Therefore, starting from A_6, we retrace the segment A_5A_6 in the reverse direction until we encounter either A_5 or A_7 (the vertices adjacent to A_6). In the case shown in Figure 7.5, the vertex $C = A_7$ is encountered first. The closed polygonal curve consisting of the segment $CA_6 \subset A_5A_6$ followed by the segment $A_6C \subset A_6A_7$ is a "2-gon," along which the integral of $f(z)$ vanishes (see Step 1). Therefore, removing the polygonal curve CA_6C from L', we obtain another closed polygonal curve

$$L'' = A_0BA_3A_4A_5CA_0$$

such that

$$\int_{L''} f(z)\,dz = \int_{L'} f(z)\,dz.$$

The curve L'' has one less vertex than L', since one vertex (A_6) is lost along with the 2-gon, and no new vertex replaces it, since $C = A_7$ (if $C = A_5$, we also lose one vertex, but if $C = A_5 = A_7$, we lose two vertices).

This argument is of a perfectly general character. After a finite number of steps, we obtain either a closed polygonal curve which is a Jordan curve (L'' is such a curve in our example), or else we obtain a 2-gon. Thus, in either case, we find that $\int_L f(z)\,dz$ vanishes, thereby proving Cauchy's integral theorem for arbitrary closed polygonal curves. It should be noted that in extending the theorem from the case of triangles to the case of arbitrary closed polygonal curves, we used no function-theoretic considerations, but only geometric arguments of an elementary character.

STEP 6 (*Closed rectifiable curves*). Finally let L be an arbitrary closed rectifiable curve contained in the domain G. According to the key lemma (see p. 147), given any $\varepsilon > 0$, we can find a closed polygonal curve Λ inscribed in L and contained in G, such that

$$\left| \int_L f(z)\,dz - \int_\Lambda f(z)\,dz \right| < \varepsilon. \tag{7.33}$$

But as we have just seen,

$$\int_\Lambda f(z)\,dz = 0.$$

It follows that

$$\left| \int_L f(z)\,dz \right| < \varepsilon,$$

and since ε is arbitrarily small,

$$\int_L f(z)\,dz = 0.$$

The proof of Theorem 7.1 is now complete.

In proving Theorem 7.1 we have assumed that G is simply connected. It is easy to see that this fact has only been used to guarantee that for sufficiently small ε, G contains the closed polygonal domains bounded by the segments of the polygonal curve Λ inscribed in L and satisfying (7.33). This observation leads to the following variants of Cauchy's integral theorem, where we drop the requirement that G itself be simply connected:

THEOREM 7.1′. *Let G be an arbitrary domain, and let $f(z)$ be a single-valued analytic function on G. Then*

$$\int_L f(z)\,dz = 0,$$

where L is any closed rectifiable curve contained in a simply connected subdomain of G.

THEOREM 7.1″. *Let G be an arbitrary domain, and let $f(z)$ be a single-valued analytic function on G. Then*

$$\int_L f(z)\,dz = 0,$$

where L is any closed rectifiable Jordan curve such that G contains both L and its interior $I(L)$.

Remark. The conditions given in Theorems 7.1′ and 7.1″ are sufficient but not necessary for $\int_L f(z)\,dz$ to vanish. For example, the functions $1/z$ and $1/z^2$ are both analytic on the domain G consisting of the whole plane with the single point $z = 0$ deleted, and G contains any circle $\gamma: |z| = r > 0$ but not $I(\gamma)$. However,

$$\int_\gamma \frac{dz}{z} = \int_0^{2\pi} \frac{ire^{i\theta}\,d\theta}{re^{i\theta}} = i\int_0^{2\pi} d\theta = 2\pi i,$$

(recall the example on p. 145), while, on the other hand,

$$\int_\gamma \frac{dz}{z^2} = \int_0^{2\pi} \frac{ire^{i\theta}\,d\theta}{r^2e^{2i\theta}} = \frac{i}{r}\int_0^{2\pi} e^{-i\theta}\,d\theta = 0.$$

Furthermore, Figure 7.6 shows a doubly connected domain G and a curve $L \subset G$ such that

$$\int_L f(z)\,dz = 0,$$

where $f(z)$ is an arbitrary single-valued analytic function on G, even though L is not contained in any simply connected subdomain of G.

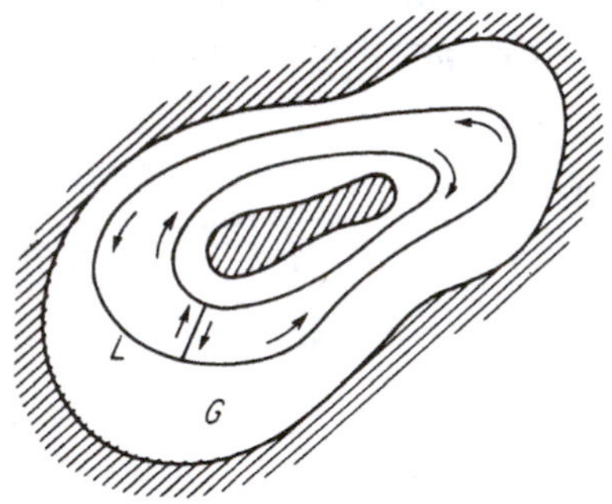

FIGURE 7.6

Problem 1. Prove that the intersection of two or more convex sets is also convex.

Problem 2. Let $L = A_0A_1 \cdots A_{n-1}A_0$ be a closed polygonal Jordan curve, with interior $I(L)$. Prove that the polygon $\overline{I(L)}$ is convex if and only if none of the extensions of its sides $A_{k-1}A_k$ ($k = 1, \ldots, n$; $A_n = A_0$) intersects $I(L)$.

Problem 3. Prove that every convex polygon is the intersection of a finite number of (closed) half-planes.

Problem 4. Prove that every convex domain is simply connected.

Problem 5. Prove Theorem 7.1″ in detail.

Hint. Recall the proof of the key lemma.

Problem 6 (M1, p. 270). Prove the following generalization of Cauchy's integral theorem: Let G be the interior of a closed rectifiable Jordan curve L, and let $f(z)$ be a single-valued function which is analytic on G and continuous on $\bar{G}$. Then

$$\int_L f(z)\, dz = 0$$

if G is *starlike*, i.e., if G contains a point ζ such that every ray emanating from ζ intersects L in only one point.

Comment. It turns out that the restriction to starlike domains is actually unnecessary (see M3, Sec. 12).

38. APPLICATION TO THE EVALUATION OF DEFINITE INTEGRALS

Cauchy's integral theorem can be used as a tool for evaluating various definite integrals of functions of a real variable, especially improper integrals.[11]

Example. *Evaluate the Fresnel integrals*

$$\int_0^\infty \cos x^2\, dx, \qquad \int_0^\infty \sin x^2\, dx, \tag{7.34}$$

encountered in the theory of diffraction.

We begin by deriving two preliminary results which will be needed to calculate (7.34), i.e., the formula

$$\int_0^\infty e^{-x^2} = \frac{\sqrt{\pi}}{2} \tag{7.35}$$

and the inequality

$$\sin \theta \geqslant \frac{2}{\pi}\theta \qquad \left(0 \leqslant \theta \leqslant \frac{\pi}{2}\right). \tag{7.36}$$

[11] In Sec. 60 we will systematically exploit a more powerful method for evaluating definite integrals, of which the present method is merely a special case.

To prove (7.35), let Q be the square $-R \leqslant x \leqslant R$, $-R \leqslant y \leqslant R$, let $C_1: |z| = R$ be the circle inscribed in Q, and let $C_2: |z| = \sqrt{2}\,R$ be the circle circumscribed about Q. Then, if $\bar{K}_1$ and $\bar{K}_2$ are the closed disks with boundaries C_1 and C_2, respectively, we have

$$\begin{aligned}\iint_{\bar{K}_1} e^{-x^2-y^2}\,dx\,dy &< \iint_Q e^{-x^2-y^2}\,dx\,dy \\ &= \left(\int_{-R}^{R} e^{-x^2}\,dx\right)^2 < \iint_{\bar{K}_2} e^{-x^2-y^2}\,dx\,dy,\end{aligned} \tag{7.37}$$

since the exponential is positive. If we introduce polar coordinates r and θ, (7.37) becomes

$$\int_0^{2\pi}\int_0^R e^{-r^2} r\,dr\,d\theta < 4\left(\int_0^R e^{-x^2}\,dx\right)^2 < \int_0^{2\pi}\int_0^{\sqrt{2}R} e^{-r^2} r\,dr\,d\theta.$$

Evaluating the integrals, and taking the square roots of all terms of the resulting inequalities, we find that

$$\sqrt{\pi(1 - e^{-R^2})} < 2\int_0^R e^{-x^2}\,dx < \sqrt{\pi(1 - e^{-2R^2})}.$$

Thus

$$\lim_{R\to\infty} 2\int_0^R e^{-x^2}\,dx = \sqrt{\pi},$$

which implies (7.35).

To prove (7.36), we note that

$$f(\theta) = \frac{\sin\theta}{\theta}$$

is a decreasing function in the interval $(0, \pi/2)$, since

$$f'(\theta) = \frac{\theta\cos\theta - \sin\theta}{\theta^2} = \frac{\cos\theta(\theta - \tan\theta)}{\theta^2} < 0$$

if $0 < \theta < \pi/2$ (why?). Therefore $f(\theta) > f(\pi/2)$ if $0 < \theta < \pi/2$, i.e.,

$$\frac{\sin\theta}{\theta} > \frac{2}{\pi}$$

or

$$\sin\theta > \frac{2}{\pi}\theta \qquad \left(0 < \theta < \frac{\pi}{2}\right).$$

Moreover, it is obvious that $\sin\theta = 2\theta/\pi$ if $\theta = 0$ or $\theta = \pi/2$, and hence (7.36) is proved.

Turning to the evaluation of the Fresnel integrals (7.34), we consider the function

$$f(z) = e^{iz^2}$$

(the reason for this choice will soon be apparent). Since both e^z and iz^2 are entire functions, it follows from the rule for differentiating a composite function (see p. 46) that $f(z)$ is entire, with derivative $f'(z) = 2ize^{iz^2}$. Therefore we can apply Cauchy's theorem to $f(z)$, with any closed rectifiable curve as the contour L. For L we choose the contour shown in Figure 7.7, consisting of the segment $OA: 0 \leqslant x \leqslant R$ of the nonnegative real axis, the arc AB of the circle of radius R with center at the origin, and the segment BO of the line bisecting the angle formed by the nonnegative real and imaginary axes. According to Cauchy's integral theorem,

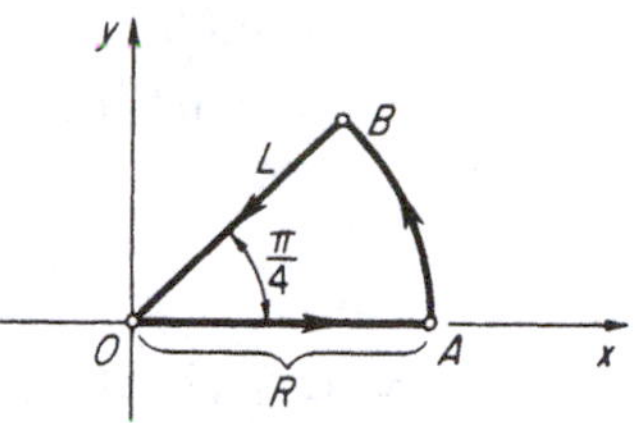

FIGURE 7.7

$$\int_L e^{iz^2}\,dz = \int_{OA} e^{iz^2}\,dz + \int_{AB} e^{iz^2}\,dz + \int_{BO} e^{iz^2}\,dz = 0. \tag{7.38}$$

On the segment OA, z reduces to a real number x, so that $dz = dx$, and

$$\mathscr{I}_1(R) = \int_{OA} e^{iz^2}\,dz = \int_0^R e^{ix^2}\,dx.$$

On the arc AB

$$z = Re^{i\theta} \quad (0 \leqslant \theta \leqslant \pi/4), \quad z^2 = R^2e^{2i\theta}, \quad dz = iRe^{i\theta}\,d\theta,$$

so that

$$\mathscr{I}_2(R) = \int_{AB} e^{iz^2}\,dz = \int_0^{\pi/4} \exp\,(iR^2e^{2i\theta})\,iRe^{i\theta}\,d\theta.$$

Finally, on the segment BO

$$z = re^{i\pi/4} \quad (R \geqslant r \geqslant 0), \quad z^2 = r^2e^{i\pi/2} = ir^2, \quad dz = e^{i\pi/4}\,dr,$$

and consequently

$$\mathscr{I}_3(R) = \int_{BO} e^{iz^2}\,dz = \int_R^0 e^{-r^2}\,e^{i\pi/4}\,dr = -e^{i\pi/4}\int_0^R e^{-r^2}\,dr$$

$$= -\frac{\sqrt{2}}{2}(1 + i)\int_0^R e^{-r^2}\,dr.$$

Next let $R \to \infty$. Then $\mathscr{I}_3(R)$ approaches the limit

$$-\frac{\sqrt{2}}{2}(1 + i)\int_0^\infty e^{-r^2}\,dr = -\frac{\sqrt{2\pi}}{4}(1 + i),$$

where we have used formula (7.35). Moreover, $\mathscr{I}_2(R) \to 0$ as $R \to \infty$. To show this, we estimate the modulus of $I_2(R)$, obtaining

$$|\mathscr{I}_2(R)| \leqslant R\int_0^{\pi/4} |\exp\,[iR^2(\cos 2\theta + i \sin 2\theta)]|\;|e^{i\theta}|\,d\theta.$$

Since $|e^{i\theta}| = 1$ and

$$|\exp[iR^2(\cos 2\theta + i \sin 2\theta)]| = \exp(-R^2 \sin 2\theta),$$

we have

$$|\mathscr{I}_2(R)| \leqslant R \int_0^{\pi/4} \exp(-R^2 \sin 2\theta)\, d\theta.$$

But according to (7.36), $\sin 2\theta \geqslant 4\theta/\pi$ if $0 \leqslant 2\theta \leqslant \pi/2$. Therefore

$$|\mathscr{I}_2(R)| \leqslant R \int_0^{\pi/4} \exp(-4R^2\theta/\pi)\, d\theta = R \left. \frac{\exp(-4R^2\theta/\pi)}{-4R^2/\pi} \right|_{\theta=0}^{\theta=\pi/4}$$

$$= \frac{\pi}{4} \frac{1 - e^{-R^2}}{R},$$

so that

$$\lim_{R\to\infty} \mathscr{I}_2(R) = 0,$$

as asserted. Finally, we consider the integral

$$\mathscr{I}_1(R) = \int_0^R e^{ix^2}\, dx = \int_0^R \cos x^2\, dx + i \int_0^R \sin x^2\, dx.$$

Since

$$\mathscr{I}_1(R) + \mathscr{I}_2(R) + \mathscr{I}_3(R) = 0$$

[cf. (7.38)], it follows that

$$\lim_{R\to\infty} \mathscr{I}_1(R) = -\lim_{R\to\infty} \mathscr{I}_2(R) - \lim_{R\to\infty} \mathscr{I}_3(R) = \frac{\sqrt{2\pi}}{4}(1 + i),$$

i.e.,

$$\lim_{R\to\infty} \int_0^R \cos x^2\, dx + i \int_0^R \sin x^2\, dx = \frac{\sqrt{2\pi}}{4}(1 + i),$$

which implies that the improper integrals

$$\int_0^\infty \cos x^2\, dx = \lim_{R\to\infty} \int_0^R \cos x^2\, dx$$

and

$$\int_0^\infty \sin x^2\, dx = \lim_{R\to\infty} \int_0^R \sin x^2\, dx$$

exist, and have the same value

$$\int_0^\infty \cos x^2\, dx = \int_0^\infty \sin x^2\, dx = \frac{\sqrt{2\pi}}{4}.$$

Problem 1 (M1, p. 275). Prove that

$$\int_{-\infty}^\infty e^{-\lambda x^2} \cos(2\lambda a x)\, dx = \sqrt{\frac{\pi}{\lambda}}\, e^{-\lambda a^2} \qquad (a > 0, \lambda > 0).$$

Problem 2 (M1, p. 277). Prove that

$$\int_0^\infty \frac{\sin x}{x}\, dx = \frac{\pi}{2}.$$

Problem 3. Prove the inequality

$$\cos \theta \geqslant 1 - \frac{2\theta}{\pi} \qquad \left(0 \leqslant \theta \leqslant \frac{\pi}{2}\right).$$

39. CAUCHY'S INTEGRAL THEOREM FOR A SYSTEM OF CONTOURS

We now prove a result which will be used repeatedly later on:

THEOREM 7.3 (*Cauchy's integral theorem for a system of contours*). *Let G be an arbitrary domain, and let $f(z)$ be a single-valued analytic function on G. Moreover, let $\Gamma, \gamma_1, \ldots, \gamma_n$ be a system of $n + 1$ closed rectifiable Jordan curves contained in G which satisfy the following conditions:*

1. *$I(\Gamma)$ contains every curve $\gamma_1, \ldots, \gamma_n$;*
2. *For every $k = 1, \ldots, n$, $E(\gamma_k)$ contains γ_j $(j \neq k)$;*
3. *G contains the multiply connected domain*

$$D = I(\Gamma) - \overline{I(\gamma_1)} - \cdots - \overline{I(\gamma_n)},$$

with boundary $\Gamma \cup \gamma_1 \cup \cdots \cup \gamma_n$.

Then

$$\int_\Gamma f(z)\, dz = \int_{\gamma_1} f(z)\, dz + \cdots + \int_{\gamma_n} f(z)\, dz, \qquad (7.39)$$

where all the integrals are taken in the same direction, e.g., in the positive (counterclockwise) direction.[12]

Proof. First we note that the theorem is trivial in the special case where G is simply connected, since then all the integrals in both sides of (7.39) vanish. To prove the theorem in the general case, we draw $n + 1$ rectifiable Jordan arcs $\delta_1', \delta_2', \ldots, \delta_{n+1}'$ contained in G such that δ_1' joins a point $z_0 \in \Gamma$ to a point $\zeta_1 \in \gamma_1$, δ_2' joins a point $z_1 \in \gamma_1$ $(z_1 \neq \zeta_1)$ to a point $\zeta_2 \in \gamma_2$, and so on, until finally δ_{n+1}' joins a point $z_n \in \gamma_n$ to a point $\zeta_0 \in \Gamma$ $(\zeta_0 \neq z_0)$. In general, these arcs may partially leave the domain D, but they can always be replaced by $n + 1$ new arcs $\delta_1, \delta_2, \ldots, \delta_{n+1}$ which, except for their end points, are contained in D. In fact, as we go along δ_1' from the point $z_0 \in \Gamma$ to the point $\zeta_1 \in \gamma_1$, we need only

[12] The positive direction, which is easily determined in the simplest cases (e.g., when the curves are circles or polygons), can be defined in complete generality by methods beyond the scope of this course (see e.g., P. S. Aleksandrov, *op. cit.*, Chap. 2).

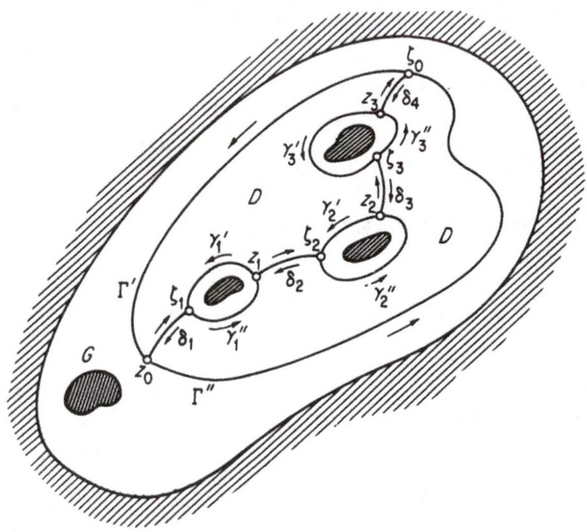

FIGURE 7.8

note the last point of intersection of δ_1' with Γ and the first point of intersection of δ_1' with γ_1. Then the part of δ_1' lying between these two points of intersection gives the required arc δ_1, and similarly we can construct the rest of the arcs $\delta_2, \ldots, \delta_{n+1}$.[13] The initial and final points of the arcs $\delta_1, \delta_2, \ldots, \delta_{n+1}$ divide each of the closed curves $\Gamma, \gamma_1, \ldots, \gamma_n$ into two arcs, which we denote by adding one or two primes to the symbols denoting the original curves:[14]

$$\Gamma = \Gamma'' - \Gamma', \quad \gamma_1 = \gamma_1'' - \gamma_1', \ldots, \gamma_n = \gamma_n'' - \gamma_n'.$$

As the common initial point of the arcs Γ', Γ'' we choose the initial point z_0 of the arc δ_1, and as the common final point of Γ', Γ'' we choose the final point ζ_0 of δ_{n+1}. Similarly, the arcs γ_1', γ_1'' have initial point ζ_1 (the final point of δ_1) and final point z_1 (the initial point of δ_2) and so on, with ζ_n (the final point of δ_n) and z_n (the initial point of δ_{n+1}) as the initial and final points, respectively, of the arcs γ_n', γ_n''. The whole construction is illustrated by Figure 7.8, which corresponds to the case $n = 3$.

[13] In general, the arcs $\delta_1, \delta_2, \ldots, \delta_{n+1}$ so obtained have points of intersection, but they can always be replaced by *disjoint* arcs (which we denote by the same symbols) satisfying the same conditions (see Prob. 1).

[14] By $\gamma - \delta$ is meant the curve $\gamma + (-\delta)$, where γ and δ are two curves (arcs) with a common final point. The context precludes any possibility of confusion with the set-theoretic difference between γ and δ.

Now consider the two closed rectifiable Jordan curves

$$L' = -\Gamma' + \delta_1 + \gamma_1' + \cdots + \gamma_n' + \delta_{n+1}$$

and

$$L'' = \Gamma'' - \delta_{n+1} - \gamma_n'' - \cdots - \gamma_1'' - \delta_1.$$

Clearly L' and L'' are both traversed in the same (positive) direction, and

$$L' \subset G, \quad L'' \subset G, \quad I(L') \subset D \subset G, \quad I(L'') \subset D \subset G.$$

Therefore, by Theorem 7.1″,

$$\int_{L'} f(z)\,dz = \int_{L''} f(z)\,dz = 0,$$

or

$$\begin{aligned} 0 &= \left(\int_{L'} + \int_{L''}\right) f(z)\,dz \\ &= \left(\int_{-\Gamma'} + \int_{\Gamma''} + \int_{\delta_1} + \int_{-\delta_1} + \int_{\gamma_1'} + \int_{-\gamma_1''} + \cdots \right. \\ &\qquad \left. + \int_{\gamma_n'} + \int_{-\gamma_n''} + \int_{\delta_{n+1}} + \int_{-\delta_{n+1}}\right) f(z)\,dz. \end{aligned} \tag{7.40}$$

But it is easy to see that

$$\left(\int_{\delta_j} + \int_{-\delta_j}\right) f(z)\,dz = 0 \qquad (j = 1, \ldots, n + 1),$$

and

$$\left(\int_{-\Gamma'} + \int_{\Gamma''}\right) f(z)\,dz = \int_{\Gamma} f(z)\,dz,$$

while

$$\left(\int_{\gamma_k'} + \int_{-\gamma_k''}\right) f(z)\,dz = \int_{-\gamma_k} f(z)\,dz = -\int_{\gamma_k} f(z)\,dz \qquad (k = 1, \ldots, n).$$

Therefore (7.40) implies (7.39), and the proof of Theorem 7.3 is complete.

Problem 1. Prove the assertion made in footnote 13, p. 164.

Hint. The new disjoint arcs need not connect the points $z_0, \zeta_1, \ldots, z_n, \zeta_0$ in the same order as the old intersecting ones.

Problem 2. Prove that

$$\int_L \frac{dz}{z - a} = \pm 2\pi i,$$

if L is any closed rectifiable Jordan curve whose interior contains the point a, where we choose the plus sign if L is traversed in the positive direction and the minus sign if L is traversed in the negative direction.

CHAPTER 8

CAUCHY'S INTEGRAL FORMULA AND ITS IMPLICATIONS

40. INDEFINITE INTEGRALS

Let G be a simply connected domain, and let $f(z)$ be a single-valued analytic function on G. Suppose z_0 and z are any two points of G, while L_1 and L_2 are any two rectifiable curves joining z_0 to z and contained in G. Then it is an immediate consequence of Cauchy's integral theorem that

$$\int_{L_1} f(z)\,dz = \int_{L_2} f(z)\,dz, \tag{8.1}$$

since

$$0 = \int_{L_1 - L_2} f(z)\,dz = \int_{L_1} f(z)\,dz - \int_{L_2} f(z)\,dz.$$

Therefore the integral of the function $f(z)$ along any rectifiable curve joining z_0 to z is *path-independent*, i.e., depends only on its end points z_0 and z, a fact which can be emphasized by writing (8.1) as

$$\int_{z_0}^{z} f(\zeta)\,d\zeta.$$

If the lower limit of integration z_0 is held fixed, while the upper limit of integration z is allowed to vary, we obtain a single-valued function

$$F(z) = \int_{z_0}^{z} f(\zeta)\, d\zeta \qquad (z_0, z \in G). \tag{8.2}$$

We now show that $F(z)$ is itself an analytic function on G, with derivative $f(z)$.

THEOREM 8.1. *Let G be a simply connected domain, and let $f(z)$ be a single-valued analytic function on G. Then the single-valued function $F(z)$ defined by* (8.2) *is analytic on G and*

$$F'(z) = f(z).$$

Proof. Given a point $z \in G$, let the modulus of Δz be so small that G contains $\overline{I(\gamma)}$, where γ is the circle with center z and radius $|\Delta z|$. Then

$$\Delta F \equiv \int_{z_0}^{z+\Delta z} f(\zeta)\, d\zeta - \int_{z_0}^{z} f(\zeta)\, d\zeta = \int_{z}^{z+\Delta z} f(\zeta)\, d\zeta,$$

where the last integral can be evaluated along the line segment joining z to $z + \Delta z$. It follows that

$$\begin{aligned}\Delta F &= \int_{z}^{z+\Delta z} f(z)\, d\zeta + \int_{z}^{z+\Delta z} [f(\zeta) - f(z)]\, d\zeta \\ &= \Delta z f(z) + \int_{z}^{z+\Delta z} [f(\zeta) - f(z)]\, d\zeta,\end{aligned}$$

and hence

$$\frac{\Delta F}{\Delta z} = f(z) + \frac{1}{\Delta z}\int_{z}^{z+\Delta z} [f(\zeta) - f(z)]\, d\zeta$$

or

$$\left|\frac{\Delta F}{\Delta z} - f(z)\right| \leqslant \frac{1}{|\Delta z|}\left|\int_{z}^{z+\Delta z} [f(\zeta) - f(z)]\, d\zeta\right| \leqslant \frac{1}{|\Delta z|} M(\Delta z)\,|\Delta z| = M(\Delta z),$$

where

$$M(\Delta z) = \max_{|\zeta - z| \leqslant |\Delta z|} |f(\zeta) - f(z)|.$$

[The existence of $M(\Delta z)$ follows from Theorem 2.7 and the continuity of $f(\zeta) - f(z)$ on the disk $\overline{I(\gamma)}$.] Since $f(\zeta)$ is continuous at z,

$$\lim_{\Delta z \to 0} \left|\frac{\Delta F}{\Delta z} - f(z)\right| = \lim_{\Delta z \to 0} M(\Delta z) = 0,$$

i.e.,

$$\lim_{\Delta z \to 0} \frac{\Delta F}{\Delta z} \equiv F'(z) = f(z),$$

as asserted.

Let $f(z)$ be a single-valued analytic function on a domain G. Then a function $\Phi(z)$ is said to be an *indefinite integral* (synonymously, a *primitive* or *antiderivative*) of $f(z)$ on G if $\Phi(z)$ is single-valued and analytic on G, and

$$\Phi'(z) = f(z) \qquad (z \in G).$$

Thus, according to Theorem 8.1, the function (8.2) is an indefinite integral of $f(z)$ on G. We now show that (8.2) is essentially the most general indefinite integral of $f(z)$ on G, if G is simply connected.

THEOREM 8.2. *Let G be a simply connected domain, and let $f(z)$ be a single-valued analytic function on G. Then any indefinite integral of $f(z)$ on G can be represented in the form*

$$\Phi(z) = \int_{z_0}^{z} f(\zeta)\, d\zeta + C, \tag{8.3}$$

where $z_0 \in G$ and C is a constant.

Proof. Writing

$$\varphi(z) = \Phi(z) - \int_{z_0}^{z} f(\zeta)\, d\zeta = u(x, y) + iv(x, y),$$

we have

$$\varphi'(z) = \Phi'(z) - f(z) = 0 \qquad (z \in G),$$

where $\varphi(z)$ is analytic on G. But according to Theorem 3.2,

$$\varphi'(z) = \frac{\partial u}{\partial x} + i\frac{\partial v}{\partial x} = \frac{\partial v}{\partial y} - i\frac{\partial u}{\partial y},$$

so that

$$\frac{\partial u}{\partial x} = \frac{\partial u}{\partial y} = \frac{\partial v}{\partial x} = \frac{\partial v}{\partial y} = 0$$

everywhere in G. Since $u(x, y)$ and $v(x, y)$ are differentiable functions of x and y, it follows that

$$u(x, y) = C_1, \qquad v(x, y) = C_2,$$

where C_1 and C_2 are real constants (why?). Therefore

$$\varphi(z) = C_1 + iC_2 = C \qquad (z \in G),$$

which implies (8.3).

COROLLARY (*Fundamental theorem of integral calculus for complex functions*). *Let G be a simply connected domain, and let $f(z)$ be a single-valued analytic function on G. Then if $z_0, z \in G$,*

$$\int_{z_0}^{z} f(\zeta)\, d\zeta = \Phi(z) - \Phi(z_0),$$

where $\Phi(z)$ is any indefinite integral of $f(z)$ on G.

Proof. We merely use the fact that

$$\Phi(z_0) = C.$$

Remark. According to the corollary to Theorem 8.2, we can calculate integrals of elementary functions of a complex variable by using formulas resembling those for elementary functions of a real variable. For example,

$$\int_{z_0}^{z} \zeta^n \, d\zeta = \frac{z^{n+1} - z_0^{n+1}}{n+1}$$

if n is an integer other than -1, and

$$\int_{z_0}^{z} e^{\zeta} \, d\zeta = e^z - e^{z_0},$$

$$\int_{z_0}^{z} \cos \zeta \, d\zeta = \sin z - \sin z_0.$$

$$\int_{z_0}^{z} \sin \zeta \, d\zeta = \cos z_0 - \cos z.$$

Next let G be a multiply connected domain, and let $f(z)$ be a single-valued analytic function on G. If

$$\int_L f(z) \, dz = 0$$

for every closed rectifiable curve contained in G, then (8.2) again turns out to be a single-valued analytic function on G, in fact an indefinite integral of $f(z)$ on G. For example, this is the case if

$$f(z) = \frac{1}{z^2 + 1} \tag{8.4}$$

and G is the ring-shaped domain or *annulus* $1 < |z| < R$, where $R > 1$ is arbitrary (see Prob. 1). However, suppose G contains at least one closed rectifiable curve $L = BCDE$ such that[1]

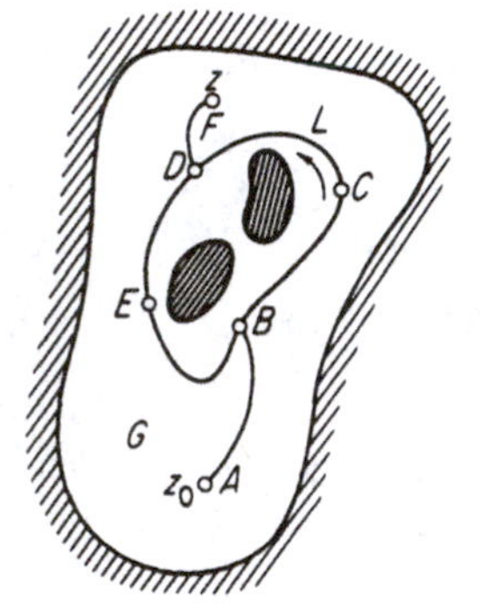

FIGURE 8.1

$$\int_L f(z) \, dz \neq 0,$$

and let $A = z_0$, $F = z$ be arbitrary points of G (see Figure 8.1). Then

$$\begin{aligned}\int_{ABCDF} f(z) \, dz - \int_{ABEDF} f(z) \, dz &= \int_{BCD} f(z) \, dz - \int_{BED} f(z) \, dz \\ &= \int_L f(z) \, dz \neq 0,\end{aligned}$$

and therefore

$$\int_{z_0}^{z} f(\zeta) \, d\zeta$$

---

[1] E.g., suppose $f(z) = 1/z$ and G is the domain consisting of all points $z \neq 0$ (recall Example 3, p. 140), or suppose $f(z)$ is given by (8.4) and G is the set of all points z satisfying $z^2 + 1 \neq 0$ (see Prob. 1).

has at least two different values at z, i.e., we can continue to write (8.2), but now the function $F(z)$ is multiple-valued on G. The appropriate generalization of Theorem 8.1 to the case of multiply connected domains is given by

THEOREM 8.3. *Let G be a multiply connected domain, and let $f(z)$ be a single-valued analytic function on G. Then the function* (8.2), *which is in general multiple-valued, has single-valued analytic branches on any simply connected subdomain $D \subset G$, and*

$$F'(z) = f(z) \qquad (z \in D),$$

where $F'(z)$ denotes the derivative of any of these single-valued branches.

Proof. If G contains no closed rectifiable curve L such that

$$\int_L f(z)\,dz \neq 0,$$

then $F(z)$ is single-valued and the proof of the theorem is identical with that of Theorem 8.1. Otherwise, let

$$F(z) = \int_{z_0}^{z_1} f(\zeta)\,d\zeta + \int_{z_1}^{z} f(\zeta)\,d\zeta \equiv \int_{z_0}^{z_1} f(\zeta)\,d\zeta + \varphi(z), \qquad (8.5)$$

where z_1 is any fixed point of D, and the path of integration $\widehat{z_1 z}$ for the second integral lies in D. Then, according to Theorem 8.1, $\varphi(z)$ is a single-valued analytic function on D such that $\varphi'(z) = f(z)$, while by the argument just given, the integral

$$\int_{z_0}^{z_1} f(\zeta)\,d\zeta \qquad (\widehat{z_0 z_1} \subset G)$$

takes at least two different values, thereby generating at least two different single-valued branches of $F(z)$ on D, whose values differ by a constant. Obviously, every such branch is an indefinite integral of $f(z)$ on D, and the proof is complete.

Remark. Actually, *every* value of the integral

$$\int_{z_0}^{z} f(\zeta)\,d\zeta$$

is of the form (8.5), and therefore belongs to one of the single-valued branches constructed in the proof of Theorem 8.3. In fact, given a curve $\Gamma \subset G$ joining z_0 to z, let $\gamma \subset D$ be any curve joining z_1 to z. Then

$$\int_\Gamma f(\zeta)\,d\zeta = \int_{\Gamma-\gamma} f(\zeta)\,d\zeta + \int_\gamma f(\zeta)\,d\zeta = \int_{\Gamma-\gamma} f(\zeta)\,d\zeta + \varphi(z),$$

where $\Gamma - \gamma \subset G$ is a curve joining z_0 to z_1.

Problem 1. Prove that

$$\int_L \frac{dz}{z^2+1} = 0$$

if L is an arbitrary closed rectifiable curve contained in the annulus $1 < |z| < R$ $(R > 1)$, but not if L is an arbitrary closed rectifiable curve contained in the domain consisting of all points z such that $z^2 + 1 \neq 0$.

Hint. Use

$$\frac{1}{z^2+1} = \frac{1}{2i}\left(\frac{1}{z-i} - \frac{1}{z+i}\right)$$

and Sec. 39, Prob. 2.

Problem 2 (M1, p. 286). Let

$$F(z) = \int_1^z \frac{d\zeta}{\zeta} \qquad (z \in G),$$

where G is the finite plane minus the origin. Prove that $F(z) = \operatorname{Ln} z$. Define single-valued branches of $F(z)$ on appropriate simply connected subdomains of G.

41. CAUCHY'S INTEGRAL FORMULA

We now prove a result of fundamental importance in the theory of analytic functions:

THEOREM 8.4 (*Cauchy's integral formula*). *If $f(z)$ is analytic on a domain G and if G contains a closed rectifiable Jordan curve L and its interior $I(L)$, then*

$$\frac{1}{2\pi i}\int_L \frac{f(z)}{z-z_0}\,dz = f(z_0) \qquad \text{if} \quad z_0 \in I(L), \tag{8.6}$$

where the integral on the left is called Cauchy's integral.[2] *On the other hand,*

$$\frac{1}{2\pi i}\int_L \frac{f(z)}{z-z_0}\,dz = 0 \qquad \text{if} \quad z_0 \in E(L), \tag{8.7}$$

where $E(L)$ denotes the exterior of L.

Proof. First let z_0 be an arbitrary point of the domain $I(L)$. Then the function

$$g(z) = \frac{f(z)}{z-z_0} \tag{8.8}$$

[2] It is assumed that L is traversed in the positive direction (cf. footnote 12, p. 163). This can be thought of as the counterclockwise direction or the direction such that $I(L)$ lies to the left of an observer moving along L.

is defined and analytic on the domain $G' = G - \{z_0\}$ (why?). Let ρ be so small that $I(L)$ contains $\overline{I(\gamma_\rho)}$, where γ_ρ is the circle $|z - z_0| = \rho$ traversed in the positive direction (see Figure 8.2). Then Theorem 7.3, applied to the contours L and γ_ρ, implies

$$\int_L g(z)\,dz = \int_{\gamma_\rho} g(z)\,dz$$

or

$$\int_L \frac{f(z)}{z - z_0}\,dz = \int_{\gamma_\rho} \frac{f(z)}{z - z_0}\,dz. \qquad (8.9)$$

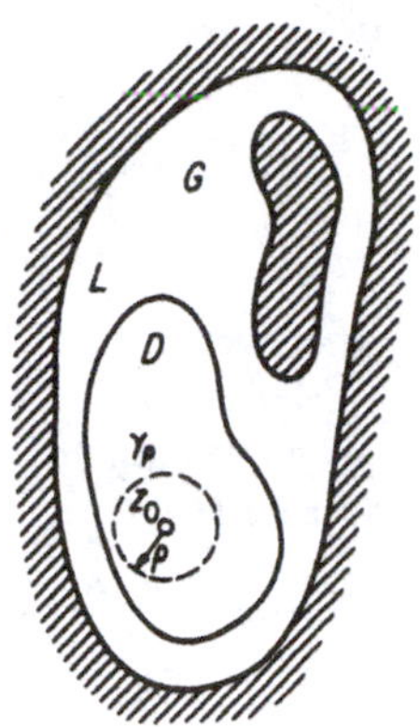

FIGURE 8.2

According to (8.9) and the choice of γ_ρ, the value of the integral

$$\int_{\gamma_\rho} \frac{f(z)}{z - z_0}\,dz$$

does not change if we decrease the radius ρ, and hence

$$\int_L \frac{f(z)}{z - z_0}\,dz = \lim_{\rho \to 0} \int_{\gamma_\rho} \frac{f(z)}{z - z_0}\,dz.$$

Therefore, to prove (8.6), we need only show that

$$\lim_{\rho \to 0} \int_{\gamma_\rho} \frac{f(z)}{z - z_0}\,dz = 2\pi i f(z_0),$$

i.e., that given any $\varepsilon > 0$, there exists a $\delta = \delta(\varepsilon) > 0$ such that

$$\left| \int_{\gamma_\rho} \frac{f(z)}{z - z_0}\,dz - 2\pi i f(z_0) \right| < \varepsilon \qquad (8.10)$$

whenever $\rho < \delta$. Since

$$\int_{\gamma_\rho} \frac{dz}{z - z_0} = 2\pi i$$

(see Example 3, p. 140), we can write the left-hand side of (8.10) in the form

$$\left| \int_{\gamma_\rho} \frac{f(z)}{z - z_0}\,dz - 2\pi i f(z_0) \right| = \left| \int_{\gamma_\rho} \frac{f(z)}{z - z_0}\,dz - f(z_0) \int_{\gamma_\rho} \frac{dz}{z - z_0} \right|$$
$$= \left| \int_{\gamma_\rho} \frac{f(z) - f(z_0)}{z - z_0}\,dz \right|.$$

Since $f(z)$ is continuous at z_0, given any $\varepsilon > 0$, we can find a $\delta = \delta(\varepsilon) > 0$ such that

$$|f(z) - f(z_0)| < \frac{\varepsilon}{2\pi}$$

whenever

$$|z - z_0| = \rho < \delta.$$

Therefore

$$\left| \int_{\gamma_\rho} \frac{f(z)}{z - z_0}\, dz - 2\pi i f(z_0) \right| < \frac{\varepsilon/2\pi}{\rho} 2\pi\rho = \varepsilon$$

whenever $\rho < \delta$. This proves (8.10) and hence (8.6).

To prove (8.7), we use Theorem 7.1″, p. 158 and the fact that (8.8) is analytic on the domain $G' = G - \{z_0\}$ which now contains both L and $I(L)$.

COROLLARY. *If $f(z)$ is analytic on a domain G and if G contains the circle γ_ρ: $|z - z_0| = \rho$ and its interior, then*

$$f(z_0) = \frac{1}{2\pi} \int_0^{2\pi} f(z_0 + \rho e^{i\theta})\, d\theta, \tag{8.11}$$

i.e., the value of $f(z)$ at the point z_0 equals the average of its values on the circle γ_ρ (with center z_0).

Proof. The equation of γ_ρ is

$$z = z_0 + \rho e^{i\theta} \qquad (0 \leqslant \theta \leqslant 2\pi). \tag{8.12}$$

It follows from (8.12) and Cauchy's integral formula (8.6) that

$$f(z_0) = \frac{1}{2\pi i} \int_0^{2\pi} \frac{f(z_0 + \rho e^{i\theta}) i \rho e^{i\theta}}{\rho e^{i\theta}}\, d\theta = \frac{1}{2\pi} \int_0^{2\pi} f(z_0 + \rho e^{i\theta})\, d\theta,$$

which proves (8.11).

A particularly important consequence of Theorem 8.4 is

THEOREM 8.5. *If $f(z)$ is analytic on a domain G, then $f(z)$ has derivatives of all orders on G, and in fact, given any $z_0 \in G$,*

$$f^{(n)}(z_0) = \frac{n!}{2\pi i} \int_L \frac{f(z)}{(z - z_0)^{n+1}}\, dz \qquad (n = 0, 1, 2, \ldots), \tag{8.13}$$

where L is any closed rectifiable Jordan curve such that $z_0 \in I(L)$, $\overline{I(L)} \subset G$.

Proof. We will prove (8.13) by mathematical induction, noting first that (8.13) reduces to Cauchy's integral formula for $n = 0$.[3] Assuming that (8.13) holds for a nonnegative integer $n - 1$, we now show that (8.13) also holds for n. This will be accomplished by direct calculation of the quantity

$$f^{(n)}(z_0) = \lim_{h \to 0} \frac{f^{(n-1)}(z_0 + h) - f^{(n-1)}(z_0)}{h},$$

[3] By definition, $f^{(0)}(z) = f(z)$ and $0! = 1$.

where z_0 is an arbitrary point of $I(L)$ and hence of G (since $L \subset G$ is itself arbitrary). Using Theorem 7.3, we can replace (8.13) by

$$f^{(n)}(z_0) = \frac{n!}{2\pi i}\int_{\gamma_\rho} \frac{f(z)}{(z-z_0)^{n+1}}\,dz, \tag{8.14}$$

where γ_ρ: $|z - z_0| = \rho$ is a circle so small that $\gamma_\rho \subset I(L)$. Therefore

$$\begin{aligned}\frac{f^{(n-1)}(z_0+h) - f^{(n-1)}(z_0)}{h} &= \frac{(n-1)!}{2\pi ih}\int_{\gamma_\rho} f(z)\left[\frac{1}{(z-z_0-h)^n} - \frac{1}{(z-z_0)^n}\right]dz \\ &= \frac{(n-1)!}{2\pi ih}\int_{\gamma_\rho} f(z)\frac{(z-z_0)^n - (z-z_0-h)^n}{(z-z_0-h)^n(z-z_0)^n}\,dz,\end{aligned}$$

provided that $|h| < \rho$. Writing $z - z_0 = t$, $z - z_0 - h = t - h$ and using the algebraic identity

$$a^n - b^n = (a-b)(a^{n-1} + a^{n-2}b + \cdots + b^{n-1}),$$

we find that

$$\begin{aligned}&\frac{f^{(n-1)}(z_0+h) - f^{(n-1)}(z_0)}{h} \\ &= \frac{(n-1)!}{2\pi i}\int_{\gamma_\rho} f(z)\,\frac{t^{n-1} + t^{n-2}(t-h) + \cdots + (t-h)^{n-1}}{t^n(t-h)^n}\,dz.\end{aligned} \tag{8.15}$$

In the new variables, the right-hand side of (8.14) becomes

$$\frac{n!}{2\pi i}\int_{\gamma_\rho} \frac{f(z)}{t^{n+1}}\,dz. \tag{8.16}$$

It follows from (8.15) and (8.16) that

$$\begin{aligned}&\frac{f^{(n-1)}(z_0+h) - f^{(n-1)}(z_0)}{h} - \frac{n!}{2\pi i}\int_{\gamma_\rho} \frac{f(z)}{t^{n+1}}\,dz \\ &= \frac{(n-1)!}{2\pi i}\int_{\gamma_\rho} f(z)\,\frac{t^n + t^{n-1}(t-h) + \cdots + t(t-h)^{n-1} - n(t-h)^n}{t^{n+1}(t-h)^n}\,dz\end{aligned}$$

Therefore, by (7.13),

$$\begin{aligned}&\left|\frac{f^{(n-1)}(z_0+h) - f^{(n-1)}(z_0)}{h} - \frac{n!}{2\pi i}\int_{\gamma_\rho} \frac{f(z)}{t^{n+1}}\,dz\right| \\ &\leqslant (n-1)!\,\rho|h|M(\rho)\,\frac{(2\rho)^{n-1} + 2(2\rho)^{n-1} + \cdots + n(2\rho)^{n-1}}{\rho^{n+1}(\rho-|h|)^n},\end{aligned} \tag{8.17}$$

where

$$M(\rho) = \max_{z\in\gamma_\rho} |f(z)|.$$

But the right-hand side of (8.17) approaches zero as $h \to 0$, and hence

$$f^{(n)}(z_0) = \lim_{h\to 0} \frac{f^{(n-1)}(z_0+h) - f^{(n-1)}(z_0)}{h} = \frac{n!}{2\pi i}\int_{\gamma_\rho} \frac{f(z)}{t^{n+1}}\,dz,$$

which is equivalent to (8.14) or (8.13). This completes the proof.

COROLLARY. *If $f(z)$ is analytic on a domain G, then every derivative*

$$f^{(n)}(z) = \frac{d^n f(z)}{dz^n} \quad (n = 0, 1, 2, \ldots)$$

is analytic on G.

Problem 1. Evaluate the integral

$$\int_L \frac{dz}{z^2+9},$$

given that

a) $I(L)$ contains the point $3i$, while $E(L)$ contains the point $-3i$;
b) $I(L)$ contains the point $-3i$, while $E(L)$ contains the point $3i$;
c) $I(L)$ contains both points $\pm 3i$.

Problem 2. Assuming that none of the points 0, 1 and -1 lies on L, calculate all possible values of the integral

$$\int_L \frac{dz}{z(z^2-1)}$$

for various positions of L.

Problem 3. Evaluate the integral

$$\int_{|z-a|=a} \frac{z}{z^4-1}\,dz \qquad (a > 1).$$

Problem 4. Evaluate the integral

$$\frac{1}{2\pi i}\int_L \frac{e^z}{z^2+a^2}\,dz$$

if $I(L)$ contains the closed disk $|z| \leqslant a$.

Problem 5. Evaluate the integral

$$\frac{1}{2\pi i}\int_L \frac{ze^z}{(z-a)^3}\,dz,$$

given that $a \in I(L)$.

Hint. Use formula (8.13).

Problem 6. Prove that

$$\int_0^{2\pi} \cos(\cos\theta)\cosh(\sin\theta)\,d\theta = 2\pi.$$

Hint. Apply formula (8.11) to the function $f(z) = \cos z$.

Problem 7. Prove that if $f(z)$ is a nonvanishing analytic function on a domain G and if G contains the circle $\gamma_\rho : |z - z_0| = \rho$ and its interior $I(\gamma_\rho)$, then

$$\ln |f(z_0)| = \frac{1}{2\pi} \int_0^{2\pi} \ln |f(z_0 + \rho e^{i\theta})| \, d\theta.$$

Problem 8. Prove that

$$\int_0^{2\pi} \ln [\cosh^2 (\sin \theta) - \sin^2 (\cos \theta)] \, d\theta = 0.$$

Hint. Apply the preceding problem to the function $f(z) = \cos z$.

42. MORERA'S THEOREM. CAUCHY'S INEQUALITIES

The next result serves as the converse of Cauchy's integral theorem:

THEOREM 8.6 (*Morera's theorem*). *Let $f(z)$ be continuous on a domain G, and suppose that*

$$\int_L f(z) \, dz = 0 \tag{8.18}$$

for every closed rectifiable curve L contained in G. Then $f(z)$ is analytic on G.

Proof. If z_0 and z are any two points in G, it follows from (8.18) and the considerations given in Sec. 40 that the integral

$$F(z) = \int_{z_0}^{z} f(\zeta) \, d\zeta$$

is path-independent. Therefore $F(z)$ is single-valued, and hence, by the same argument as in the proof of Theorem 8.1, $F(z)$ is analytic on G with derivative $F'(z) = f(z)$. But then the corollary to Theorem 8.5 implies that $f(z)$ itself is analytic on G.

THEOREM 8.7. *Let $f(z)$ be analytic on a domain G, and suppose G contains the circle γ_ρ: $|z - z_0| = \rho$ and its interior. Then*

$$|f^{(n)}(z_0)| \leqslant n! \frac{M(\rho)}{\rho^n} \qquad (n = 0, 1, 2, \ldots), \tag{8.19}$$

where

$$M(\rho) = \max_{z \in \gamma_\rho} |f(z)|. \tag{8.20}$$

Proof. The inequalities (8.19), known as the *Cauchy inequalities*, are an immediate consequence of the representation (8.13) and the estimate (7.13), which together imply

$$|f^{(n)}(z_0)| \leqslant \frac{n!}{2\pi} \frac{M(\rho)}{\rho^{n+1}} 2\pi\rho = n! \frac{M(\rho)}{\rho^n}.$$

It should be noted that knowledge of the single *maximum modulus function* (8.20) is sufficient to write estimates for *all* the derivatives of $f(z)$.

Problem 1. Prove (8.19) for the case $n = 0$ directly from the corollary to Theorem 8.4.

Problem 2. A point ζ is said to be a limit point of a *sequence* $\{z_n\}$ if given any neighborhood $\mathscr{N}(\zeta)$, there is a subsequence $\{z_{k_n}\}$ all of whose terms belong to $\mathscr{N}(\zeta)$. Equivalently, ζ is said to be a limit point of $\{z_n\}$ if every neighborhood $\mathscr{N}(\zeta)$ contains infinitely many terms of $\{z_n\}$. Prove that every bounded sequence $\{z_n\}$ has at least one limit point.

Hint. Use Theorem 2.2.

Comment. Note that the *set* E consisting of the distinct terms of $\{z_n\}$ need not have a limit point as defined on p. 25. In other words, when referring to a sequence, the phrase "infinitely many terms" means infinitely many terms with distinct indices, but only finitely many of these terms need be distinct complex numbers.

Problem 3. If $\{a_n\}$ is a bounded sequence of real numbers, then, according to the preceding problem, $\{a_n\}$ has at least one limit point. An unbounded sequence $\{a_n\}$ also has at least one limit point if we allow the values $+\infty$ and $-\infty$.[4] The largest limit point of the (bounded or unbounded) sequence $\{a_n\}$ is called the *limit superior* or *upper limit* of the sequence, denoted by[5]

$$\overline{\lim_{n\to\infty}}\, a_n.$$

Starting from (8.19), prove the following *Cauchy-Hadamard inequality*: If $f(z)$ is analytic on a domain G, then

$$\overline{\lim_{n\to\infty}} \sqrt[n]{\frac{|f^{(n)}(z_0)|}{n!}} \leqslant \frac{1}{\Delta}$$

for every $z_0 \in G$, where Δ is the distance between z_0 and the boundary of G ($\Delta = \infty$ if G is the whole finite plane).

Problem 4. Prove that if $f(z)$ is an entire function, then

$$\lim_{n\to\infty} \sqrt[n]{\frac{|f^{(n)}(z_0)|}{n!}} = 0$$

for every finite z_0. Verify this formula for the function $f(z) = e^z$ (see M1, p. 304).

---

[4] In fact, $\{a_n\}$ has $+\infty$ as a limit point if it contains arbitrarily large positive terms and $-\infty$ as a limit point if it contains negative terms of arbitrarily large absolute value.

[5] Obviously

$$\overline{\lim_{n\to\infty}}\, a_n = \lim_{n\to\infty} a_n$$

in the case where $\{a_n\}$ has a limit (infinite limits are defined in the natural way).

Problem 5. The smallest limit point (possibly $\pm\infty$) of a sequence of real numbers $\{a_n\}$ is called the *limit inferior* or *lower limit* of the sequence, denoted by

$$\varliminf_{n\to\infty} a_n.$$

Prove that

a) $\displaystyle \varliminf_{n\to\infty} a_n = -\varlimsup_{n\to\infty} (-a_n);$

b) $\displaystyle \varliminf_{n\to\infty} a_n = \frac{1}{\varlimsup\limits_{n\to\infty} \dfrac{1}{a_n}}$ if $a_n > 0$.

CHAPTER 9

COMPLEX SERIES. UNIFORM CONVERGENCE

43. COMPLEX SERIES

Let

$$\sum_{n=1}^{\infty} z_n = z_1 + \cdots + z_n + \cdots \tag{9.1}$$

be an *infinite series* (briefly, a *series*) with complex terms z_n, and let

$$\{s_n\} = \{z_1 + \cdots + z_n\} \tag{9.2}$$

be the sequence of *partial sums* of (9.2). Then the series (9.1) is said to *converge* or to be *convergent* if the sequence (9.2) converges, and the limit of the sequence is called the *sum* of the series. Moreover, if

$$\lim_{n\to\infty} s_n = s,$$

we write

$$\sum_{n=1}^{\infty} z_n = z_1 + \cdots + z_n + \cdots = s.$$

A series which does not converge is said to be *divergent*.

If we set

$$z_n = x_n + iy_n, \qquad s_n = \sigma_n + i\tau_n = (x_1 + \cdots + x_n) + i(y_1 + \cdots + y_n),$$

it follows from Theorem 1.1 that the series (9.1) converges if and only if both of the sequences $\{\sigma_n\}$ and $\{\tau_n\}$ converge. But these are the sequences of partial sums of the following series with real terms:

$$\sum_{n=1}^{\infty} x_n = x_1 + \cdots + x_n + \cdots, \tag{9.3}$$

$$\sum_{n=1}^{\infty} y_n = y_1 + \cdots + y_n + \cdots. \tag{9.4}$$

Therefore a series with complex terms converges if and only if the series formed from the real and imaginary parts of its terms converge. Moreover, if

$$\sum_{n=1}^{\infty} x_n = \lim_{n\to\infty} (x_1 + \cdots + x_n) = \sigma,$$

$$\sum_{n=1}^{\infty} y_n = \lim_{n\to\infty} (y_1 + \cdots + y_n) = \tau,$$

then

$$\lim_{n\to\infty} (z_1 + \cdots + z_n) = \sigma + i\tau.$$

Thus the sum of the series (9.1) equals $\sigma + i\tau$, where σ and τ are the sums of the series (9.3) and (9.4), and conversely, the sums of (9.3) and (9.4) are the real and imaginary parts of the sum of (9.1).

Applying the Cauchy convergence criterion (Theorem 1.2) to the sequence of partial sums (9.2), and noting that

$$s_{n+p} - s_n = z_{n+1} + \cdots + z_{n+p},$$

we find that the series (9.1) converges if and only if given any $\varepsilon > 0$, there exists an integer $N(\varepsilon) > 0$ such that

$$|z_{n+1} + \cdots + z_{n+p}| < \varepsilon$$

for all $n > N(\varepsilon)$ and $p > 0$. In particular, a necessary condition for the series (9.1) to converge is that

$$\lim_{n\to\infty} z_n = 0.$$

The series (9.1) is said to be *absolutely convergent* if the series

$$\sum_{n=1}^{\infty} |z_n| = |z_1| + \cdots + |z_n| + \cdots \tag{9.5}$$

formed from the moduli of its terms converges. Since

$$|z_{n+1} + \cdots + z_{n+p}| \leqslant |z_{n+1}| + \cdots + |z_{n+p}|,$$

a series with complex terms is convergent if it is absolutely convergent. The converse is not true, as shown by the familiar example

$$1 - \frac{1}{2} + \frac{1}{3} - \frac{1}{4} + \cdots.$$

It follows from the inequalities

$$|x_n| \leqslant |z_n|, \quad |y_n| \leqslant |z_n| \quad \text{and} \quad |z_n| \leqslant |x_n| + |y_n|$$

that the series (9.1) is absolutely convergent if and only if the series (9.3) and (9.4) are absolutely convergent.

Let the series (9.1) be absolutely convergent with sum $\sigma + i\tau$. Then the series (9.3) and (9.4) are also absolutely convergent with sums σ and τ, respectively. Suppose we make an arbitrary rearrangement of the terms of (9.1), thereby obtaining a new series

$$\sum_{n=1}^{\infty} z_{k_n} = z_{k_1} + \cdots + z_{k_n} + \cdots, \tag{9.6}$$

where the sequence $k_1, \ldots, k_n, \ldots$ contains all the positive integers, each just once. Correspondingly, the series formed from the real and imaginary parts of (9.6) have the form

$$\sum_{n=1}^{\infty} x_{k_n} = x_{k_1} + \cdots + x_{k_n} + \cdots,$$

$$\sum_{n=1}^{\infty} y_{k_n} = y_{k_1} + \cdots + y_{k_n} + \cdots.$$

These series are obtained by rearranging the terms of the absolutely convergent series (9.3) and (9.4) and therefore, by a familiar property of real series, they converge to their previous sums σ and τ. It follows that the series (9.6) with complex terms also converges to its original sum $\sigma + i\tau$.

Problem 1. Prove the following properties of complex series:

a) For any integer $N > 1$, the two series

$$\sum_{n=1}^{\infty} z_n, \quad \sum_{n=N}^{\infty} z_n$$

converge and diverge together;

b) If the series $z_1 + \cdots + z_n + \cdots$ converges and its sum is s, then the series $\lambda z_1 + \cdots + \lambda z_n + \cdots$ converges and its sum is λs;

c) If the series $z_1 + \cdots + z_n + \cdots$ converges and its sum is s, then

$$(z_1 + \cdots + z_{k_1}) + (z_{k_1+1} + \cdots + z_{k_2}) + \cdots + (z_{k_n+1} + \cdots + z_{k_{n+1}}) + \cdots = s,$$

where $0 < k_1 < k_2 < \cdots < k_n < k_{n+1} < \cdots$;

d) If

$$\sum_{n=1}^{\infty} z_n = s, \quad \sum_{n=1}^{\infty} z'_n = s',$$

then

$$\sum_{n=1}^{\infty} (z_n \pm z'_n) = s \pm s'.$$

Problem 2. What is the analogue for complex series of the familiar ratio and root tests for real series? State the analogue of the comparison test.

Problem 3 (M1, p. 39). Prove that if the series

$$\sum_{n=1}^{\infty} z_n, \quad \sum_{n=1}^{\infty} z'_n$$

are absolutely convergent, with sums s and s' respectively, then the series

$$\sum_{n=1}^{\infty} (z_1 z'_n + z_2 z'_{n-1} + \cdots + z_n z'_1)$$

is also absolutely convergent, with sum ss'.

Problem 4. Investigate the convergence of the following series:

a) $\displaystyle\sum_{n=0}^{\infty} \frac{1}{(1+i)^n}$; b) $\displaystyle\sum_{n=0}^{\infty} \left(\cos\frac{n\pi}{4} + i\sin\frac{n\pi}{4}\right)$;

c) $\displaystyle\sum_{n=0}^{\infty} \frac{1}{(1+i)^n}\left(\cos\frac{n\pi}{4} + i\sin\frac{n\pi}{4}\right)$; d) $\displaystyle\sum_{n=0}^{\infty} \frac{(1+i)^n}{i(1+i)^n + 2}$.

44. UNIFORMLY CONVERGENT SERIES AND SEQUENCES

Having discussed series whose terms are complex *numbers*, we now consider series whose terms are complex *functions*. Thus let

$$\sum_{n=1}^{\infty} f_n(z) = f_1(z) + \cdots + f_n(z) + \cdots \tag{9.7}$$

be an (infinite) series whose terms are functions of a complex variable defined on a set E in the complex plane, and let

$$s_n(z) = f_1(z) + \cdots + f_n(z) \tag{9.8}$$

be the nth partial sum of (9.7). Then the series (9.7) is said to be *uniformly convergent* on E, with *sum* (*function*) $f(z)$, if given any $\varepsilon > 0$, there exists an integer $N(\varepsilon) > 0$ such that

$$|s_n(z) - f(z)| < \varepsilon$$

for all $n > N(\varepsilon)$ and all $z \in E$.

Remark 1. Obviously, if a series is uniformly convergent on E, it is also convergent on E (i.e., at every point of E). However, the converse is not true, as shown by the example below.

Remark 2. In some problems, especially those involving power series, we will find it convenient to introduce an extra "zeroth" term $f_0(z)$, writing

$$\sum_{n=0}^{\infty} f_n(z) = f_0(z) + f_1(z) + \cdots + f_n(z) + \cdots$$

and

$$s_n(z) = f_0(z) + f_1(z) + \cdots + f_n(z),$$

instead of (9.7) and (9.8).

Remark 3. It follows from the Cauchy convergence criterion (Theorem 1.2) that a necessary and sufficient condition for (9.7) to converge uniformly on E to $f(z)$ is that there exist an integer $N(\varepsilon) > 0$ such that

$$|s_{n+p}(z) - s_n(z)| < \varepsilon$$

for all $n > N(\varepsilon)$, all $p > 0$ and all $z \in E$.[1]

Example. The *geometric series*

$$\sum_{n=0}^{\infty} z^n = 1 + z + \cdots + z^n + \cdots \tag{9.9}$$

converges on the open unit disk K: $|z| < 1$, since

$$1 + z + \cdots + z^n = \frac{1 - z^{n+1}}{1 - z} \to \frac{1}{1 - z} \quad \text{as} \quad n \to \infty$$

if $|z| < 1$. Moreover,

$$|s_{n+p}(z) - s_n(z)| = |z^{n+1}(1 + z + \cdots + z^{p-1})|$$

$$= \frac{|z|^{n+1}\,|1 - z^p|}{|1 - z|} \geqslant \frac{|z|^{n+1}(1 - |z|^p)}{|1 - z|},$$

and hence choosing

$$p = n, \quad z_n = \frac{n - 1}{n} \qquad (n > 0),$$

we find that[2]

$$|s_{2n}(z_n) - s_n(z_n)| \geqslant n\left(1 - \frac{1}{n}\right)^{n+1}\left[1 - \left(1 - \frac{1}{n}\right)^n\right] \to \infty \qquad \text{as} \quad n \to \infty.$$

---

[1] For brevity, we usually just say "for all $n > N(\varepsilon)$, $p > 0$ and $z \in E$." It may sometimes be appropriate to write $N(E, \varepsilon)$ instead of $N(\varepsilon)$, to emphasize the fact that $N(E, \varepsilon)$ depends on the set E (which may not be fixed), as well as on the number ε (see the remark on p. 188).

[2] Note that $\lim\limits_{n\to\infty}\left(1 - \dfrac{1}{n}\right)^n = \dfrac{1}{e}$.

Thus, for sufficiently large n and appropriate p and $z \in K$, we can make the difference $|s_{n+p}(z) - s_n(z)|$ as large as we please. It follows that the series (9.9) is not uniformly convergent on K.

Next we give an important sufficient condition for uniform convergence:

THEOREM 9.1 (*Weierstrass' M-test*). *Given a convergent series*

$$\sum_{n=1}^{\infty} M_n, \tag{9.10}$$

whose terms are nonnegative constants, suppose the functions $f_1(z), \ldots, f_n(z), \ldots$ are such that

$$|f_n(z)| \leqslant M_n \tag{9.11}$$

for all $z \in E$ and all n exceeding a certain integer $N > 0$. Then the series

$$\sum_{n=1}^{\infty} f_n(z) = f(z) \qquad (z \in E) \tag{9.12}$$

is uniformly convergent on E.

Proof. Since (9.10) converges, given any $\varepsilon = 0$, there is an integer $N_1(\varepsilon) > 0$ such that

$$M_{n+1} + \cdots + M_{n+p} < \varepsilon$$

for all $n > N_1(\varepsilon)$ and $p > 0$. If $N_2(\varepsilon) = \max\{N_1(\varepsilon), N\}$, then, according to (9.11),

$$\begin{aligned} |s_{n+p}(z) - s_n(z)| &\leqslant |f_{n+1}(z)| + \cdots + |f_{n+p}(z)| \\ &\leqslant M_{n+1} + \cdots + M_{n+p} < \varepsilon \end{aligned}$$

for all $n > N_2(\varepsilon)$, $p > 0$ and $z \in E$, i.e., (9.12) is uniformly convergent on E (recall Remark 3). We note, in passing, that (9.12) is also absolutely convergent on E.

The sum of a convergent series of continuous functions need not be continuous, as the reader will recall from advanced calculus.[3] The situation is different if the series is *uniformly* convergent:

THEOREM 9.2. *If the series* (9.12) *converges uniformly on a set E with a limit point $z_0 \in E$ and if every term $f_n(z)$ is continuous at z_0, then the sum $f(z)$ is also continuous at z_0.*

Proof. If z ($\neq z_0$) is any point of E, then

$$\begin{aligned} |f(z) - f(z_0)| &= |[f(z) - s_n(z)] + [s_n(z) - s_n(z_0)] + [s_n(z_0) - f(z_0)]| \\ &\leqslant |s_n(z) - f(z)| + |s_n(z) - s_n(z_0)| + |s_n(z_0) - f(z_0)|. \end{aligned}$$

---

[3] See e.g., D. V. Widder, *op. cit.*, p. 300.

Since (9.12) is uniformly convergent on E, given any $\varepsilon > 0$, there is an integer $N(\varepsilon) > 0$ such that

$$|s_n(z) - f(z)| < \frac{\varepsilon}{3},$$

for all $n > N(\varepsilon)$ and $z \in E$. In particular,

$$|s_n(z_0) - f(z_0)| < \frac{\varepsilon}{3}$$

for all $n > N(\varepsilon)$. Therefore, if n_0 is any integer greater than $N(\varepsilon)$, we can set $n = n_0$ and then combine the last three inequalities, obtaining

$$|f(z) - f(z_0)| < |s_{n_0}(z) - s_{n_0}(z_0)| + \frac{2\varepsilon}{3}. \tag{9.13}$$

Moreover, since $s_{n_0}(z)$ is continuous at $z_0 \in E$, being the sum of a *finite* number of functions continuous at z_0, we can choose $\delta(\varepsilon) > 0$ such that

$$|s_{n_0}(z) - s_{n_0}(z_0)| < \frac{\varepsilon}{3}$$

if $|z - z_0| < \delta$, and hence, according to (9.13),

$$|f(z) - f(z_0)| < \varepsilon$$

if $|z - z_0| < \delta$. But this is precisely what is meant by saying that $f(z)$ is continuous at z_0.

A related result is

THEOREM 9.3. *Given a rectifiable curve L, suppose the series*

$$\sum_{n=1}^{\infty} f_n(z) = f(z) \tag{9.14}$$

is uniformly convergent on L, and suppose every term $f_n(z)$ is continuous on L. Then (9.14) *can be integrated term by term along L, i.e.,*

$$\sum_{n=1}^{\infty} \int_L f_n(z)\,dz = \int_L f(z)\,dz. \tag{9.15}$$

Proof. According to Theorem 9.2, $f(z)$ is continuous and hence integrable along L. Since (9.14) is uniformly convergent on L, given any $\varepsilon > 0$, there is an integer $N(\varepsilon) > 0$ such that

$$|s_n(z) - f(z)| < \frac{\varepsilon}{l}$$

for all $n > N(\varepsilon)$ and $z \in L$, where l is the length of L. Then obviously

$$\left| \sum_{k=1}^{n} \int_L f_k(z)\,dz - \int_L f(z)\,dz \right| = \left| \int_L [s_n(z) - f(z)]\,dz \right| < \frac{\varepsilon}{l} l = \varepsilon,$$

and since ε is arbitrary, the theorem is proved.

COROLLARY. *Given a rectifiable curve L contained in a domain G, suppose the series* (9.14) *is uniformly convergent on every compact subset of G, and suppose every term $f_n(z)$ is continuous on G. Then* (9.14) *can be integrated term by term along L.*

Proof. According to Theorem 2.6, any continuous curve (rectifiable or not) contained in G is a compact subset of G.

Remark. Obviously, every series which is uniformly convergent on a domain G is uniformly convergent on every compact subset of G. However, the converse is not true. For example, consider the geometric series

$$\sum_{n=0}^{\infty} z^n = 1 + z + \cdots + z^n + \cdots,$$

which, as we have already shown (see p. 185) is convergent on the disk K: $|z| < 1$. Let F be any compact subset of K, and let $\delta > 0$ be the distance between F and the circle $|z| = 1$ (the boundary of K). Then for any point $z \in F$, we have $|z| \leqslant 1 - \delta$, and hence

$$\begin{aligned} |z^{n+1} + \cdots + z^{n+p}| &= \left| z^{n+1} \frac{1 - z^p}{1 - z} \right| \\ &\leqslant |z|^{n+1} \frac{1 + |z|^p}{1 - |z|} \leqslant (1 - \delta)^{n+1} \frac{2}{\delta}. \end{aligned} \tag{9.16}$$

Since $(1 - \delta)^{n+1}(2/\delta) \to 0$ as $n \to \infty$, then, given any $\varepsilon > 0$, there is an integer $N(F, \varepsilon) > 0$ such that the left-hand side of (9.16) is less than ε for all $n > N(F, \varepsilon)$ and $z \in F$. It follows that the geometric series is uniformly convergent on every compact subset of K (but not on K itself, as already noted).

Finally we turn our attention to (infinite) sequences of complex functions, defining uniform convergence in the natural way. Thus let $\{f_n(z)\}$ be a sequence whose terms are functions of a complex variable defined on a set E in the complex plane. Then the sequence $\{f_n(z)\}$ is said to be *uniformly convergent* on E, with *limit (function)* $f(z)$, if given any $\varepsilon > 0$, there exists an integer $N(\varepsilon) > 0$ such that

$$|f_n(z) - f(z)| < \varepsilon$$

for all $n > N(\varepsilon)$ and $z \in E$. Since $\{f_n(z)\}$ can be regarded as the sequence of partial sums of the series

$$f_1(z) + [f_2(z) - f_1(z)] + \cdots + [f_n(z) - f_{n-1}(z)] + \cdots, \tag{9.17}$$

$\{f_n(z)\}$ is a uniformly convergent sequence on E if and only if (9.17) is a uniformly convergent series on E. Moreover, the terms of the sequence $\{f_n(z)\}$ are continuous if and only if the terms of the series (9.17) are continuous. It follows that every theorem of this section has an obvious counterpart for sequences.

Problem 1. Find the set on which each of the following series converges:

a) $\displaystyle\sum_{n=1}^{\infty} \frac{z^n}{1 - z^n}$; b) $\displaystyle\sum_{n=1}^{\infty} \frac{\sin nz}{n}$; c) $\displaystyle\sum_{n=1}^{\infty} \frac{\cos nz}{n^2}$; d) $\displaystyle\sum_{n=1}^{\infty} \frac{(-1)^n}{z + n}$;

e) $\displaystyle\sum_{n=1}^{\infty} \frac{2^n}{z^{2^n} + 1}$.

Hint. In part e) use the fact that

$$\frac{2^n}{z^{2^n} - 1} = \frac{2}{z^2 - 1} - \frac{2}{z^2 + 1} - \cdots - \frac{2^{n-1}}{z^{2^{n-1}} + 1}.$$

Problem 2. Where does each of the series given in the preceding problem converge *uniformly*?

Problem 3. Prove that uniform convergence of the series

$$\sum_{n=1}^{\infty} |f_n(z)|$$

on a set E implies uniform convergence of the series

$$\sum_{n=1}^{\infty} f_n(z)$$

on E, but not conversely.

Hint. Consider the series

$$\sum_{n=1}^{\infty} \frac{z^n}{n}$$

in the interval $(-1, 0)$.

Problem 4. A set E is said to be *dense in itself* if every point of E is a limit point of E. For example, any nonempty open set (in particular, a domain) or any continuous curve is dense in itself. Prove the following variant of Theorem 9.2: If the series (9.12) converges on a set E which is dense in itself and if every term $f_n(z)$ is continuous on E, then the sum $f(z)$ is also continuous on E.

Problem 5. Prove that if the series

$$\sum_{n=1}^{\infty} f_n(z) = f(z)$$

is uniformly convergent on a set E, and if $|\varphi(z)| \leqslant$ const on E, then the series

$$\sum_{n=1}^{\infty} \varphi(z) f_n(z)$$

is uniformly convergent on E, with sum $\varphi(z)f(z)$.

Problem 6. State the analogues of Theorems 9.1–9.3 for sequences.

45. SERIES AND SEQUENCES OF ANALYTIC FUNCTIONS

We begin by proving the following

LEMMA. *A series*

$$\sum_{n=1}^{\infty} f_n(z) = f(z) \tag{9.18}$$

which is convergent on a domain G is uniformly convergent on every compact subset of G if and only if, given any point $z_0 \in G$, there is a neighborhood $\mathscr{N}(z_0) \subset G$ in which (9.18) *is uniformly convergent.*

Proof. The condition is obviously necessary, since given any $z_0 \in G$, (9.18) is uniformly convergent in some closed neighborhood

$$\overline{\mathscr{N}(z_0)}: |z - z_0| \leqslant \rho$$

contained in G, and hence uniformly convergent on $\mathscr{N}(z_0)$. To prove that the condition is sufficient, suppose there is a compact set $F \subset G$ on which (9.18) is not uniformly convergent, although the condition of the theorem is satisfied. Then there must exist a number $\varepsilon > 0$, a sequence $\{z_n\}$ of points in F, and a sequence $\{k_n\}$ of positive integers, where $k_1 < \cdots < k_n < \cdots$, such that

$$|s_{k_n}(z_n) - f(z_n)| \geqslant \varepsilon \tag{9.19}$$

for all $n = 1, 2, \ldots$. [This is the negative of the assertion that (9.18) is uniformly convergent on F.] The sequence $\{z_n\}$, being bounded, has at least one limit point ζ (see Sec. 42, Prob. 2), and moreover contains a subsequence $\{z_n'\}$ converging to ζ. Since F is closed, $\zeta \in F \subset G$ and hence, by hypothesis, there is a neighborhood $\mathscr{N}(\zeta) \subset G$ in which the series (9.18) is uniformly convergent, i.e.,

$$|s_n(z) - f(z)| < \varepsilon$$

for sufficiently large n and all $z \in \mathcal{N}(\zeta)$. On the other hand, for sufficiently large n, $\mathcal{N}(\zeta)$ must contain points (in fact, all the points of $\{z'_n\}$ starting from a certain value of n) satisfying the opposite inequality (9.19). This contradiction completes the proof.

The next theorem is basic to the whole subject:

THEOREM 9.4 (*Weierstrass' theorem on uniformly convergent series of analytic functions*). *If the series* (9.18) *is uniformly convergent on every compact subset of a domain G, and if every term $f_n(z)$ is analytic on G, then the sum $f(z)$ of the series is also analytic on G. Moreover, the series* (9.18) *can be differentiated term by term any number of times, i.e.,*[4]

$$\sum_{n=1}^{\infty} f_n^{(k)}(z) = f^{(k)}(z) \qquad (k = 1, 2, \ldots) \tag{9.20}$$

for all $z \in G$, and each differentiated series is uniformly convergent on every compact subset of G.

Proof. Let z_0 be an arbitrary point of G, and choose $\rho > 0$ such that G contains the circle γ_ρ: $|z - z_0| = \rho$ and its interior $I(\gamma_\rho)$. By hypothesis, the series (9.18) is uniformly convergent on γ_ρ, and hence so is each of the series

$$\frac{k!}{2\pi i} \sum_{n=1}^{\infty} \frac{f_n(z)}{(z - z_0)^{k+1}} = \frac{k!}{2\pi i} \frac{f(z)}{(z - z_0)^{k+1}} \qquad (k = 0, 1, 2, \ldots) \tag{9.21}$$

$(0! = 1)$, since

$$\left| \frac{k!}{2\pi i} \frac{1}{(z - z_0)^{k+1}} \right| = \frac{k!}{2\pi \rho^{k+1}}$$

for all $z \in \gamma_\rho$ (see Sec. 44, Prob. 5). Therefore, by the corollary to Theorem 9.3, we can integrate (9.21) term by term along γ_ρ, obtaining

$$\frac{k!}{2\pi i} \sum_{n=1}^{\infty} \int_{\gamma_\rho} \frac{f_n(z)}{(z - z_0)^{k+1}}\, dz = \frac{k!}{2\pi i} \int_{\gamma_\rho} \frac{f(z)}{(z - z_0)^{k+1}}\, dz \qquad (k = 0, 1, 2, \ldots). \tag{9.22}$$

For $k = 0$, (9.22) reduces to

$$\frac{1}{2\pi i} \sum_{n=1}^{\infty} \int_{\gamma_\rho} \frac{f_n(z)}{z - z_0}\, dz = \frac{1}{2\pi i} \int_{\gamma_\rho} \frac{f(z)}{z - z_0}\, dz$$

or

$$f(z_0) = \sum_{n=1}^{\infty} f_n(z_0) = \frac{1}{2\pi i} \int_{\gamma_\rho} \frac{f(z)}{z - z_0}\, dz, \tag{9.23}$$

[4] As always, we use the notation

$$f_n^{(0)}(z) = f_n(z), f_n^{(1)}(z) = \frac{df_n(z)}{dz}, \ldots, f_n^{(k)}(z) = \frac{d^k f_n(z)}{dz^k}, \ldots$$

where we use Theorem 8.4 and the fact that every $f_n(z)$ is analytic on $\overline{I(\gamma_\rho)}$. Therefore $f(z)$ is represented by Cauchy's integral at every point $z_0 \in G$. But in proving Theorem 8.5, we used the analyticity of $f(z)$ only to establish (8.14) for $n = 0$ and hence if $f(z)$ satisfies (9.23) it must have derivatives of all orders on G. In particular, $f(z)$ is analytic on G. For $k > 0$, (9.22) reduces to (9.20), with z replaced by z_0, where we now use formula (8.14) and the fact that $f(z)$ itself, as well as $f_n(z)$ is analytic on $\overline{I(\gamma_\rho)}$.

Finally we must prove that each differentiated series

$$\sum_{n=1}^{\infty} f_n^{(k)}(z) = \frac{k!}{2\pi i} \sum_{n=1}^{\infty} \int_{\gamma_\rho} \frac{f_n(\zeta)}{(\zeta - z)^{k+1}}\, d\zeta = \frac{k!}{2\pi i} \int_{\gamma_\rho} \frac{f(\zeta)}{(\zeta - z)^{k+1}}\, d\zeta = f^{(k)}(z) \tag{9.24}$$

is uniformly convergent on every compact subset of G. To show this, we use the lemma, choosing the neighborhood $\mathcal{N}(z_0)$ to be the open disk $|z - z_0| < \frac{1}{2}\rho$. Since the original series (9.18) is uniformly convergent on γ_ρ, there is an integer $N(\varepsilon) > 0$ such that

$$|s_n(z) - f(z)| < \varepsilon$$

for all $n > N(\varepsilon)$ and $z \in \gamma_\rho$, where $s_n(z)$ denotes the nth partial sum of (9.18). It follows that

$$\begin{aligned} &\left| \frac{k!}{2\pi i} \sum_{j=1}^{n} \int_{\gamma_\rho} \frac{f_j(\zeta)}{(\zeta - z)^{k+1}}\, d\zeta - \frac{k!}{2\pi i} \int_{\gamma_\rho} \frac{f(\zeta)}{(\zeta - z)^{k+1}}\, d\zeta \right| \\ &\qquad = \left| \frac{k!}{2\pi i} \int_{\gamma_\rho} \frac{s_n(\zeta) - f(\zeta)}{(\zeta - z)^{k+1}}\, d\zeta \right| < \frac{k!}{2\pi} \frac{\varepsilon}{(\rho/2)^{k+1}} 2\pi\rho \end{aligned} \tag{9.25}$$

for all $n > N(\varepsilon)$ and $z \in \mathcal{N}(z_0)$. The right-hand side of (9.25) obviously goes to zero with ε itself. Therefore the series (9.24) is uniformly convergent on $\mathcal{N}(z_0)$, and the proof is complete.

Corollary (*Weierstrass' theorem on uniformly convergent sequences of analytic functions*). *If the sequence $\{f_n(z)\}$ is uniformly convergent on every compact subset of a domain G, and if every term $f_n(z)$ is analytic on G, then the limit function*

$$f(z) = \lim_{n \to \infty} f_n(z)$$

is also analytic on G. Moreover, as $n \to \infty$ each differentiated sequence $\{f_n^{(k)}(z)\}$, $k = 1, 2, \ldots$, converges uniformly to $f^{(k)}(z)$ on every compact subset of G.

Problem 1 (M1, p. 329). Prove that Theorem 9.4 does not hold if G is assumed to be an arbitrary set instead of a domain.

Hint. The series

$$\sin x + \left(\frac{\sin 2x}{2} - \sin x\right) + \left(\frac{\sin 3x}{3} - \frac{\sin 2x}{2}\right) + \cdots \tag{9.26}$$

converges uniformly to zero on the real axis. Examine the result of differentiating (9.26) term by term.

Problem 2. Prove that (9.26) cannot converge uniformly on any domain containing points of the real axis.

CHAPTER 10

POWER SERIES

46. THE CAUCHY-HADAMARD THEOREM

In the preceding chapter, we considered series whose terms are analytic functions. Of all such series, the simplest, and at the same time, the most important, are the power series. By a *power series* we mean a series of the form

$$a_0 + a_1(z - z_0) + \cdots + a_n(z - z_0)^n + \cdots, \tag{10.1}$$

where $a_0, a_1, \ldots, a_n, \ldots$ and z_0 are given complex numbers.

THEOREM 10.1 (*Cauchy-Hadamard theorem*). *Given the power series* (10.1), *let*[1]

$$R = \frac{1}{\Lambda}, \quad \textit{where} \quad \Lambda = \overline{\lim_{n\to\infty}} \sqrt[n]{|a_n|}, \tag{10.2}$$

and let γ *be the circle* $|z - z_0| = R$, *with interior* $I(\gamma)$ *and exterior* $E(\gamma)$. *Then there are three possibilities:*

1. *If* $R = 0$, *the series is divergent for all* $z \neq z_0$;
2. *If* $0 < R < \infty$, *the series is absolutely convergent for all* $z \in I(\gamma)$ *and divergent for all* $z \in E(\gamma)$;
3. *If* $R = \infty$, *the series is absolutely convergent for all finite* z.

[1] Concerning the meaning of $\overline{\lim_{n\to\infty}}$, see Sec. 42, Prob. 3.

Proof. We examine each of the three possibilities in turn (obviously, in every case, the series is absolutely convergent for $z = z_0$):

1. $R = 0$, $\Lambda = \infty$. In this case,

$$\Lambda > \frac{1}{|z - z_0|}$$

for any $z \neq z_0$, i.e., for any $z \neq z_0$ we can find an infinite sequence $\{n_k\}$ of positive integers, where $n_1 < \cdots < n_k < \cdots$ (recall the remark on p. 17) such that

$$\sqrt[n_k]{|a_{n_k}|} > \frac{1}{|z - z_0|} \qquad (k = 1, 2, \ldots). \tag{10.3}$$

But this implies

$$|a_{n_k}(z - z_0)^{n_k}| > 1 \qquad (k = 1, 2, \ldots), \tag{10.4}$$

and hence the necessary condition

$$\lim_{n\to\infty} a_n(z - z_0)^n = 0 \tag{10.5}$$

(recall p. 182) for the convergence of the series (10.1) cannot be satisfied for any $z \neq z_0$. Therefore, if $R = 0$, the series diverges for all $z \neq z_0$, and converges only at the single point $z = z_0$.

2. $0 < R < \infty$, $0 < \Lambda < \infty$. In this case, if $z \in I(\gamma)$, we can write

$$|z - z_0| = \frac{\theta^2}{\Lambda},$$

where $0 < \theta < 1$ (except for the trivial case $z = z_0$). Then

$$\Lambda < \frac{\Lambda}{\theta} = \frac{\theta}{|z - z_0|},$$

and hence all the values of $\sqrt[n]{|a_n|}$ starting from some value of n must be less than

$$\frac{\theta}{|z - z_0|},$$

i.e., the inequality

$$|a_n(z - z_0)^n| < \theta^n$$

holds, starting from some value of n. Therefore, since the geometric series

$$1 + \theta + \cdots + \theta^n + \cdots$$

is convergent, the series (10.1) is absolutely convergent if $z \in I(\gamma)$. On the other hand, if $z \in E(\gamma)$, then

$$\Lambda > \frac{1}{|z - z_0|},$$

and just as in the case $R = 0$, we can find an infinite sequence $\{n_k\}$ of positive integers, where $n_1 < \cdots < n_k < \cdots$, such that (10.3) and (10.4) hold. Therefore the necessary condition (10.5) for convergence of the series cannot be satisfied, and the series diverges.

3. $R = \infty$, $\Lambda = 0$. In this case,

$$\Lambda < \frac{\theta}{|z - z_0|}$$

for any $z \neq z_0$ and any θ $(0 < \theta < 1)$. Therefore the series (10.1) converges absolutely for any $z \neq z_0$ by exactly the same argument as when $0 < R < \infty$, $z \in I(\gamma)$.

Remark 1. The circle γ is called the *circle of convergence* and R is called the *radius of convergence* [of the series (10.1)]. The formula (10.2) is called the *Cauchy-Hadamard formula*, and expresses the radius of convergence in terms of the coefficients a_n of the power series. Obviously, the behavior of the series for $R = 0$ and $R = \infty$ can be regarded as the appropriate limiting case of its behavior for $0 < R < \infty$. It should be noted that Theorem 10.1 says nothing about the behavior of the series on its circle of convergence, i.e., for $z \in \gamma$. From now on, we will only consider power series for which $R > 0$.

Remark 2. In applying the Cauchy-Hadamard formula, the following relation often turns out to be useful:

$$\lim_{n\to\infty} \sqrt[n]{\frac{n!}{n^n}} = \frac{1}{e}. \tag{10.6}$$

To prove (10.6), we first note that

$$e^n = 1 + \frac{n}{1!} + \cdots + \frac{n^{n-1}}{(n-1)!} + \frac{n^n}{n!}\left[1 + \frac{n}{n+1} + \frac{n^2}{(n+1)(n+2)} + \cdots\right],$$

which implies

$$e^n < n\frac{n^n}{n!} + \frac{n^n}{n!}\left[1 + \frac{n}{n+1} + \left(\frac{n}{n+1}\right)^2 + \cdots\right] = (2n+1)\frac{n^n}{n!}.$$

On the other hand, it is obvious that

$$e^n > \frac{n^n}{n!},$$

and hence

$$\frac{1}{e^n} < \frac{n!}{n^n} < \frac{2n+1}{e^n},$$

from which (10.6) follows immediately.[2]

[2] Actually, it can be shown that

$$\frac{n!}{n^n} = \frac{\sqrt{2\pi n}}{e^n}(1 + \varepsilon),$$

where $\varepsilon \to 0$ as $n \to \infty$, a result known as *Stirling's formula*. See e.g., D. V. Widder, *op. cit.*, p. 386.

Example 1. The radius of convergence of the series

$$z + \frac{2^2}{2!} z^2 + \cdots + \frac{n^n}{n!} z^n + \cdots \tag{10.7}$$

equals $1/e$, as we see at once from (10.2) and (10.6).

Example 2. Consider the series

$$z + \frac{z^{2^2}}{2!} + \cdots + \frac{z^{n^2}}{n!} + \cdots. \tag{10.8}$$

In this case,

$$\overline{\lim_{n\to\infty}} \sqrt[n]{|a_n|} = \lim_{n\to\infty} \sqrt[n^2]{\frac{1}{n!}} = \lim_{n\to\infty} \sqrt[n^2]{\frac{n^n}{n!}\frac{1}{n^n}} = 1$$

(supply the details), and the radius of convergence equals 1.

THEOREM 10.2. *Let γ: $|z - z_0| = R$ be the circle of convergence of the power series* (10.1). *Then the series is uniformly convergent on every compact subset of $I(\gamma)$.*

Proof. Since any compact set $F \subset I(\gamma)$ is contained in some closed disk $|z - z_0| \leqslant r < R$, if r is near enough to R, we need only prove that the series is uniformly convergent on every such closed disk. Thus, given a closed disk $|z - z_0| \leqslant r < R$, let ζ be a point such that

$$r < |\zeta - z_0| = \rho < R.$$

Since the series (10.1) is absolutely convergent for $z = \zeta$, the series

$$\sum_{n=0}^{\infty} |a_n|\, |\zeta - z_0|^n = \sum_{n=0}^{\infty} |a_n|\, \rho^n$$

converges. Therefore, by Weierstrass' M-test (Theorem 9.1), the series (10.1) is uniformly convergent on the disk $|z - z_0| \leqslant r$, since

$$|a_n(z - z_0)^n| < |a_n|\, \rho^n$$

for all z in the disk $|z - z_0| \leqslant r$.

Remark. The power series (10.1) need not converge uniformly on $I(\gamma)$ itself. In fact, as we already know (see the remark on p. 188), the geometric series

$$\sum_{n=0}^{\infty} z^n,$$

which is a power series with radius of convergence 1, is not uniformly convergent on the unit disk K: $|z| < 1$, but only on every compact subset of K.

Problem 1. Deduce the results of Examples 1 and 2 above by direct use of the ratio test (cf. Sec. 43, Prob. 2).

Problem 2. Prove that if a power series is absolutely convergent at a single point of its circle of convergence, then it is absolutely convergent at every point of its circle of convergence.

Problem 3. Find the radius of convergence of each of the following power series:

a) $\sum_{n=1}^{\infty} n^k z^n \ (k = 0, 1, 2, \ldots)$; b) $\sum_{n=1}^{\infty} n^n z^n$; c) $\sum_{n=1}^{\infty} \frac{n^k}{n!} z^n \ (k = 0, 1, 2, \ldots)$;

d) $\sum_{n=1}^{\infty} \frac{n!}{n^n} z^n$; e) $\sum_{n=0}^{\infty} \frac{2^n}{n!} z^n$; f) $\sum_{n=1}^{\infty} n!\, z^n$; g) $\sum_{n=1}^{\infty} \frac{z^n}{n}$; h) $\sum_{n=1}^{\infty} \frac{n}{2^n} z^n$; i) $\sum_{n=0}^{\infty} z^{n!}$;

j) $\sum_{n=0}^{\infty} 2^n z^{n!}$; k) $\sum_{n=0}^{\infty} z^{2n}$; l) $\sum_{n=0}^{\infty} [3 + (-1)^n]^n z^n$; m) $\sum_{n=0}^{\infty} (\cos in) z^n$;

n) $\sum_{n=0}^{\infty} (n + a^n) z^n$.

Ans. b) 0; d) e; f) 0; h) 2; j) 1;
l) $\frac{1}{4}$; n) 1 if $|a| \leqslant 1$, $1/|a|$ if $|a| > 1$.

Problem 4. Given that the radius of convergence of the power series

$$\sum_{n=0}^{\infty} c_n z^n$$

is R $(0 < R < \infty)$, what is the radius of convergence of each of the following series:

a) $\sum_{n=0}^{\infty} n^k c_n z^n \ (k = 0, 1, 2, \ldots)$; b) $\sum_{n=0}^{\infty} (2^n - 1) c_n z^n$; c) $\sum_{n=0}^{\infty} \frac{c_n}{n!} z^n$;

d) $\sum_{n=1}^{\infty} n^n c_n z^n$; e) $\sum_{n=0}^{\infty} c_n^k z^n \ (k = 0, 1, 2, \ldots)$; f) $\sum_{n=0}^{\infty} (1 + z_0^n) c_n z^n$?

Ans. a) R; c) ∞; e) R^k.

Problem 5. Given that the radii of convergence of the power series

$$\sum_{n=0}^{\infty} a_n z^n \quad \text{and} \quad \sum_{n=0}^{\infty} b_n z^n$$

equal R_1 and R_2, respectively, what can be said about the radius of convergence R of each of the following series:

a) $\sum_{n=0}^{\infty} (a_n + b_n) z^n$; b) $\sum_{n=0}^{\infty} (a_n - b_n) z^n$; c) $\sum_{n=0}^{\infty} a_n b_n z^n$; $\sum_{n=0}^{\infty} \frac{a_n}{b_n} z^n \ (b_n \neq 0)$?

Ans. a) $R \geqslant \min(R_1, R_2)$; c) $R \geqslant R_1 R_2$.

Problem 6. Give an example of two power series

$$\sum_{n=0}^{\infty} a_n z^n, \qquad \sum_{n=0}^{\infty} b_n z^n$$

with the same finite radius of convergence, such that the radius of convergence of the series

$$\sum_{n=0}^{\infty} (a_n + b_n) z^n$$

is infinite.

47. TAYLOR SERIES. THE UNIQUENESS THEOREM FOR POWER SERIES

According to Theorem 10.2, the power series

$$f(z) = a_0 + a_1(z - z_0) + \cdots + a_n(z - z_0)^n + \cdots, \tag{10.9}$$

with radius of convergence R given by formula (10.2), is a series of analytic functions (in fact, entire functions) which converges uniformly on every compact subset of the disk $K: |z - z_0| < R$. It follows from Theorem 9.4 that the sum function $f(z)$ is analytic on K, and that $f^{(k)}(z)$ can be calculated by differentiating (10.9) term by term, i.e.,

$$f^{(k)}(z) = k!\, a_k + \frac{(k+1)!\, a_{k+1}}{1!}(z - z_0) + \cdots + \frac{n!\, a_n}{(n-k)!}(z - z_0)^{n-k} + \cdots, \tag{10.10}$$

where the series (10.10) is uniformly convergent on every compact subset of K. Setting $z = z_0$ in (10.10), we obtain

$$f^{(k)}(z_0) = k!\, a_k,$$

and hence

$$a_k = \frac{f^{(k)}(z_0)}{k!} \qquad (k = 0, 1, 2, \ldots). \tag{10.11}$$

[The fact that (10.11) holds for $k = 0$ can be seen by setting $z = z_0$ in (10.9) itself.] Substituting (10.11) into (10.9), we find that

$$f(z) = f(z_0) + \frac{f'(z_0)}{1!}(z - z_0) + \cdots + \frac{f^{(n)}(z_0)}{n!}(z - z_0)^n + \cdots,$$

where the series on the right is known as the *Taylor series* (*expansion*) of the function $f(z)$ at the point z_0.[3] Thus we have proved that every power series is the Taylor series of its sum $f(z)$. These results are summarized in

THEOREM 10.3. *The power series* (10.9), *with radius of convergence R, converges to an analytic function $f(z)$ on the disk $K: |z - z_0| < R$, and*

[3] Similarly, the numbers $a_0, a_1, \ldots, a_n, \ldots$ are often called the *Taylor coefficients* of the series (10.9).

the coefficients $a_0, a_1, \ldots, a_n, \ldots$ *are given by the formula* (10.11). *Moreover, the series* (10.9) *can be differentiated term by term any number of times, and each differentiated series* (10.10) *is uniformly convergent on every compact subset of K.*

Remark. Together, Theorems 8.7 and 10.3 imply

$$|a_n| \leqslant \frac{M(\rho)}{\rho^n} \qquad (n = 0, 1, 2, \ldots) \tag{10.12}$$

for any $0 < \rho < R$, where

$$M(\rho) = \max_{z \in \gamma_\rho} |f(z)|$$

is the maximum modulus function. The formulas (10.12) are called the *Cauchy inequalities* for the Taylor coefficients.

Now suppose two power series

$$a_0 + a_1(z - z_0) + \cdots + a_n(z - z_0)^n + \cdots \tag{10.13}$$

and

$$b_0 + b_1(z - z_0) + \cdots + b_n(z - z_0)^n + \cdots, \tag{10.14}$$

with radii of convergence R_1 and R_2, respectively, have the same sum in a neighborhood of the point z_0, i.e., suppose

$$a_0 + a_1(z - z_0) + \cdots + a_n(z - z_0)^n + \cdots = b_0 + b_1(z - z_0) + \cdots + b_n(z - z_0)^n + \cdots$$

if $|z - z_0| < r$, where $0 < r \leqslant \min (R_1, R_2)$. Then, according to (10.11),

$$a_0 = b_0 = f(z_0), \quad a_1 = b_1 = \frac{f'(z_0)}{1!}, \ldots, \quad a_n = b_n = \frac{f^{(n)}(z_0)}{n!}, \ldots,$$

where $f(z)$ denotes the common sum of the two series. Therefore coefficients of the two series with identical indices are identical, and it follows that the two series have the same radius of convergence $R_1 = R_2$. (Note that it is vital that the point z_0 be the same for both series.) Thus we have proved the following theorem, typical of a class of results known as *uniqueness theorems*:

THEOREM 10.4. *If the sums of two power series in the variable* $z - z_0$ *coincide in a neighborhood of the point* z_0, *then identical powers of* $z - z_0$ *have identical coefficients, i.e., there is a unique power series in the variable* $z - z_0$ *which has a given sum in a neighborhood of* z_0.

In the simplest case, where the power series (10.13) and (10.14) reduce to polynomials

$$a_0 + a_1(z - z_0) + \cdots + a_m(z - z_0)^m \qquad (a_m \neq 0)$$

and

$$b_0 + b_1(z - z_0) + \cdots + b_n(z - z_0)^n \qquad (b_n \neq 0),$$

the requirement that the values of these two polynomials coincide for all z belonging to a certain neighborhood of z_0, as assumed in Theorem 10.4, is unnecessary. In fact, if $N = \max(m, n)$, and if the values of the two polynomials are the same for more than N different values of z, then $m = n$ and the polynomials are identically equal, i.e.,

$$a_0 = b_0, \quad a_1 = b_1, \ldots, \quad a_n = b_n$$

(recall Sec. 18, Prob. 3). In the case of power series, which in many respects can be regarded as "polynomials of infinitely high degree," one can deduce that two power series have identical coefficients only if it is known that the two series have the same sum on an infinite set. However, even in this case, it is not necessary to assume that the sums of the two series coincide at all points of some neighborhood of z_0. Actually, as shown by the following generalization of Theorem 10.4, it is only necessary that the sums of the series (10.13) and (10.14) coincide on a set which has z_0 as a limit point, for example, on the set of points

$$z_n = z_0 + \frac{1}{n} \qquad (n = 1, 2, \ldots).$$

THEOREM 10.5 (*Uniqueness theorem for power series*). *If the sums of two power series in the variable $z - z_0$ coincide on a set E with z_0 as a limit point, then identical powers of $z - z_0$ have identical coefficients, i.e., there is a unique power series in the variable $z - z_0$ which has a given sum on E.*

Proof. Let the two power series be of the form

$$a_0 + a_1(z - z_0) + a_2(z - z_0)^2 + \cdots,$$
$$b_0 + b_1(z - z_0) + b_2(z - z_0)^2 + \cdots,$$

and suppose that

$$\lim_{n\to\infty} z_n = z_0,$$

where $z_n \in E$, $z_n \neq z_0$ $(n = 1, 2, \ldots)$. Then, since

$$\begin{aligned} a_0 + a_1(z_n - z_0) + a_2(z_n - z_0)^2 + \cdots \\ = b_0 + b_1(z_n - z_0) + b_2(z_n - z_0)^2 + \cdots \end{aligned} \tag{10.15}$$

for all $n = 1, 2, \ldots$, and since the sum of a power series is continuous on the interior of its circle of convergence (why?), we have

$$\begin{aligned} a_0 &= \lim_{n\to\infty} [a_0 + a_1(z_n - z_0) + a_2(z_n - z_0)^2 + \cdots] \\ &= \lim_{n\to\infty} [b_0 + b_1(z_n - z_0) + b_2(z_n - z_0)^2 + \cdots] = b_0, \end{aligned}$$

i.e.,

$$a_0 = b_0.$$

Suppose it is known that

$$a_0 = b_0, \quad a_1 = b_1, \ldots, \quad a_k = b_k.$$

Then (10.15) implies that

$$\begin{aligned} a_{k+1}(z_n - z_0)^{k+1} + a_{k+2}(z_n - z_0)^{k+2} + \cdots \\ = b_{k+1}(z_n - z_0)^{k+1} + b_{k+2}(z_n - z_0)^{k+2} + \cdots. \end{aligned} \tag{10.16}$$

Dividing (10.16) by $(z_n - z_0)^{k+1} \neq 0$, and passing to the limit as $n \to \infty$, we find that

$$a_{k+1} = b_{k+1}.$$

The proof now follows at once by induction.

Problem 1. Given a power series

$$f(z) = a_0 + a_1 z + a_2 z^2 + \cdots + a_n z^n + \cdots \qquad (|z| < R),$$

prove that

a) The coefficients of odd powers of z vanish if $f(z)$ is even, i.e., if

$$f(-z) \equiv f(z);$$

b) The coefficients of even powers of z vanish if $f(z)$ is odd, i.e., if

$$f(-z) \equiv -f(z).$$

Problem 2 (M1, p. 351). Series expansions of the form

$$a_0\varphi_0(z) + a_1\varphi_1(z) + \cdots + a_n\varphi_n(z) + \cdots \tag{10.17}$$

with respect to given functions

$$\varphi_0(z), \varphi_1(z), \ldots, \varphi_n(z), \ldots \tag{10.18}$$

are said to have the *uniqueness property* on a set E if, whenever two such series (10.17) and

$$b_0\varphi_0(z) + b_1\varphi_1(z) + \cdots + b_n\varphi_n(z) + \cdots$$

coincide for all $z \in E$, the corresponding coefficients coincide:

$$a_0 = b_0, a_1 = b_1, \ldots, a_n = b_n, \ldots$$

In this language, Theorem 10.4 states that series expansions with respect to the functions

$$\varphi_n(z) = (z - z_0)^n \qquad (n = 0, 1, 2, \ldots)$$

have the uniqueness property in a neighborhood of the point z_0. Give an example of a system of functions (10.18) and a set E such that expansions with respect to the given functions fail to have the uniqueness property on E.

Ans. Let

$$\varphi_0(z) = 1,\ \varphi_1(z) = z - 1, \ldots, \varphi_n(z) = z^n - z^{n-1}, \ldots,$$

with $E: |z| < 1$.

48. EXPANSION OF AN ANALYTIC FUNCTION IN A POWER SERIES

According to Theorem 10.3, every power series represents an analytic function inside its circle of convergence. Conversely, every analytic function can be represented by a power series, in the following sense:

THEOREM 10.6. *Let $f(z)$ be analytic on a domain G, let z_0 be an arbitrary (finite) point of G, and let $\Delta = \Delta(z_0)$ be the distance between z_0 and the boundary of G.*[4] *Then there exists a power series*

$$f(z) = \sum_{n=0}^{\infty} a_n(z - z_0)^n \quad (10.19)$$

converging to $f(z)$ on the disk $K: |z - z_0| < \Delta$.

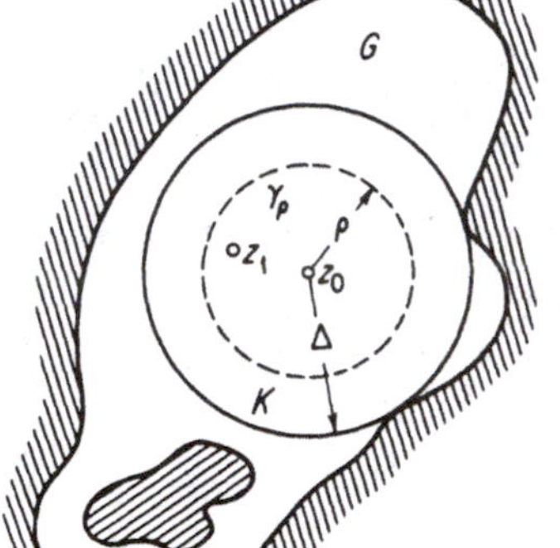

FIGURE 10.1

Proof. Let z_1 be an arbitrary point of K, and let γ_ρ be any circle with center z_0 and radius ρ $(0 < \rho < \Delta)$ such that $z_1 \in I(\gamma_\rho)$ [see Figure 10.1]. Then, according to Cauchy's integral formula,

$$f(z_1) = \frac{1}{2\pi i} \int_{\gamma_\rho} \frac{f(z)}{z - z_1}\, dz.$$

To convert this relation into a power series of the form (10.19), we represent

$$\frac{1}{z - z_1}$$

as the sum of a geometric series with ratio

$$\frac{z_1 - z_0}{z - z_0},$$

[4] If the boundary of G consists of the single point ∞, we write $\Delta = \infty$. Thus, if $f(z)$ is an entire function, the series (10.19) converges for all finite z.

where

$$\left|\frac{z_1 - z_0}{z - z_0}\right| = \frac{|z_1 - z_0|}{\rho} < 1,$$

i.e.,

$$\frac{1}{z - z_1} = \frac{1}{(z - z_0) - (z_1 - z_0)} = \frac{1}{z - z_0}\frac{1}{1 - \dfrac{z_1 - z_0}{z - z_0}}$$

$$= \frac{1}{z - z_0} + \frac{z_1 - z_0}{(z - z_0)^2} + \cdots + \frac{(z_1 - z_0)^n}{(z - z_0)^{n+1}} + \cdots \qquad (10.20)$$

$$= \sum_{n=0}^{\infty} \frac{(z_1 - z_0)^n}{(z - z_0)^{n+1}}.$$

Then, multiplying (10.20) by the function

$$\frac{1}{2\pi i} f(z),$$

we obtain

$$\frac{1}{2\pi i}\frac{f(z)}{z - z_1} = \sum_{n=0}^{\infty} \frac{1}{2\pi i}\frac{f(z)}{(z - z_0)^{n+1}}(z_1 - z_0)^n. \qquad (10.20')$$

If $z \in \gamma_\rho$, then

$$\left|\frac{1}{2\pi i}\frac{f(z)}{(z - z_0)^{n+1}}(z_1 - z_0)^n\right| \leqslant \frac{1}{2\pi}\frac{M(\rho)}{\rho}\left(\frac{|z_1 - z_0|}{\rho}\right)^n, \qquad (10.21)$$

where, as usual,

$$M(\rho) = \max_{z \in \gamma_\rho} |f(z)|.$$

The right-hand side of (10.20′) is uniformly convergent on γ_ρ by Weierstrass' M-test (Theorem 9.1).[5] Therefore, according to Theorem 9.3, we can integrate (10.20) term by term along γ_ρ. The result is

$$f(z_1) = \frac{1}{2\pi i}\int_{\gamma_\rho} \frac{f(z)}{z - z_1}\, dz = \sum_{n=0}^{\infty} a_n(z_1 - z_0)^n,$$

where

$$a_n = \frac{1}{2\pi i}\int_{\gamma_\rho} \frac{f(z)}{(z - z_0)^{n+1}}\, dz = \frac{f^{(n)}(z_0)}{n!} \qquad (n = 0, 1, 2, \ldots).$$

Since the point $z_1 \in K$ is arbitrary, this (Taylor) series expansion is just the desired result (10.19).

---

[5] Note that the right-hand side of (10.21) is the general term of a convergent numerical series, i.e., a geometric series with ratio $|z_1 - z_0|/\rho < 1$.

Remark. According to Theorem 10.6, the Taylor series

$$f(z) = \sum_{n=0}^{\infty} a_n(z - z_0)^n = \sum_{n=0}^{\infty} \frac{f^{(n)}(z_0)}{n!}(z - z_0)^n \tag{10.22}$$

representing the function $f(z)$ in a neighborhood of the point z_0 converges on the disk $|z - z_0| < \Delta$. Therefore the radius of convergence R of the series (10.22) can be no smaller than Δ, i.e.,

$$R \geqslant \Delta.$$

Writing this inequality in the form

$$\frac{1}{R} \leqslant \frac{1}{\Delta},$$

and substituting from the Cauchy-Hadamard formula (10.2), we find that

$$\overline{\lim_{n\to\infty}} \sqrt[n]{|a_n|} = \overline{\lim_{n\to\infty}} \sqrt[n]{\frac{|f^{(n)}(z_0)|}{n!}} \leqslant \frac{1}{\Delta}.$$

This is the *Cauchy-Hadamard inequality*, already encountered in Sec. 42, Prob. 3. In particular, $\lim\limits_{n\to\infty} \sqrt[n]{|a_n|} = 0$ if (10.22) is an entire function.

Example 1. Choosing $z_0 = 0$ in (10.22), and calculating the derivatives of all orders of the entire functions e^z, $\cos z$, $\sin z$, $\cosh z$ and $\sinh z$, we immediately obtain the following power series expansions, valid for all finite z:

$$e^z = \sum_{n=0}^{\infty} \frac{z^n}{n!}, \quad \cos z = \sum_{n=0}^{\infty} (-1)^n \frac{z^{2n}}{(2n)!}, \quad \sin z = \sum_{n=0}^{\infty} (-1)^n \frac{z^{2n+1}}{(2n+1)!},$$

$$\cosh z = \sum_{n=0}^{\infty} \frac{z^{2n}}{(2n)!}, \quad \sinh z = \sum_{n=0}^{\infty} \frac{z^{2n+1}}{(2n+1)!}.$$

Example 2. Theorem 10.6 applies equally well to single-valued branches of multiple-valued functions. For example, let G be the domain with the half-line $x \leqslant 0$, $y = 0$ as its boundary (i.e., the whole plane "cut along" the negative real axis, including the origin), and let $\ln z$ denote the single-valued branch of the multiple-valued function $\operatorname{Ln} z$ which is defined on G and satisfies the condition $\ln 1 = 0$.[6] Then, since

$$\frac{d^n}{dz^n} \ln z = (-1)^{n-1} \frac{(n-1)!}{z^n} \qquad (n = 1, 2, \ldots),$$

[6] This branch can be represented in the form

$$\ln z = \ln |z| + i \arg z,$$

where $|\arg z| < \pi$.

it follows from (10.22), with $z_0 = 1$ that

$$\ln z = \sum_{n=1}^{\infty}(-1)^{n-1}\frac{(z-1)^n}{n}. \tag{10.23}$$

The distance between the point $z_0 = 1$ and the boundary of G equals 1, and therefore the expansion (10.23) is valid for $|z - 1| < 1$. Replacing $z - 1$ by z, we obtain the following power series for $\ln(1 + z)$, valid for $|z| < 1$:

$$\ln(1+z) = \sum_{n=1}^{\infty}(-1)^{n-1}\frac{z^n}{n}.$$

Here $\ln(1 + z)$ is the single-valued branch of the function $\text{Ln}(1 + z)$ which is defined on the domain G and satisfies the condition

$$\ln(1+z)|_{z=0} = \ln 1 = 0.$$

Problem 1 (M1, p. 365). Verify the *binomial expansion*

$$(1+z)^\alpha = 1 + \sum_{n=1}^{\infty}\frac{\alpha(\alpha-1)\cdots(\alpha-n+1)}{n!}z^n,$$

valid for $|z| < 1$ and arbitrary complex α.

Problem 2. Sum the following power series for $|z| < 1$:

a) $\displaystyle\sum_{n=1}^{\infty}\frac{z^n}{n}$; b) $\displaystyle\sum_{n=0}^{\infty}\frac{z^{2n+1}}{2n+1}$.

Ans. a) $-\ln(1 - z)$.

Problem 3. Expand each of the following functions in a Taylor series at the point $z = 0$:

a) $\sin^2 z$; b) $\cosh^2 z$; c) $\sqrt{z+i}$, where $\sqrt{i} = \dfrac{1+i}{\sqrt{2}}$;

d) $\dfrac{1}{az+b}$, where $b \neq 0$; e) $\dfrac{z}{z^2-4z+13}$; f) $\dfrac{z^2}{(z+1)^2}$;

g) $\arctan z$, where $\arctan 0 = 0$; h) $\displaystyle\int_0^z e^{\zeta^2}\,d\zeta$; i) $\displaystyle\int_0^z \frac{\sin\zeta}{\zeta}\,d\zeta$.

In each case, find the radius of convergence of the Taylor series.

Problem 4. Expand each of the following functions in a Taylor series at the point $z = 1$:

a) $\dfrac{z}{z+2}$; b) $\dfrac{z^2}{(z+1)^2}$; c) $\sqrt[3]{z}$, where $\sqrt[3]{1} = \dfrac{-1+i\sqrt{3}}{2}$;

d) $\dfrac{z}{z^2-2z+5}$.

In each case, find the radius of convergence of the Taylor series.

49. LIOUVILLE'S THEOREM. THE UNIQUENESS THEOREM FOR ANALYTIC FUNCTIONS

THEOREM 10.7 (*Liouville's theorem*). *If $f(z)$ is an entire function and if $|f(z)| \leqslant M$ for all finite z, then $f(z) \equiv$ const, i.e., every bounded entire function is a constant.*

Proof. Since $f(z)$ is entire, the expansion

$$f(z) = \sum_{n=0}^{\infty} a_n (z - z_0)^n \tag{10.24}$$

implied by Theorem 10.6 is valid for arbitrary z. Writing Cauchy's inequalities (10.12) for $n > 0$ only, we have

$$|a_n| \leqslant \frac{M(\rho)}{\rho^n} \leqslant \frac{M}{\rho^n} \qquad (n = 1, 2, \ldots) \tag{10.25}$$

for any $\rho > 0$. Taking the limit of (10.25) as $\rho \to \infty$, we find that $|a_n| \leqslant 0$ and hence $a_n = 0$ for all $n = 1, 2, \ldots$. Therefore the power series (10.24) reduces to

$$f(z) \equiv a_0,$$

and the theorem is proved.

COROLLARY 1. *Every function analytic in the extended plane must be a constant.*

Proof. If $f(z)$ is analytic at infinity (see Remark 1, p. 63) as well as for all finite z, then $f(z)$ is bounded (why?).

COROLLARY 2 (*Fundamental theorem of algebra*). *Every polynomial*

$$P_n(z) = a_0 + a_1 z + \cdots + a_n z^n \qquad (a_n \neq 0, n > 0)$$

has a zero in the complex plane.

Proof. If $P_n(z)$ had no zeros, then $1/P_n(z)$ would be analytic in the extended plane and hence a constant, which is impossible,

The next theorem, of basic importance in complex analysis, is a far-reaching generalization of the uniqueness theorem for power series (Theorem 10.5):

THEOREM 10.8 (*Uniqueness theorem for analytic functions*). *If two single-valued analytic functions $f(z)$ and $g(z)$ defined on a domain G coincide on a set $E \subset G$ with a limit point $z_0 \in G$, then $f(z)$ and $g(z)$ coincide on the whole domain G.*

Proof. Given any point $z' \in G$, $z' \neq z_0$, we join z_0 to z' by a continuous curve $L \subset G$, with equation $z = \lambda(t)$, $a \leqslant t \leqslant b$. Let ρ be the distance between L and the boundary of G, and use the uniform continuity of $\lambda(t)$ on $[a, b]$ to choose a partition $\mathscr{P} = \{t_0, t_1, \ldots, t_n\}$ of the interval $[a, b]$ such that

$$|z_j - z_{j-1}| < \rho \qquad (j = 1, \ldots, n),$$

where

$$z_0 = \lambda(t_0) = \lambda(a), \ldots,$$
$$z_j = \lambda(t_j), \ldots,$$
$$z_n = \lambda(t_n) = \lambda(b) = z'.$$

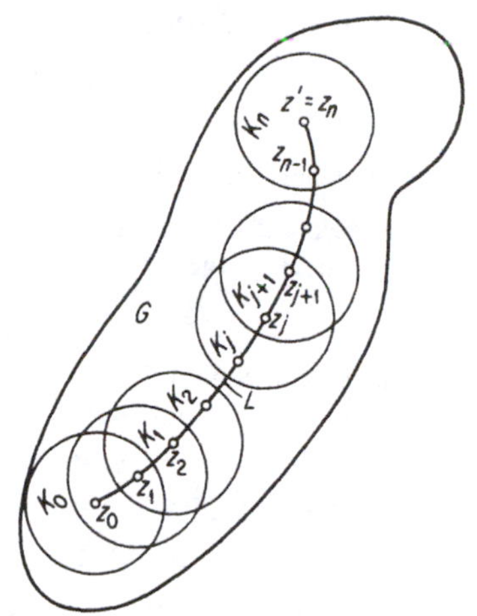

FIGURE 10.2

Then, as shown in Figure 10.2, construct the disks K_j: $|z - z_j| < \rho$. Clearly each disk K_j contains the center z_{j+1} of the "next" disk K_{j+1}. According to Theorem 10.6, both $f(z)$ and $g(z)$ have convergent power series expansions on every K_j $(j = 0, 1, \ldots, n)$. Since $f(z)$ and $g(z)$ coincide on the set $E \cap K_0$ with limit point z_0, it follows from Theorem 10.5 that they coincide on the whole disk K_0. But then they coincide on the set $K_0 \cap K_1$ which has z_1 as a limit point (why?), and hence, using Theorem 10.5 again, we find that $f(z)$ and $g(z)$ coincide on K_1. Repeating this argument $n - 1$ more times, we eventually find that $f(z)$ and $g(z)$ coincide on K_n and hence at the point z'. Since z' is an arbitrary point of G, the proof is now complete.

Remark. It follows from Theorem 10.8 that if $f(z)$ and $g(z)$ coincide on a subdomain $D \subset G$ (in particular, in a neighborhood of a point $z_0 \in G$), or on an arc of some continuous curve contained in G, then $f(z)$ and $g(z)$ coincide on G itself. In particular, if $f(z)$ is analytic on a domain G and if $f(z)$ is not identically equal to a constant on G, then $f(z)$ cannot have a constant value in any neighborhood of a point $z_0 \in G$ or on any arc of a curve $\gamma \subset G$. Moreover, if $f(z) \not\equiv$ const on G, and if z_0 is an arbitrary point of G, then at least one of the derivatives $f'(z_0), \ldots, f^{(n)}(z_0), \ldots$ must be nonzero, since otherwise the Taylor series expansion of $f(z)$ would have the form

$$f(z) = a_0 + 0 \cdot (z - z_0) + \cdots + 0 \cdot (z - z_0)^n + \cdots$$

in a neighborhood of z_0, which implies $f(z) \equiv f(z_0)$ for all $z \in G$.

Problem 1. Give an alternative proof of Liouville's theorem, based on the use of Cauchy's integral formula.

Hint. Consider the integral

$$\int_{|z|=R} \frac{f(z)}{(z-a)(z-b)}\,dz,$$

where $a \neq b$, $|a| < R$, $|b| < R$, and take the limit as $R \to \infty$.

Problem 2. Prove the following generalization of Liouville's theorem: If $f(z)$ is an entire function, and if the function

$$M(\rho) = \max_{|z|=\rho} |f(z)|$$

satisfies the inequality

$$M(\rho) \leqslant M\rho^k,$$

where M is a positive constant and k is a fixed positive integer, then $f(z)$ is a polynomial of degree no higher than k.

Problem 3. Verify that Theorem 10.8 reduces to Theorem 10.5 if G is a disk.

Problem 4. Give an alternative derivation of the addition theorem for $\cos z$ and $\sin z$ (see Sec. 31).

Hint. Use Theorem 10.8, regarding the addition theorems as known for real arguments.

Problem 5. Does there exist a function which is analytic at the point $z = 0$ and takes the following values at the points $z = 1/n$ $(n = 1, 2, \ldots)$:

a) $0, 1, 0, 1, 0, 1, \ldots 0, 1, \ldots;$

b) $0, \frac{1}{2}, 0, \frac{1}{4}, 0, \frac{1}{6}, \ldots, 0, \frac{1}{2k}, \ldots;$

c) $\frac{1}{2}, \frac{1}{2}, \frac{1}{4}, \frac{1}{4}, \frac{1}{6}, \frac{1}{6}, \ldots, \frac{1}{2k}, \frac{1}{2k}, \ldots;$

d) $\frac{1}{2}, \frac{2}{3}, \frac{3}{4}, \frac{4}{5}, \frac{5}{6}, \frac{6}{7}, \ldots, \frac{n}{n+1}, \ldots$?

Problem 6. Does there exist a function which is analytic at the point $z = 0$ and satisfies the following condition for every positive integer n:

a) $f(1/n) = f(-1/n) = 1/n^2$; b) $f(1/n) = f(-1/n) = 1/n^3$?

Ans. a) Yes; b) no.

50. A-POINTS AND ZEROS

By an *A-point* of a function $f(z)$, we mean a root of the equation $f(z) = A$, where A is an arbitrary finite complex number. It follows from Theorem 10.8 that *if $f(z) \not\equiv A$ is analytic on a domain G, the set of A-points of $f(z)$ cannot have a limit point in G.* Moreover, any compact subset $F \subset G$ can only contain a finite number of A-points (for fixed A), since if F contained infinitely many A-points, these points would have a limit point belonging to

F and hence to G. If $z_0 \in G$ is any A-point of $f(z)$, then $f(z_0) = A$, and hence the power series expansion of $f(z)$ in a neighborhood of z_0 is of the form

$$f(z) = A + f'(z_0)(z - z_0) + \frac{f''(z_0)}{2!}(z - z_0)^2 + \cdots$$

or

$$f(z) - A = f'(z_0)(z - z_0) + \frac{f''(z_0)}{2!}(z - z_0)^2 + \cdots. \tag{10.26}$$

If $f(z) \not\equiv$ const, then, according to the remark on p. 209, at least one of the coefficients $f'(z_0), f''(z_0), \ldots$ is nonzero. Let $(z - z_0)^k$ be the smallest power of $z - z_0$ with a nonzero coefficient. Then (10.26) becomes

$$f(z) - A = (z - z_0)^k\left[\frac{f^{(k)}(z_0)}{k!} + \frac{f^{(k+1)}(z_0)}{(k+1)!}(z - z_0) + \cdots\right], \tag{10.27}$$

where $f^{(k)}(z_0) \neq 0$. The integer $k \geqslant 1$ is called the *order* of the A-point z_0, and z_0 is called a *simple* A-point if $k = 1$ and a *multiple* A-point if $k > 1$. It follows from (10.27) and this definition that

$$f(z_0) = A, \qquad f'(z_0) \neq 0$$

if z_0 is a simple A-point, whereas

$$f(z_0) = A, \quad f'(z_0) = 0, \ldots, \quad f^{(k-1)}(z_0) = 0, \quad f^{(k)}(z_0) \neq 0 \tag{10.28}$$

if z_0 is an A-point of order k. From now on, in determining the number of A-points in a compact subset $F \subset G$, we shall count each A-point a number of times equal to its order. Then the number of A-points in F will still be finite, since F contains only a finite number of multiple A-points and each is of finite order.

THEOREM 10.9. *If $f(z)$ is analytic at z_0, then $f(z)$ has an A-point of order k at z_0 if and only if the function*

$$\varphi(z) = \frac{f(z) - A}{(z - z_0)^k} \tag{10.29}$$

is analytic at z_0 and

$$\varphi(z_0) = \lim_{z \to z_0} \varphi(z) \neq 0. \tag{10.30}$$

Proof. If $f(z)$ has an A-point of order k at z_0, then, according to (10.27) and (10.29), $\varphi(z)$ has the power series expansion

$$\varphi(z) = \frac{f^{(k)}(z_0)}{k!} + \frac{f^{(k+1)}(z_0)}{(k+1)!}(z - z_0) + \cdots$$

in a neighborhood of z_0, where

$$\varphi(z_0) = \frac{f^{(k)}(z_0)}{k!} \neq 0$$

[cf. (10.28)] Conversely, if $\varphi(z)$ is analytic at z_0 and the condition (10.30) holds, then

$$\varphi(z) = a_k + a_{k+1}(z - z_0) + \cdots$$

in a neighborhood of z_0, where $a_k = \varphi(z_0) \neq 0$. Substituting for $\varphi(z)$ from (10.29) and solving for $f(z)$, we find that

$$f(z) = A + a_k(z - z_0)^k + a_{k+1}(z - z_0)^{k+1} + \cdots.$$

Then obviously the conditions (10.28) hold, i.e., $f(z)$ has an A-point of order k at z_0.

Everything just said about A-points applies equally well to the special case where $A = 0$. For simplicity, *zero-points* are called *zeros*. The reader can easily verify that our present definition of the order of a zero is consistent with the definition already given in the case where $f(z)$ is a polynomial (see p. 66) or a rational function (see p. 79). Since the A-points of the function $f(z)$ are the zeros of the function $f(z) - A$, any general property which we prove for the zeros of an analytic function (e.g., Theorem 10.10) is also valid for the A-points of an analytic function.

Remark. If the function $f(z)$ is analytic at ∞ (see Remark 1, p. 63) and if $f(\infty) = A$, we say that $f(z)$ has an A-point of order k at ∞ if $f^*(\zeta) = f(1/\zeta)$ has an A-point of order k at $\zeta = 0$.

THEOREM 10.10. *If $f(z)$ is analytic on a domain G, then the set of all zeros of $f(z)$ in G is countable.*

Proof. Since the union of countably many countable sets is countable (see Sec. 5, Prob. 5d), and since any compact set contains only finitely many zeros of $f(z)$, the theorem will be proved if we show that G can always be represented as a union of countably many compact sets. If G is the whole plane, G is (trivially) the union of the closed disks

$$|z| \leqslant 1, \quad |z| \leqslant 2, \ldots, \quad |z| \leqslant n, \ldots$$

On the other hand, if Γ, the boundary of G, contains at least one finite po .t, let $\rho(z, \Gamma)$ denote the distance between an arbitrary point z and Γ (see p. 28). Then let F_n denote the set of all points in G such that

$$|z| \leqslant n, \qquad \rho(z, \Gamma) \geqslant \frac{1}{n}.$$

Clearly all the F_n starting from some value of n are nonempty, and every point of G belongs to all the F_n starting from some value of n. It follows that

$$G = F_1 \cup \cdots \cup F_n \cup \cdots.$$

Since the sets F_n are obviously bounded, it only remains to show that they are closed. If F_n is empty, F_n is closed by definition (see p. 25). Every nonempty F_n is also closed, since if ζ is a limit point of F_n, then ζ is also a limit point of G, and moreover

$$\rho(\zeta, \Gamma) = \lim_{\substack{z \to \zeta \\ z \in F_n}} \rho(z, \Gamma) \geqslant \frac{1}{n},$$

which implies $\zeta \in G$, $\zeta \in F_n$. Here we use the fact that $\rho(z, \Gamma)$ is a continuous function of z (see Sec. 36, Prob. 3). The proof is now complete.

Problem 1. Prove that if $f(z)$ and $g(z)$ are analytic at a point z_0 and have zeros at z_0 of orders p and q, respectively, then $f(z)g(z)$ is analytic at z_0 and has a zero at z_0 of order $p + q$.

Problem 2. Find the order of the zero at $z = 0$ of the function

a) $z^2(e^{z^2} - 1)$; b) $6 \sin z^3 + z^3(z^6 - 6)$; c) $e^{\sin z} - e^{\tan z}$.

Ans. b) 15.

Problem 3. Find the zeros (possibly including ∞), together with their orders, of each of the following functions:

a) $z^2 + 9$; b) $\dfrac{z^2 + 9}{z^4}$; c) $z \sin z$; d) $(1 - e^z)(z^2 - 4)^3$;

e) $1 - \cos z$; f) $\dfrac{(z^2 - \pi^2)^2 \sin z}{z^7}$; g) $\dfrac{1 - \cot z}{z}$; h) $e^{\tan z}$;

i) $\sin^3 z$; j) $\dfrac{\sin^3 z}{z}$; k) $\sin z^3$; l) $\cos^3 z$; m) $\cos z^3$; n) $(\sqrt{z} - 2)^3$;

o) $(1 - \sqrt{2 - 2 \cos z})^2$.

51. WEIERSTRASS' DOUBLE SERIES THEOREM

In principle, the problem of finding the Taylor series expansion of an analytic function $f(z)$ in a neighborhood of a point z_0 is solved by using the formula

$$a_n = \frac{f^{(n)}(z_0)}{n!} \qquad (n = 0, 1, 2, \ldots) \tag{10.31}$$

to find the coefficients of the series

$$f(z) = \sum_{n=0}^{\infty} a_n (z - z_0)^n. \tag{10.32}$$

However, it often turns out to be a formidable task to calculate the coefficients (10.31) directly. Fortunately, in many practical problems, the Taylor series (10.32) can readily be deduced by starting from other series expansions with known coefficients.

THEOREM 10.11 (*Weierstrass' double series theorem*). *Suppose the series*

$$f(z) = \sum_{k=1}^{\infty} f_k(z) \tag{10.33}$$

is uniformly convergent on every compact subset of the disk $K: |z - z_0| < R$, *and suppose every function* $f_k(z)$ *is analytic on* K. *Then* $f(z)$ *has the Taylor series expansion* (10.32) *on* K, *where the coefficients* (10.31) *are related to the derivatives of the original series* (10.33) *by the formula*

$$a_n = \sum_{k=1}^{\infty} \frac{f_k^{(n)}(z_0)}{n!} \qquad (n = 0, 1, 2, \ldots).$$

Proof. According to Theorem 9.4, $f(z)$ is analytic on K, and hence, according to Theorem 10.6, $f(z)$ has the Taylor series expansion (10.32) on K, with coefficients (10.31). The rest of the proof follows at once by combining (10.31) and the formula

$$f^{(n)}(z) = \sum_{k=1}^{\infty} f_k^{(n)}(z) \qquad (n = 0, 1, 2, \ldots)$$

[cf. (9.20)].

Remark. Since each of the functions $f_k(z)$ in the right-hand side of (10.33) has its own Taylor series expansion on K, i.e.,

$$f_k(z) = \sum_{l=0}^{\infty} a_{kl} (z - z_0)^l \qquad (k = 1, 2, \ldots),$$

where

$$a_{kl} = \frac{f_k^{(l)}(z_0)}{l!} \qquad (l = 0, 1, 2, \ldots),$$

it follows from Theorem 10.10 that

$$a_n = \sum_{k=1}^{\infty} a_{kn}.$$

In other words, the coefficient of a given power of $z - z_0$ in the right-hand side of (10.32) is obtained by adding all the coefficients of the same power of $z - z_0$ in the separate Taylor series expansions of $f_k(z)$, $k = 1, 2, \ldots$.

Example. If

$$f_k(z) = \frac{z^k}{1 - z^k},$$

(10.33) becomes

$$f(z) = \sum_{k=1}^{\infty} \frac{z^k}{1 - z^k}. \tag{10.34}$$

Every term $f_k(z)$ is analytic on the unit disk $K: |z| < 1$, and the series (10.34) is uniformly convergent on every compact subset of K. In fact, let $F \subset K$ be compact, and let $\delta > 0$ be the distance between F and the unit circle $|z| = 1$. Then $|z| \leqslant 1 - \delta = \rho < 1$ for all $z \in F$, and hence

$$\left|\frac{z^k}{1 - z^k}\right| \leqslant \frac{\rho^k}{1 - \rho^k} \leqslant \frac{\rho^k}{1 - \rho}.$$

Since the series

$$\sum_{k=1}^{\infty} \frac{\rho^k}{1 - \rho}$$

is convergent (being a geometric series with ratio $\rho < 1$), it follows from Weierstrass' M-test (Theorem 9.1) that the series (10.34) is uniformly convergent on F, as asserted.

According to Theorem 10.11, to find the coefficient of z^n in the Taylor series of the analytic function $f(z)$, we need only add all the coefficients of z^n in the separate Taylor series

$$f_k(z) = \frac{z^k}{1 - z^k} = z^k + z^{2k} + \cdots.$$

In each such expansion, the coefficient of z^n equals 1 if n is divisible by k, and 0 otherwise. Therefore the coefficient of z^n in the expansion of $f(z)$ equals the number of positive integers which divide n (including 1 and n). Denoting this number by $\tau(n)$,[7] we have

$$f(z) = \sum_{n=1}^{\infty} \tau(n) z^n \qquad (|z| < 1).$$

Problem 1. The function

$$\zeta(z) = \sum_{k=1}^{\infty} \frac{1}{k^z} = \sum_{k=1}^{\infty} e^{-z \ln k} \tag{10.35}$$

is known as the *Riemann zeta function.* Prove that the series (10.35) is uniformly convergent on every compact subset of the half-plane $\operatorname{Re} z > 1$, so that $\zeta(z)$ is analytic for $\operatorname{Re} z > 1$. Find the Taylor series expansion of $\zeta(z)$ at the point $z = 2$. What is the radius of convergence of this Taylor series?

---

[7] Thus $\tau(1) = 1$, $\tau(2) = 2$, $\tau(3) = 2$, $\tau(4) = 3$, $\tau(5) = 2, \ldots$

Problem 2 (M1, p. 431). Find the Taylor series expansion of

$$f(z) = \sum_{k=1}^{\infty} \frac{2z}{z^2 - k^2\pi^2}.$$

Ans. $f(z) = -2 \sum_{m=1}^{\infty} \left[\sum_{k=1}^{\infty} \frac{1}{(k\pi)^{2m}} \right] z^{2m-1} \qquad (|z| < \pi).$

52. SUBSTITUTION OF ONE POWER SERIES INTO ANOTHER

Given a Taylor series

$$f(z) = a_0 + a_1(z - z_0) + \cdots + a_n(z - z_0)^n + \cdots \tag{10.36}$$

at the point z_0 in the z-plane, and a second Taylor series

$$\varphi(w) = A_0 + A_1(w - w_0) + \cdots + A_n(w - w_0)^n + \cdots \tag{10.37}$$

at the point $w_0 = f(z_0) = a_0$ in the w-plane, suppose we replace w by $f(z)$ in (10.37). Then we obtain a new series expansion[8]

$$F(z) = A_0 + A_1[f(z) - a_0] + \cdots + A_n[f(z) - a_0]^n + \cdots, \tag{10.38}$$

corresponding to the *composite function* $F(z) = \varphi[f(z)]$. Suppose the algebraic operations called for in the right-hand side of (10.38) are carried out formally: This involves raising $f(z) - a_0$ to various powers after replacing $f(z)$ by the series (10.36) and then grouping together and adding all the coefficients of each power $(z - z_0)^n$, $n = 0, 1, 2, \ldots$ in the expression so obtained. The result is a Taylor series of the form

$$F(z) = b_0 + b_1(z - z_0) + \cdots + b_n(z - z_0)^n + \cdots, \tag{10.39}$$

which is said to be obtained by *formal substitution* of the series (10.36) into the series (10.37).

THEOREM 10.12. *If $f(z)$ is analytic at the point $z = z_0$, and if $\varphi(w)$ is analytic at the point $w_0 = f(z_0)$, then the composite function $F(z) = \varphi[f(z)]$ is analytic at the point $z = z_0$. Moreover, the Taylor series of $F(z)$ is obtained by formal substitution of the Taylor series of $w = f(z)$ into the Taylor series of $\varphi(w)$.*

Proof. Let (10.36) be the Taylor series of $f(z)$, valid for $|z - z_0| < r$, and let (10.37) be the Taylor series of $\varphi(w)$, valid for $|w - w_0| < R$, where $w_0 = f(z_0) = a_0$. (By hypothesis, such positive numbers r and R exist.) Since $f(z) \to f(z_0)$ as $z \to z_0$, there is a number r' $(0 < r' \leqslant r)$ such that

[8] As it stands, (10.38) is not a Taylor series in the z-plane. Note that $F(z_0) = \varphi(w_0) = A_0$, as required. Thus b_0 must equal A_0 in (10.39).

$|f(z) - w_0| < R' < R$ if $|z - z_0| < r'$. Clearly the expansion (10.38) holds for all $|z - z_0| < r'$. This expansion is of the form

$$F(z) = \sum_{k=0}^{\infty} F_k(z), \tag{10.40}$$

where every function

$$F_k(z) = A_k[f(z) - a_0]^k \qquad (k = 0, 1, 2, \ldots)$$

is analytic on the disk $K: |z - z_0| < r'$. Moreover, since (10.37) is uniformly convergent for $|w - w_0| < R'$ (recall Theorem 10.2), (10.40) is uniformly convergent on K, and hence, *a fortiori*, on every compact subset of K. It follows from Theorem 10.11 and the remark on p. 214 that $F(z)$ is analytic on K (in particular at z_0), and that the coefficient of a given power $(z - z_0)^n$ in the series (10.39) is obtained by adding all the coefficients of the same power $(z - z_0)^n$ in the separate Taylor series of $F_k(z)$, $k = 0, 1, 2, \ldots$. All that remains to be proved is that the Taylor series of $F_k(z)$ is obtained by raising the Taylor series of $f(z) - a_0$ to the kth power and multiplying the result by A_k. But this is an immediate consequence of Sec. 43, Prob. 3 and the fact that the Taylor series of $f(z) - a_0$ is absolutely convergent on K.

Remark. The fact that $F(z)$ is differentiable for $|z - z_0| < r'$, and hence analytic at z_0, also follows at once from the rule for differentiation of a composite function (see Rule 4, p. 46).

Example. Let

$$F(z) = +\sqrt{\cos z}, \tag{10.41}$$

where $+\sqrt{\cos z}$ denotes the single-valued branch of the double-valued function $\sqrt{\cos z}$ which is analytic at $z = 0$ and takes the value $+1$ at $z = 0$. With this understanding, we shall henceforth omit the plus sign in (10.41). To apply Theorem 10.12, we write (10.41) in the form

$$F(z) = [1 - (1 - \cos z)]^{1/2}.$$

Then

$$w = f(z) = 1 - \cos z = \frac{z^2}{2!} - \frac{z^4}{4!} + \frac{z^6}{6!} - \cdots \qquad (|z| < \infty)$$

and

$$\varphi(w) = (1 - w)^{1/2} = 1 - \frac{w}{2} - \frac{w^2}{8} - \frac{w^3}{16} - \cdots \qquad (|w| < 1)$$

(recall Sec. 48, Prob. 1), so that

$$F(z) = \varphi[f(z)] = 1 - \frac{1}{2}\left(\frac{z^2}{2} - \frac{z^4}{24} + \frac{z^6}{720} - \cdots\right) - \frac{1}{8}\left(\frac{z^2}{2} - \frac{z^4}{24} + \cdots\right)^2 - \frac{1}{16}\left(\frac{z^2}{2} - \cdots\right)^3 - \cdots,$$

where the omitted terms contribute only to the coefficients of $z^8, z^{10}, \ldots$ in the Taylor series of $F(z)$. Thus, writing

$$\left(\frac{z^2}{2}-\frac{z^4}{24}+\cdots\right)^2=\frac{z^4}{4}-\frac{z^6}{24}+\cdots, \qquad \left(\frac{z^2}{2}-\cdots\right)^3=\frac{z^6}{8}-\cdots,$$

we find that

$$(z)=1-\frac{1}{2}\left(\frac{z^2}{2}-\frac{z^4}{24}+\frac{z^6}{720}-\cdots\right)-\frac{1}{8}\left(\frac{z^4}{4}-\frac{z^6}{24}+\cdots\right)-\frac{1}{16}\left(\frac{z^6}{8}-\cdots\right)$$

$$=1-\frac{z^2}{4}-\frac{z^4}{96}-\frac{19z^6}{5760}-\cdots. \tag{10.42}$$

The series (10.42) converges on the disk $K:|z|<\pi/2$, since $F(z)=\sqrt{\cos z}$ is analytic on K, with derivative

$$F'(z)=-\frac{\sin z}{2\sqrt{\cos z}} \qquad (z\in K).$$

Moreover, the fact that $F'(z)$ becomes infinite for $z=\pi/2$ implies that K is the largest disk on which (10.42) converges (why?).

Problem 1 (M1, p. 435). Find the power series expansion at $z=0$ of the function $\exp\,[1/(1-z)]$.

Problem 2. Find the terms up to degree 5 in z in the power series expansions of the functions

a) $e^{z\sin z}$; b) $(1+z)^z=e^{z\ln(1+z)}$; c) e^{e^z}.

53. DIVISION OF POWER SERIES

Let

$$a_0+a_1(z-z_0)+\cdots+a_n(z-z_0)^n+\cdots \tag{10.43}$$

be a power series with radius of convergence $r>0$, and let

$$b_0+b_1(z-z_0)+\cdots+b_n(z-z_0)^n+\cdots \qquad (b_0\neq 0) \tag{10.44}$$

be a second power series (in the same variable $z-z_0$) with radius of convergence $\rho>0$. Since we assume $b_0\neq 0$, there is a disk $K:|z-z_0|<R$ such that both series converge on K, while the second series has no zeros in K. In fact, we need only choose

$$R=\min\,(r,\rho,|z_1-z_0|),$$

where z_1 is the zero of (10.44) which lies closest to z_0.[9] It follows from the rule for differentiating a quotient (see Rule 3, p. 46) that

$$f(z) = \frac{a_0 + a_1(z - z_0) + \cdots + a_n(z - z_0)^n + \cdots}{b_0 + b_1(z - z_0) + \cdots + b_n(z - z_0)^n + \cdots} \tag{10.45}$$

is an analytic function on K. Therefore, according to Theorem 10.6, $f(z)$ has a power series representation

$$f(z) = c_0 + c_1(z - z_0) + \cdots + c_n(z - z_0)^n + \cdots \tag{10.46}$$

on K. By *division of power series* we mean the process of finding the "quotient series" (10.46) from a knowledge of the "dividend series" (10.43) and the "divisor series" (10.44). For further justification of this terminology, see Prob. 1.

To find (10.46), we first combine (10.45) and (10.46), obtaining

$$\begin{aligned} &[c_0 + c_1(z - z_0) + \cdots + c_n(z - z_0)^n + \cdots] \\ &\qquad \times [b_0 + b_1(z - z_0) + \cdots + b_n(z - z_0)^n + \cdots] \\ &\qquad\qquad = a_0 + a_1(z - z_0) + \cdots + a_n(z - z_0)^n + \cdots. \end{aligned} \tag{10.47}$$

Since both series in the left-hand side are absolutely convergent on K, they can be multiplied term by term (see Sec. 43, Prob. 3). As a result, (10.47) becomes

$$\begin{aligned} &c_0b_0 + (c_0b_1 + c_1b_0)(z - z_0) + (c_0b_2 + c_1b_1 + c_2b_0)(z - z_0)^2 + \cdots \\ &\qquad + (c_0b_n + c_1b_{n-1} + \cdots + c_nb_0)(z - z_0)^n + \cdots \\ &\qquad = a_0 + a_1(z - z_0) + a_2(z - z_0)^2 + \cdots + a_n(z - z_0)^n + \cdots, \end{aligned}$$

where both sides are convergent power series on K. It follows from Theorem 10.4 that identical powers of $z - z_0$ have identical coefficients. Thus we obtain an infinite system of linear equations

$$\begin{aligned} &c_0b_0 = a_0, \\ &c_0b_1 + c_1b_0 = a_1, \\ &c_0b_2 + c_1b_1 + c_2b_0 = a_2, \\ &\cdots\cdots\cdots\cdots \\ &c_0b_n + c_1b_{n-1} + \cdots + c_nb_0 = a_n, \\ &\cdots\cdots\cdots\cdots \end{aligned} \tag{10.48}$$

which determine the unknown coefficients c_n $(n = 0, 1, 2, \ldots)$.

---

[9] If there is no unique z_1, we choose any of the (finitely many) zeros which lie at the same distance from z_0. If (10.44) has no zeros at all, we set $|z_1 - z_0| = \rho$. [As always, $\min(\alpha, \alpha, \beta) = \min(\alpha, \beta)$ and $\min(\alpha, \alpha) = \alpha$, and similarly with min replaced by max.]

The problem of solving the system (10.48) is greatly facilitated by the fact that the first $n + 1$ equations ($n = 0, 1, 2, \ldots$) involve only the first $n + 1$ unknowns, giving the system (10.48) its characteristic "triangular shape." Solving the first equation of the system, we obtain

$$c_0 = \frac{a_0}{b_0}$$

($b_0 \neq 0$, by hypothesis). Substituting this value of c_0 into the second equation of the system, we obtain

$$c_1 = \frac{a_1 b_0 - a_0 b_1}{b_0^2},$$

and so on. Thus, having found $c_0, c_1, \ldots, c_{n-1}$, we substitute these values into the $(n + 1)$th equation of the system (10.48), obtaining

$$c_n = \frac{a_n - c_0 b_n - c_1 b_{n-1} - \cdots - c_{n-1} b_1}{b_0}.$$

Example. The function

$$f(z) = \frac{z}{e^z - 1} \tag{10.49}$$

is analytic everywhere except at those points where $e^z - 1$ vanishes and z does not, i.e., except at the points $\pm 2\pi i, \pm 4\pi i, \ldots$. Substituting the power series

$$e^z - 1 = \frac{z}{1!} + \frac{z^2}{2!} + \cdots + \frac{z^n}{n!} + \cdots$$

into (10.49), and dividing both numerator and denominator by z, we obtain

$$f(z) = \frac{1}{1 + \dfrac{z}{2!} + \cdots + \dfrac{z^n}{(n+1)!} + \cdots}$$

(note that this expression is meaningful for $z = 0$). The series in the denominator converges for all z and does not vanish for $z = 0$, but otherwise has the same zeros as $e^z - 1$. The two zeros closest to the origin are at $\pm 2\pi i$, and therefore $f(z)$ has a power series representation (10.46) on the disk $K: |z| < 2\pi$. To find this representation, we use our technique of division of power series. In the present case, the coefficients a_n, b_n appearing in the system (10.48) are just

$$a_0 = 1, \quad a_n = 0 \qquad (n = 1, 2, \ldots),$$

$$b_n = \frac{1}{(n+1)!} \qquad (n = 0, 1, 2, \ldots).$$

Therefore the first of the equations (10.48) gives

$$c_0 = 1,$$

and the rest reduce to the *recurrence relation*

$$c_0 \frac{1}{(n+1)!} + c_1 \frac{1}{n!} + \cdots + c_n = 0 \qquad (n = 1, 2, \ldots), \tag{10.50}$$

relating c_n to the values $c_0, c_1, \ldots, c_{n-1}$.

The numbers $c_n n!$ are called the *Bernoulli numbers* and are denoted by B_n. To calculate B_n, we use the recurrence relation (10.50), which now takes the form

$$B_0 \frac{1}{0!(n+1)!} + B_1 \frac{1}{1!n!} + \cdots + B_n \frac{1}{n!1!} = 0 \qquad (n = 1, 2, \ldots) \tag{10.51}$$

(obviously $B_0 = c_0 0! = 1$). Multiplying (10.51) by $(n + 1)!$ and introducing the notation

$$\frac{(n+1)!}{k!(n+1-k)!} = \binom{n+1}{k}$$

(the familiar *binomial coefficient*), we find that

$$B_0\binom{n+1}{0} + B_1\binom{n+1}{1} + \cdots + B_n\binom{n+1}{n} = 0 \quad (n = 1, 2, \ldots). \tag{10.52}$$

Equation (10.52) can be written symbolically as

$$(1 + B)^{n+1} - B^{n+1} = 0, \tag{10.53}$$

where after raising $1 + B$ to the $(n + 1)$th power, every B^k is changed to B_k $(k = 1, 2, \ldots, n + 1)$. Using (10.53) and the fact that $B_0 = 1$, we deduce step by step that

$$B_0 + 2B_1 = 0, \qquad B_1 = -\tfrac{1}{2}B_0 = -\tfrac{1}{2},$$

$$B_0 + 3B_1 + 3B_2 = 0, \qquad B_2 = -\tfrac{1}{3}B_0 - B_1 = \tfrac{1}{6},$$

$$B_0 + 4B_1 + 6B_2 + 4B_3 = 0, \qquad B_3 = -\tfrac{1}{4}B_0 - B_1 - \tfrac{3}{2}B_2 = 0,$$

$$B_0 + 5B_1 + 10B_2 + 10B_3 + 5B_4 = 0,$$

$$B_4 = -\tfrac{1}{5}B_0 - B_1 - 2B_2 - 2B_3 = -\tfrac{1}{30},$$

$$B_0 + 6B_1 + 15B_2 + 20B_3 + 15B_4 + 6B_5 = 0,$$

$$B_5 = -\tfrac{1}{6}B_0 - B_1 - \tfrac{5}{2}B_2 - \tfrac{10}{3}B_3 - \tfrac{5}{2}B_4 = 0,$$

$$B_0 + 7B_1 + 21B_2 + 35B_3 + 35B_4 + 21B_5 + 7B_6 = 0,$$

$$B_6 = -\tfrac{1}{7}B_0 - B_1 - 3B_2 - 5B_3 - 5B_4 - 3B_5 = \tfrac{1}{42},$$

$$\cdots\cdots\cdots\cdots\cdots$$

Collecting these results, we have

$$B_0 = 1, \quad B_1 = -\tfrac{1}{2}, \quad B_2 = \tfrac{1}{6}, \quad B_3 = 0, \quad B_4 = -\tfrac{1}{30}, \quad B_5 = 0, \quad B_6 = \tfrac{1}{42}, \ldots \tag{10.54}$$

As suggested by (10.54), the Bernoulli numbers with odd indices greater than 1 vanish. To see this, we write

$$\begin{aligned} f(z) = \frac{z}{e^z - 1} &= c_0 + c_1 z + c_2 z^2 + c_3 z^3 + \cdots + c_n z^n + \cdots \\ &= B_0 + \frac{B_1}{1!} z + \frac{B_2}{2!} z^2 + \frac{B_3}{3!} z^3 + \cdots + \frac{B_n}{n!} z^n + \cdots \end{aligned} \tag{10.55}$$

and then replace z by $-z$, obtaining

$$\begin{aligned} f(-z) = \frac{-z}{e^{-z} - 1} &= \frac{-ze^z}{(e^{-z} - 1)e^z} = \frac{ze^z}{e^z - 1} \\ &= B_0 - \frac{B_1}{1!} z + \frac{B_2}{2!} z^2 - \frac{B_3}{3!} z^3 + \cdots + \frac{B_n}{n!} (-1)^n z^n + \cdots. \end{aligned} \tag{10.56}$$

Subtraction of (10.56) from (10.55) gives

$$\begin{aligned} f(z) - f(-z) &= \frac{z}{e^z - 1} - \frac{ze^z}{e^z - 1} = -z \\ &= 2\frac{B_1}{1!} z + 2\frac{B_3}{3!} z^3 + \cdots + 2\frac{B_{2m+1}}{(2m+1)!} z^{2m+1} + \cdots, \end{aligned}$$

and hence, by the uniqueness of power series expansions,

$$2B_1 = -1, \quad B_3 = B_5 = \cdots = B_{2m+1} = \cdots = 0,$$

as asserted. Using this fact, we finally write the expansion (10.55) in the form

$$f(z) = \frac{z}{e^z - 1} = 1 - \frac{z}{2} + \sum_{m=1}^{\infty} \frac{B_{2m}}{(2m)!} z^{2m}, \tag{10.57}$$

where the series converges on the disk $K: |z| < 2\pi$. Moreover, the fact that $f(z)$ becomes infinite for $z = \pm 2\pi i$ implies that K is the largest disk on which (10.57) converges (why?).

Problem 1. Verify that formal "long division" of the series (10.43) by the series (10.44) gives a series (10.46) whose coefficients $c_0, c_1, \ldots, c_n, \ldots$ satisfy the system (10.48).

Problem 2 (M1, p. 442). Verify the Taylor series expansions

a) $z \cot z = 1 + \sum_{m=1}^{\infty} (-1)^m \frac{2^{2m} B_{2m}}{(2m)!} z^{2m} \qquad (|z| < \pi),$

b) $\tan z = \sum_{m=1}^{\infty} (-1)^{m-1} \frac{2^{2m}(2^{2m} - 1)B_{2m}}{(2m)!} z^{2m-1} \qquad (|z| < \pi/2),$

c) $z \sec z = 1 + \sum_{m=1}^{\infty} (-1)^{m-1} \frac{(2^{2m} - 2)B_{2m}}{(2m)!} z^{2m} \qquad (|z| < \pi),$

involving the Bernoulli numbers B_{2m}.

Hint. Start from the example on pp. 220–222.

Problem 3. Prove that the coefficients c_n in the expansion

$$\frac{1}{1 - z - z^2} = \sum_{n=0}^{\infty} c_n z^n \tag{10.58}$$

satisfy the recurrence relation

$$c_0 = c_1 = 1, \quad c_n = c_{n-1} + c_{n-2} \qquad (n \geqslant 2).$$

What is the radius of convergence of the series (10.58)?

Comment. The numbers c_n are called the *Fibonacci numbers.*

CHAPTER 11

LAURENT SERIES. SINGULAR POINTS

54. LAURENT SERIES

In the preceding chapter, we devoted considerable attention to the subject of power series, i.e., expansions in *nonnegative* integral powers of the variable $z - z_0$. We now consider series of a related type, which involve *arbitrary* integral powers of $z - z_0$. These series closely resemble power series both in structure and basic properties. Our first result is analogous to the Cauchy-Hadamard theorem (Theorem 10.1):

THEOREM 11.1. *Given a series*

$$A_0 + A_1(z - z_0)^{-1} + A_2(z - z_0)^{-2} + \cdots + A_n(z - z_0)^{-n} + \cdots \quad (11.1)$$

of negative powers of $z - z_0$, *let*

$$r = \overline{\lim_{n\to\infty}} \sqrt[n]{|A_n|},$$

and let γ *be the circle* $|z - z_0| = r$, *with interior* $I(\gamma)$ *and exterior* $E(\gamma)$. *Then there are three possibilities*:

1. *If* $r = 0$, *the series is absolutely convergent for all* z *in the extended plane except* $z = z_0$;
2. *If* $0 < r < \infty$, *the series is absolutely convergent for all* $z \in E(\gamma)$ *and divergent for all* $z \in I(\gamma)$;
3. *If* $r = \infty$, *the series is divergent for all finite* z.

Proof. The substitution

$$Z = \frac{1}{z - z_0} \tag{11.2}$$

converts (11.1) into the power series

$$A_0 + A_1 Z + A_2 Z^2 + \cdots + A_n Z^n + \cdots.$$

with radius of convergence

$$\frac{1}{\overline{\lim\limits_{n\to\infty}} \sqrt[n]{|A_n|}} = \frac{1}{r}.$$

The rest of the proof is an immediate consequence of Theorem 10.1, and the fact that (11.2) transforms the points $z = z_0$, $z = \infty$ into the points $Z = \infty$, $Z = 0$.

Remark 1. Obviously, the behavior of the series (11.1) for $r = 0$ and $r = \infty$ can be regarded as the appropriate limiting case of its behavior for $0 < r < \infty$. We shall henceforth assume that $r < \infty$, so that (11.1) converges on a domain, i.e., the exterior of the circle $\gamma: |z - z_0| = r$.

Remark 2. It is clear that every compact subset of $E(\gamma)$ is mapped into a compact subset of $I(\gamma^*)$ by the transformation (11.2), where γ^* is the circle

$$|Z| = \frac{1}{r}.$$

Therefore, according to Theorem 10.2, the series (11.1) is uniformly convergent on every compact subset of $E(\gamma)$. It follows from Theorem 9.4 that (11.1) represents a function $f(z)$ which is analytic at every finite point of $E(\gamma)$.

Remark 3. The function $f(z)$ is also analytic at $z = \infty$, since the function $f^*(\zeta) = f(1/\zeta)$ is analytic at $\zeta = 0$ (see Prob. 1). Correspondingly, we define

$$f(\infty) = \lim_{z\to\infty} f(z) = A_0.$$

We now introduce series of the form

$$\sum_{n=-\infty}^{\infty} a_n (z - z_0)^n, \tag{11.3}$$

interpreted as the sum of the two series

$$\sum_{n=0}^{\infty} a_n (z - z_0)^n, \quad \sum_{m=1}^{\infty} a_{-m} (z - z_0)^{-m}. \tag{11.4}$$

The series (11.3) is known as a *Laurent series*, and is regarded as convergent if and only if both series (11.4) converge. In other words,

$$\sum_{n=-\infty}^{\infty} a_n(z - z_0)^n = \lim_{\nu\to\infty} \sum_{n=0}^{\nu} a_n(z - z_0)^n + \lim_{\mu\to\infty} \sum_{m=1}^{\mu} a_{-m}(z - z_0)^{-m},$$

or equivalently,

$$\sum_{n=-\infty}^{\infty} a_n(z - z_0)^n = \lim_{\substack{\mu\to\infty \\ \nu\to\infty}} \sum_{n=-\mu}^{\nu} a_n(z - z_0)^n, \tag{11.5}$$

where μ and ν approach infinity independently. Thus (11.5) means that given any $\varepsilon > 0$, there is an integer $N(\varepsilon) > 0$ such that

$$\left|\sum_{n=-\infty}^{\infty} a_n(z - z_0)^n - \sum_{n=-\mu}^{\nu} a_n(z - z_0)^n\right| < \varepsilon$$

whenever $\mu > N(\varepsilon)$ and $\nu > N(\varepsilon)$.

According to this definition, the convergence behavior of a Laurent series depends on the corresponding behavior of the series (11.4). Consider the two circles $\Gamma: |z-z_0| = R$ and $\gamma: |z - z_0| = r$, where

$$\frac{1}{R} = \overline{\lim_{n\to\infty}} \sqrt[n]{|a_n|}, \qquad r = \overline{\lim_{m\to\infty}} \sqrt[m]{|a_{-m}|}.$$

Then the first of the series (11.4) converges absolutely and uniformly on every compact subset of $I(\Gamma)$ and diverges on $E(\Gamma)$, whereas the second of the series (11.4) converges absolutely and uniformly on every compact subset of $E(\gamma)$ and diverges on $I(\gamma)$. The domains $I(\Gamma)$ and $E(\gamma)$ have a nonempty intersection if and only if $r < R$, and if this condition is satisfied, the domain $D = I(\Gamma) \cap E(\gamma)$ is a ring-shaped region or *annulus*, consisting of all points z such that

$$r < |z - z_0| < R.$$

Clearly, both series (11.4) converge absolutely and uniformly on every compact subset of D, and hence the same is true of the series (11.3). It follows from Theorem 9.4 that the function $f(z)$ defined by

$$f(z) = \sum_{n=-\infty}^{\infty} a_n(z - z_0)^n \qquad (r < |z - z_0| < R) \tag{11.6}$$

is analytic on D. Moreover, since one or the other of the series (11.4) diverges if $|z- z_0| < r$ or $|z - z_0| > R$, so does the series (11.3).

Remark. Henceforth, in talking about a Laurent series, we shall always assume that the condition $r < R$ is satisfied, so that there is an annulus on which the series converges.

THEOREM 11.2. *The coefficients of the Laurent series* (11.6) *are given by the formula*

$$a_k = \frac{1}{2\pi i}\int_{\gamma_\rho} \frac{f(z)}{(z-z_0)^{k+1}}\,dz \qquad (k = 0, \pm 1, \pm 2, \ldots), \tag{11.7}$$

where γ_ρ *is any circle* $|z - z_0| = \rho$, $r < \rho < R$.

Proof. The series (11.6) is uniformly convergent on γ_ρ, and hence the same is true of the series obtained by multiplying (11.6) by

$$\frac{1}{2\pi i}\frac{1}{(z-z_0)^{k+1}},$$

where k is an arbitrary integer (why?). It follows from Theorem 9.3 that

$$\begin{aligned}\frac{1}{2\pi i}\int_{\gamma_\rho} \frac{f(z)}{(z-z_0)^{k+1}}\,dz &= \sum_{n=-\infty}^{\infty} a_n \frac{1}{2\pi i}\int_{\gamma_\rho} (z-z_0)^{n-k-1}\,dz \\ &= \sum_{n=-\infty}^{\infty} a_n \frac{1}{2\pi}\int_0^{2\pi} \rho^{n-k} e^{i(n-k)\theta}\,d\theta = a_k,\end{aligned}$$

where we use the fact that the last integral on the right vanishes if $n \neq k$ and equals 2π if $n = k$.

COROLLARY 1. *Let*

$$f(z) = \sum_{n=-\infty}^{\infty} a_n(z-z_0)^n, \qquad g(z) = \sum_{n=-\infty}^{\infty} b_n(z-z_0)^n \tag{11.8}$$

be two Laurent series, which converge on annuli D and Δ, respectively, and suppose $f(z)$ and $g(z)$ coincide on a circle $\gamma_\rho : |z - z_0| = \rho$ contained in both D and Δ. Then

$$a_k = b_k \qquad (k = 0, \pm 1, \pm 2, \ldots),$$

i.e., the two series (11.8) *are identical.*

Proof.

$$a_k = \frac{1}{2\pi i}\int_{\gamma_\rho} \frac{f(z)}{(z-z_0)^{k+1}}\,dz = \frac{1}{2\pi i}\int_{\gamma_\rho} \frac{g(z)}{(z-z_0)^{k+1}}\,dz = b_k$$
$$(k = 0, \pm 1, \pm 2, \ldots).$$

COROLLARY 2. *There is a unique Laurent series in the variable $z - z_0$ which has a given sum on an annulus $D : r < |z - z_0| < R$.*

Proof. Choose $D = \Delta$ in Corollary 1.

Problem 1. Suppose the series (11.1) converges to $f(z)$ for all z such that $r < |z - z_0| < \infty$, and let $f^*(\zeta) = f(1/\zeta)$, $f^*(0) = A_0$. Prove that $f^*(\zeta)$ is analytic at $\zeta = 0$.

Hint. If $z_0 \neq 0$, consider the series

$$f^*(\zeta) = \sum_{n=0}^{\infty} A_n \left(\frac{\zeta}{1 - \zeta z_0}\right)^n,$$

where $|\zeta| < \min(r, |z_0|^{-1})$.

Problem 2. Prove that if the Laurent series

$$\sum_{n=-\infty}^{\infty} a_n z^n$$

represents an even function, then

$$a_{2k+1} = 0 \qquad (k = 0, \pm 1, \pm 2, \ldots),$$

while if the series represents an odd function, then

$$a_{2k} = 0 \qquad (k = 0, \pm 1, \pm 2, \ldots).$$

Problem 3. Show that the Cauchy inequalities [cf. (10.12)] continue to hold for the coefficients a_n ($n = 0, \pm 1, \pm 2, \ldots$) of a Laurent series.

55. LAURENT'S THEOREM

Our next result is the analogue of Theorem 10.6:

THEOREM 11.3 (*Laurent's theorem*). *Let $f(z)$ be an analytic function on an annulus $D: r < |z - z_0| < R$.*[1] *Then there exists a Laurent series*

$$f(z) = \sum_{n=-\infty}^{\infty} a_n (z - z_0)^n \tag{11.9}$$

converging to $f(z)$ on D.

Proof. Let z_1 be an arbitrary point of D, and choose values of r' and R' satisfying the inequalities

$$r < r' < |z_1 - z_0| < R' < R.$$

Then the annulus

$$D': r' < |z - z_0| < R'$$

is contained in D and contains the point z_1. Let γ_ρ be the circle $|z - z_0| = \rho$, $r < \rho < R$, and let Γ be a circle with center z_1 such that $\Gamma \subset D'$ (see Figure 11.1). Since the function

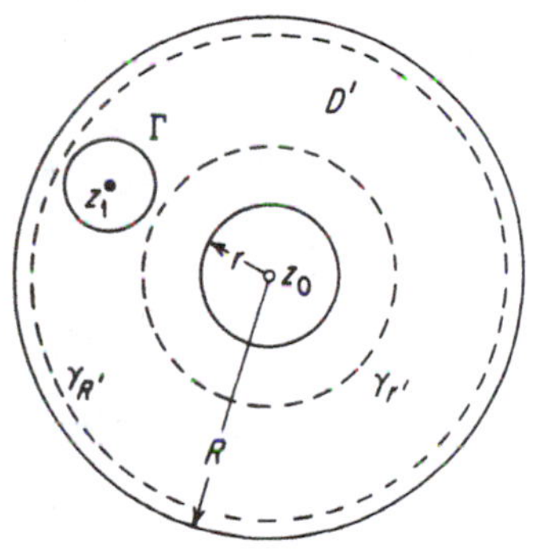

FIGURE 11.1

$$\varphi(z) = \frac{f(z)}{z - z_1}$$

[1] The annulus may reduce to a "deleted neighborhood," i.e., to a disk without its center (if $r = 0$, $R < \infty$), to the exterior of a circle with the point at infinity deleted (if $r > 0$, $R = \infty$), or to the whole plane with the points z_0 and ∞ deleted (if $r = 0$, $R = \infty$). See the remark on p. 232.

is analytic on $D - \{z_1\}$, it follows from Cauchy's integral theorem for a system of contours (Theorem 7.3) that[2]

$$\frac{1}{2\pi i}\int_{\gamma_{R'}} \frac{f(z)}{z - z_1}\,dz = \frac{1}{2\pi i}\int_{\gamma_{r'}} \frac{f(z)}{z - z_1}\,dz + \frac{1}{2\pi i}\int_{\Gamma} \frac{f(z)}{z - z_1}\,dz. \quad (11.10)$$

The last integral on the right is a Cauchy integral, and equals $f(z_1)$. Therefore (11.10) becomes

$$f(z_1) = \frac{1}{2\pi i}\int_{\gamma_{R'}} \frac{f(z)}{z - z_1}\,dz - \frac{1}{2\pi i}\int_{\gamma_{r'}} \frac{f(z)}{z - z_1}\,dz,$$

or

$$f(z_1) = \frac{1}{2\pi i}\int_{\gamma_{R'}} \frac{f(z)}{z - z_1}\,dz + \frac{1}{2\pi i}\int_{\gamma_{r'}} \frac{f(z)}{z_1 - z}\,dz. \quad (11.11)$$

As we now show, the first integral on the right in (11.11) leads to the nonnegative powers of $z - z_0$ in the Laurent series (11.9), while the second integral leads to the negative powers of $z - z_0$. With this in mind, we express the factor

$$\frac{1}{z - z_1} \qquad (z \in \gamma_{R'})$$

appearing in the first integral as a geometric series with ratio

$$\frac{z_1 - z_0}{z - z_0},$$

where

$$\left|\frac{z_1 - z_0}{z - z_0}\right| = \frac{|z_1 - z_0|}{R'} = \theta < 1.$$

The result is

$$\begin{aligned} \frac{1}{z - z_1} &= \frac{1}{(z - z_0) - (z_1 - z_0)} \\ &= \frac{1}{z - z_0}\,\frac{1}{1 - \dfrac{z_1 - z_0}{z - z_0}} = \sum_{n=0}^{\infty} \frac{(z_1 - z_0)^n}{(z - z_0)^{n+1}} \end{aligned} \quad (11.12)$$

[cf. (10.20)], where the series on the right is uniformly convergent on $\gamma_{R'}$, since

$$\left|\frac{(z_1 - z_0)^n}{(z - z_0)^{n+1}}\right| = \frac{\theta^n}{R'} \qquad (0 < \theta < 1).$$

---

[2] The circles $\gamma_{R'}$, $\gamma_{r'}$ and Γ are all traversed in the same (counterclockwise) direction.

Therefore the series

$$\frac{1}{2\pi i}\frac{f(z)}{z - z_1} = \sum_{n=0}^{\infty} \frac{1}{2\pi i}\frac{f(z)}{(z - z_0)^{n+1}}(z_1 - z_0)^n, \qquad (11.13)$$

obtained by multiplying (11.12) by the function

$$\frac{1}{2\pi i}f(z)$$

(which is bounded on $\gamma_{R'}$), is uniformly convergent on $\gamma_{R'}$, and we can integrate (11.13) term by term along $\gamma_{R'}$, obtaining

$$\frac{1}{2\pi i}\int_{\gamma_{R'}} \frac{f(z)}{z - z_1}\,dz = \sum_{n=0}^{\infty} a_n(z_1 - z_0)^n, \qquad (11.14)$$

where

$$a_n = \frac{1}{2\pi i}\int_{\gamma_{R'}} \frac{f(z)}{(z - z_0)^{n+1}}\,dz \qquad (n = 0, 1, 2, \ldots). \qquad (11.15)$$

Next we again consider (11.1) and express the factor

$$\frac{1}{z_1 - z} \qquad (z \in \gamma_{r'})$$

appearing in the second integral as a geometric series with ratio

$$\frac{z - z_0}{z_1 - z_0},$$

where

$$\left|\frac{z - z_0}{z_1 - z_0}\right| = \frac{r'}{|z_1 - z_0|} = \vartheta < 1.$$

The result is

$$\frac{1}{z_1 - z} = \frac{1}{(z_1 - z_0) - (z - z_0)}$$
$$= \frac{1}{z_1 - z_0}\frac{1}{1 - \dfrac{z - z_0}{z_1 - z_0}} = \sum_{n=0}^{\infty} \frac{(z - z_0)^n}{(z_1 - z_0)^{n+1}}.$$

The series on the right is uniformly convergent on $\gamma_{r'}$, since

$$\left|\frac{(z - z_0)^n}{(z_1 - z_0)^{n+1}}\right| = r'\vartheta^n \qquad (0 < \vartheta < 1),$$

and hence so is the series

$$\frac{1}{2\pi i}\frac{f(z)}{z_1 - z} = \sum_{n=0}^{\infty} \frac{1}{2\pi i}\frac{f(z)}{(z - z_0)^{-n}}(z_1 - z_0)^{-n-1}$$
$$= \sum_{n=1}^{\infty} \frac{1}{2\pi i}\frac{f(z)}{(z - z_0)^{-n+1}}(z_1 - z_0)^{-n}. \qquad (11.16)$$

Integrating (11.16) term by term along $\gamma_{r'}$, we find that

$$\frac{1}{2\pi i}\int_{\gamma_{r'}}\frac{f(z)}{z_1 - z}\,dz = \sum_{n=1}^{\infty} a_{-n}(z_1 - z_0)^{-n}, \tag{11.17}$$

where

$$a_{-n} = \frac{1}{2\pi i}\int_{\gamma_{r'}}\frac{f(z)}{(z - z_0)^{-n+1}}\,dz \qquad (n = 1, 2, \ldots). \tag{11.18}$$

Substituting (11.14) and (11.17) into (11.11), and replacing z_1 by z (z_1 is arbitrary), we finally find that

$$f(z) = \sum_{n=0}^{\infty} a_n(z - z_0)^n + \sum_{n=1}^{\infty} a_{-n}(z - z_0)^{-n} = \sum_{n=-\infty}^{\infty} a_n(z - z_0)^n,$$

which is the desired Laurent series (11.9). Moreover, using Theorem 7.3 to replace $\gamma_{r'}$ and $\gamma_{R'}$ by γ_ρ ($r < \rho < R$), we can combine (11.15) and (11.18) into a single formula

$$a_n = \frac{1}{2\pi i}\int_{\gamma_\rho}\frac{f(z)}{(z - z_0)^{n+1}}\,dz \qquad (n = 0, \pm 1, \pm 2, \ldots),$$

in complete agreement with (11.7).

Remark. If $r = 0$ and $R < \infty$, the annulus figuring in Theorem 11.3 reduces to a disk without its center. Such a set $\mathscr{N}(z_0) - \{z_0\}$ is called a *deleted neighborhood of* z_0, and will often be denoted by $\mathscr{N}'(z_0)$. A function $f(z)$ is said to have a *Laurent expansion* at $z = z_0$ if $f(z)$ has a Laurent expansion (in the variable $z - z_0$) in some deleted neighborhood of z_0. Similarly, if $r > 0$ and $R = \infty$, the annulus reduces to the exterior of a circle $\gamma : |z - z_0| = r$, with the point at infinity deleted. Such a set $E(\gamma) - \{\infty\} = \mathscr{N}'(\infty)$ is called a *deleted neighborhood of* ∞. A function $f(z)$ is said to have a *Laurent expansion at* $z = \infty$ if $f(z)$ has a Laurent expansion in some deleted neighborhood of ∞ (see Prob. 1).

Example. In the annulus $0 < |z| < 1$, and hence at $z = 0$, the function

$$f(z) = \frac{1}{z(1 - z)}$$

has the Laurent expansion

$$\frac{1}{z} + \sum_{n=0}^{\infty} z^n,$$

while in the annulus $0 < |z - 1| < 1$, and hence at $z = 1$, $f(z)$ has the Laurent expansion

$$-\frac{1}{z - 1} + \sum_{n=0}^{\infty}(-1)^n(z - 1)^n.$$

In the domain $1 < |z| < \infty$, and hence at $z = \infty$, $f(z)$ has the Laurent expansion

$$-\sum_{n=2}^{\infty} \frac{1}{z^n}.$$

The last expansion contains no positive powers of z, and is therefore valid at $z = \infty$, where $f(z)$ vanishes (see Remark 3, p. 226).

Problem 1. Suppose $f(z)$ has a Laurent expansion at $z = \infty$, so that

$$f(z) = \sum_{n=-\infty}^{\infty} a_n(z - z_0)^n \qquad (r < |z - z_0| < \infty).$$

Prove that $f(z)$ can also be represented in the form

$$f(z) = \sum_{n=-\infty}^{\infty} \tilde{a}_n z^n \qquad (\tilde{r} < |z| < \infty),$$

i.e., the circle figuring in the deleted neighborhood of infinity (see p. 232) can always be chosen with its center at the origin.

Problem 2. Expand each of the following functions $f(z)$ in a Laurent series on the indicated domain:

a) $\dfrac{1}{(z-a)(z-b)}$ $(0 < |a| < |b|)$ on the annulus $|a| < |z| < |b|$;

b) $\dfrac{z^2 - 2z + 5}{(z-2)(z^2+1)}$ on the annulus $1 < |z| < 2$;

c) $\sqrt{\dfrac{z}{(z-1)(z-2)}}$ $(\operatorname{Im} f(\tfrac{3}{2}) > 0)$ on the annulus $1 < |z| < 2$;

d) $e^{z+(1/z)}$ on the domain $0 < |z| < \infty$.

Ans.

b) $2\sum_{n=1}^{\infty}(-1)^n \dfrac{1}{z^{2n}} - \sum_{n=0}^{\infty} \dfrac{z^n}{2^{n+1}}$ for $1 < |z| < 2$;

d) $\sum_{n=0}^{\infty} c_n z^n + \sum_{n=1}^{\infty} c_{-n} z^{-n}$ for $0 < |z| < \infty$, where

$$c_n = c_{-n} = \sum_{k=0}^{\infty} \frac{1}{k!\,(n+k)!}.$$

Problem 3. Find the Laurent expansion of the function

$$\ln \frac{z-a}{z-b}$$

at $z = \infty$.

Ans.

$$\sum_{n=1}^{\infty} \frac{b^n - a^n}{nz^n} \text{ for } |z| > \max(|a|, |b|).$$

56. POLES AND ESSENTIAL SINGULAR POINTS

Let $f(z)$ be a single-valued function which is analytic in a deleted neighborhood of a point z_0, i.e., on some domain

$$D:0 < |z - z_0| < R.$$

If there exists a finite complex number a_0 such that setting $f(z_0) = a_0$ gives a function analytic on the whole disk $|z - z_0| < R$, including the point z_0, we call z_0 a *regular point* of $f(z)$ and $f(z)$ is said to be *regular at* z_0. However, if no such number a_0 exists, we call z_0 a *singular point* of $f(z)$ [more precisely, an *isolated singular point of single-valued character*] and we say that $f(z)$ is *singular at* z_0.

Suppose $f(z)$ is analytic in a deleted neighborhood $D:0 < |z - z_0| < R$ of an isolated singular point z_0. Then the basic tool for studying $f(z)$ is its Laurent series

$$f(z) = \sum_{n=-\infty}^{\infty} a_n(z - z_0)^n \qquad (z \in D), \tag{11.19}$$

where

$$a_n = \frac{1}{2\pi i} \int_{\gamma_\rho} \frac{f(z)}{(z - z_0)^{n+1}} \, dz \qquad (n = 0, \pm 1, \pm 2, \ldots) \tag{11.20}$$

and γ_ρ is the circle $|z - z_0| = \rho$ $(0 < \rho < R)$. The expansion (11.19) can also be used in the case where z_0 is a regular point of $f(z)$. Then $a_n = 0$ if $n < 0$, and the Laurent series reduces to a Taylor series.

THEOREM 11.4. *Let $f(z)$ be a single-valued analytic function defined on the domain $D:0 < |z - z_0| < R$. Then z_0 is a regular point of $f(z)$ if and only if $f(z)$ is bounded in some deleted neighborhood of z_0.*

Proof. If $f(z)$ is regular at z_0, then $f(z)$ is analytic on the whole disk $K:|z - z_0| < R$. Therefore $f(z)$ is bounded in any closed neighborhood $\overline{\mathcal{N}(z_0)} \subset K$ and hence in some deleted neighborhood of z_0.

Conversely, suppose there is a deleted neighborhood $\mathcal{N}'(z_0) \subset D$ and a finite constant $M > 0$ such that $|f(z)| \leqslant M$ for all $z \in \mathcal{N}'(z_0)$. Choosing a circle $\gamma_\rho:|z - z_0| = \rho$ $(0 < \rho < R)$ such that $\gamma_\rho \subset \mathcal{N}'(z_0)$, we use (11.20) to estimate the coefficients a_n of the Laurent series (11.19):

$$|a_n| \leqslant \frac{1}{2\pi} M \frac{2\pi\rho}{\rho^{n+1}}, \quad \text{i.e.,} \quad |a_n| \leqslant \frac{M}{\rho^n}$$

(cf. Sec. 54, Prob. 3). If $n < 0$, then, letting ρ $(0 < \rho < R)$ approach zero, we find that

$$a_n = 0 \qquad (n = -1, -2, \ldots).$$

Therefore the Laurent series (11.19) reduces to the Taylor series

$$f(z) = \sum_{n=0}^{\infty} a_n(z - z_0)^n.$$

The sum of this series is analytic on the whole disk $|z - z_0| < R$. Setting $f(z_0) = a_0$, we see that z_0 is a regular point of $f(z)$, as asserted.

COROLLARY. *The point z_0 is a singular point of $f(z)$ if and only if $f(z)$ is unbounded in every deleted neighborhood $\mathcal{N}'(z_0)$.*

According to the corollary, $f(z)$ can behave in two possible ways in a neighborhood of an isolated singular point:

1. $f(z) \to \infty$ as $z \to z_0$;
2. $f(z)$ approaches no limit (finite or infinite) as $z \to z_0$.

Simple examples show that both cases are possible. To illustrate the first case, let

$$f(z) = \frac{1}{(z - z_0)^n}, \tag{11.21}$$

where n is a positive integer. Then obviously $f(z)$ is analytic for $|z - z_0| > 0$, and

$$\lim_{z \to z_0} f(z) = \infty.$$

To illustrate the second possibility, let

$$f(z) = \exp\left(\frac{1}{z - z_0}\right).$$

This function is also analytic for $|z - z_0| > 0$, but unlike (11.21), $f(z)$ approaches no limit as $z \to z_0$. In fact, let z lie on the straight line passing through z_0 parallel to the real axis, so that $z - z_0 = x - x_0$, where $x - x_0$ is real. Then

$$\exp\left(\frac{1}{x - x_0}\right) \to \infty \qquad (x > x_0)$$

as $x \to x_0$ from the right, whereas

$$\exp\left(\frac{1}{x - x_0}\right) \to 0 \qquad (x < x_0)$$

as $x \to x_0$ from the left.

DEFINITION 1. *An isolated singular point z_0 of an analytic function $f(z)$ such that $f(z) \to \infty$ as $z \to z_0$ is called a pole of $f(z)$.*[3]

---

[3] Poles of rational functions have already been defined on p. 80, and the present considerations are the natural generalization of our previous results.

DEFINITION 2. *An isolated singular point z_0 of an analytic function $f(z)$ such that $f(z)$ approaches no limit (finite or infinite) as $z \to z_0$ is called an essential singular point of $f(z)$.*

THEOREM 11.5. *The point z_0 is a pole of the function $f(z)$ if and only if it is a zero of the function*

$$\varphi(z) = \frac{1}{f(z)}.$$

Proof. If z_0 is a pole of $f(z)$, then $f(z) \to \infty$ as $z \to z_0$, and hence there is a deleted neighborhood $\mathcal{N}'(z_0)$ in which $|f(z)| > 1$. Obviously, $\varphi(z) = 1/f(z)$ is analytic on $\mathcal{N}'(z_0)$, and moreover $|\varphi(z)| < 1$ for all $z \in \mathcal{N}'(z_0)$. It follows from Theorem 11.4 that z_0 is a regular point of $\varphi(z)$, and that

$$\varphi(z_0) = \lim_{z \to z_0} \frac{1}{f(z)} = 0$$

(why?). Therefore z_0 is a zero of $\varphi(z)$.

Conversely, let $\varphi(z)$ be single-valued and analytic in a neighborhood $\mathcal{N}(z_0)$, and suppose z_0 is a zero of $\varphi(z)$, where $\varphi(z) \not\equiv 0$. Then there is a number $\delta > 0$ such that the disk $|z - z_0| < \delta$ contains no zeros of $\varphi(z)$ other than z_0 itself (cf. p. 210). Forming the function $f(z) = 1/\varphi(z)$, we see that $f(z)$ is analytic for $0 < |z - z_0| < \delta$ and approaches infinity as $z \to z_0$. In other words, z_0 is a pole of $f(z)$.

This result, which establishes a relation between poles and zeros, allows us to introduce the concept of the *order* of a pole. We say that the point z_0 is a *pole of order* k ($k \geqslant 1$) of the function $f(z)$ if z_0 is a zero of order k of the function $1/f(z)$. The pole is said to be *simple* if $k = 1$ and *multiple* if $k > 1$ (cf. p. 80).

At a pole of order k, the Laurent expansion of $f(z)$ has the characteristic structure given by

THEOREM 11.6. *The point z_0 is a pole of order k of the function $f(z)$ if and only if the Laurent expansion of $f(z)$ at z_0 is of the form*

$$\begin{aligned} f(z) = a_{-k}(z - z_0)^{-k} + \cdots + a_{-1}(z - z_0)^{-1} \\ + a_0 + a_1(z - z_0) + \cdots, \end{aligned} \tag{11.22}$$

where $a_{-k} \neq 0$.

Proof. If $f(z)$ has a pole of order k at z_0, then $1/f(z)$ has a zero of order k at z_0, i.e.,

$$\frac{1}{f(z)} = A_k(z - z_0)^k + A_{k+1}(z - z_0)^{k+1} + \cdots,$$

where $A_k \neq 0$ (cf. p. 211), and hence

$$f(z) = \frac{1}{(z - z_0)^k} \frac{1}{A_k + A_{k+1}(z - z_0) + \cdots}. \tag{11.23}$$

The power series

$$A_k + A_{k+1}(z - z_0) + \cdots$$

represents an analytic function which is nonvanishing in some neighborhood of z_0 (since $A_k \neq 0$). It follows that the function

$$\frac{1}{A_k + A_{k+1}(z - z_0) + \cdots}$$

is analytic in a neighborhood of z_0, and hence has an expansion of the form

$$\alpha_0 + \alpha_1(z - z_0) + \cdots,$$

where $\alpha_0 = 1/A_k \neq 0$. After substitution into (11.23), this gives

$$f(z) = \alpha_0(z - z_0)^{-k} + \alpha_1(z - z_0)^{-k+1} + \cdots \qquad (\alpha_0 \neq 0), \tag{11.24}$$

which is just the Laurent expansion of $f(z)$ at z_0 (see Corollary 2, p. 228). In fact, (11.24) coincides with (11.22) if we make the substitution $\alpha_n = a_{n-k}$ ($n = 0, 1, 2, \ldots$).

Conversely, suppose $f(z)$ has an expansion of the form (11.22) at the point z_0, with $a_{-k} \neq 0$. Then

$$f(z) = \frac{a_{-k} + a_{-k+1}(z - z_0) + \cdots}{(z - z_0)^k},$$

which implies

$$\frac{1}{f(z)} = (z - z_0)^k \frac{1}{a_{-k} + a_{-k+1}(z - z_0) + \cdots}.$$

Replacing the function

$$\frac{1}{a_{-k} + a_{-k+1}(z - z_0) + \cdots}$$

by its Taylor series

$$\beta_0 + \beta_1(z - z_0) + \cdots,$$

we obtain

$$\begin{aligned}\frac{1}{f(z)} &= (z - z_0)^k[\beta_0 + \beta_1(z - z_0) + \cdots] \\ &= \beta_0(z - z_0)^k + \beta_1(z - z_0)^{k+1} + \cdots,\end{aligned}$$

where $\beta_0 = 1/a_{-k} \neq 0$. It follows that z_0 is a zero of order k of the function $1/f(z)$, and hence a pole of order k of the original function $f(z)$. This completes the proof.

COROLLARY. *The point z_0 is an essential singular point of the function $f(z)$ if and only if the Laurent expansion of $f(z)$ at z_0 has infinitely many terms of the form $a_{-k}(z - z_0)^{-k}$, where $k > 0$, $a_{-k} \neq 0$.*

Proof. Obviously, z_0 is a regular point if and only if there are no negative powers of $z - z_0$ in the Laurent expansion of $f(z)$ at z_0. The case of finitely many negative powers is covered by Theorem 11.6. The only other possibility is that z_0 is an essential singular point.

Because of Theorem 11.6 and its corollary, we see that a decisive role is played by the negative powers of $z - z_0$ in the Laurent series

$$\sum_{n=-\infty}^{\infty} a_n(z - z_0)^n. \tag{11.25}$$

The series of negative powers

$$\sum_{n=-\infty}^{-1} a_n(z - z_0)^n = \sum_{n=1}^{\infty} a_{-n}(z - z_0)^{-n}$$

is called the *principal part* of (11.25), while the series of nonnegative powers

$$\sum_{n=0}^{\infty} a_n(z - z_0)^n$$

is called the *regular part* of (11.25).

Example. Consider the function

$$f(z) = e^{1/z^4}.$$

Clearly, the expansion

$$f(z) = 1 + \frac{1}{z^4} + \frac{1}{2!z^8} + \frac{1}{3!z^{12}} + \cdots \tag{11.26}$$

converges for all $z \neq 0$ and represents the Laurent expansion of $f(z)$ at the point $z = 0$. Since (11.26) contains infinitely many negative powers of z, the point $z = 0$ must be an essential singular point of $f(z)$. This is easily verified directly. In fact, $f(z) \to \infty$ if we approach $z = 0$ along the coordinate axes, while $f(z) \to 0$ if we approach $z = 0$ along the lines $y = \pm x$ bisecting the coordinate axes. Therefore $f(z)$ approaches no limit as $z \to 0$.

Problem 1 (M2, p. 18). Show that the Laurent series

$$\cdots + \frac{1}{z^n} + \frac{1}{z^{n-1}} + \cdots + \frac{1}{z} + \frac{z}{2^2} + \cdots + \frac{z^n}{2^{n+1}} + \cdots$$

does not have an essential singular point at $z = 0$, although it contains infinitely many negative powers of z. Why doesn't this contradict the corollary to Theorem 11.6?

Hint. The series does not converge in a deleted neighborhood of $z = 0$.

Problem 2 (M2, Sec. 3). Given a domain G, let $f(z)$ be a (single-valued) function with the property that every point of G is either a regular point or an isolated singular point of $f(z)$. Let $g(z)$ be another function with the same property. Make a detailed investigation of the singular points of $f(z) \pm g(z)$, $f(z)g(z)$ and $f(z)/g(z)$ for the case where

a) $f(z)$ and $g(z)$ are both analytic on G;
b) $f(z)$ and $g(z)$ are allowed to have poles in G;
c) $f(z)$ but not $g(z)$ is allowed to have essential singular points as well as poles in G;
d) $f(z)$ and $g(z)$ are both allowed to have essential singular points as well as poles in G.

57. BEHAVIOR AT AN ESSENTIAL SINGULAR POINT. PICARD'S THEOREM

The complicated behavior of a function at an essential singular point is shown by

THEOREM 11.7.[4] *If z_0 is an essential singular point of $f(z)$, then, given any complex number A (finite or infinite), there exists a sequence of points $\{z_n\}$ converging to z_0 such that*

$$\lim_{n\to\infty} f(z_n) = A. \tag{11.27}$$

Proof. We can assume that $A \neq \infty$, since the theorem is obvious if $A = \infty$ [$f(z)$ is unbounded in any deleted neighborhood of z_0]. Suppose the theorem is false, i.e., suppose that in an arbitrary deleted neighborhood $\mathcal{N}'(z_0)$ of the essential singular point z_0 it is impossible to find points at which the values of $f(z)$ are arbitrarily close to A. Then there must exist a number $\alpha > 0$ such that $|f(z) - A| > \alpha$ for all $z \in \mathcal{N}'(z_0)$. It follows that for all $z \in \mathcal{N}'(z_0)$ the function

$$\varphi(z) = \frac{1}{f(z) - A}$$

is analytic and satisfies the inequality

$$|\varphi(z)| = \frac{1}{|f(z) - A|} < \frac{1}{\alpha}.$$

Therefore, according to Theorem 11.4, z_0 is a regular point of $\varphi(z)$, and the value of $\varphi(z)$ at z_0 must equal

$$\lim_{z\to z_0} \frac{1}{f(z) - A}.$$

[4] Often called the *Casorati-Weierstrass theorem.*

But $f(z)$ cannot be bounded in any deleted neighborhood of z_0, and therefore this limit can only equal zero, i.e., $\varphi(z)$ has a zero at z_0. Therefore $f(z) - A$ has a pole at z_0, and the same is true of $f(z)$ itself. This contradicts the assumption that z_0 is an essential singular point, thereby proving the theorem.

COROLLARY. *Let z_0 be an essential singular point of $f(z)$, and let E_δ be the set of values taken by $f(z)$ in the neighborhood $0 < |z - z_0| < \delta$. Then the closure of E_δ, i.e., the set consisting of E_δ and all limit points of E_δ, coincides with the extended complex plane.*

Example 1. The function

$$f(z) = \sin \frac{1}{z}$$

has an essential singular point at the origin. In fact, as $z \to 0$, $\sin (1/z)$ approaches no limit (finite or infinite), as can be seen at once by just considering real values of z. If $A = \infty$, the sequence $\{z_n\} = \{i/n\}$ satisfies the condition (11.27), since

$$f(z_n) = \sin \frac{1}{z_n} = \sin (-in) = -i \sinh n \to \infty \quad \text{as} \quad n \to \infty.$$

If $A \neq \infty$, we solve the equation

$$\sin \frac{1}{z} = A,$$

obtaining

$$\frac{1}{z} = \text{Arc} \sin A = \frac{1}{i} \text{Ln} (iA + \sqrt{1 - A^2})$$

(cf. Sec. 32, Prob. 3b) or

$$z = \frac{i}{\text{Ln} (iA + \sqrt{1 - A^2})} = \frac{i}{\ln (iA + \sqrt{1 - A^2}) + 2k\pi i}.$$

Setting

$$z_n = \frac{i}{\ln (iA + \sqrt{1 - A^2}) + 2n\pi i} \qquad (n = 1, 2, \ldots),$$

we obtain a sequence $\{z_n\}$ which converges to zero and satisfies the condition (11.27), in fact the much stronger condition $f(z_n) = A$ $(n = 1, 2, \ldots)$.

Example 2. The function

$$f(z) = e^{1/z}$$

has an essential singular point at the origin, since, as already noted on p. 235, $f(z)$ approaches no limit as $z \to 0$. If $A = \infty$, the sequence $\{z_n\} = \{1/n\}$ satisfies the condition (11.27), since

$$f(z_n) = e^n \to \infty \quad \text{as} \quad n \to \infty.$$

If $A = 0$, the sequence $\{z_n\} = \{-1/n\}$ is a sequence of the required type, since then

$$f(z_n) = e^{-n} \to 0 \quad \text{as} \quad n \to \infty.$$

Finally, assuming that $A \neq 0$, $A \neq \infty$, we solve the equation

$$e^{1/z} = A.$$

The result is

$$\frac{1}{z} = \operatorname{Ln} A,$$

which implies

$$z = \frac{1}{\operatorname{Ln} A} = \frac{1}{\ln A + 2k\pi i}.$$

Then, setting

$$z_n = \frac{1}{\ln A + 2n\pi i} \qquad (n = 1, 2, \ldots),$$

we obtain a sequence $\{z_n\}$ which converges to zero and satisfies the condition (11.27), in fact the much stronger condition $f(z_n) = A$ $(n = 1, 2, \ldots)$.

Remark. In each of the above examples, except for certain special values of A ($A = \infty$ in Example 1, $A = \infty$, $A = 0$ in Example 2), we can find a sequence $\{z_n\}$ which not only satisfies the requirement

$$\lim_{n \to \infty} f(z_n) = A$$

of Theorem 11.7, but also the much stronger condition of exact equality

$$f(z_n) = A \qquad (n = 1, 2, \ldots).$$

This is no accident, in view of the following result, known as *Picard's* (*great*) *theorem*, whose proof will not be given here:[5] *If z_0 is an essential singular point of $f(z)$, then, given any complex number $A \neq \infty$, with the possible exception of a single value $A = A_0$, there exists a sequence of points $\{z_n\}$ converging to z_0 such that*

$$f(z_n) = A \qquad (n = 1, 2, \ldots).$$

In other words, the set E_δ figuring in the corollary to Theorem 11.7 coincides with the finite complex plane, with the possible exception of a single point A_0 (which is independent of δ). It is easy to see that this theorem implies Theorem 11.7. In Example 1, there is no exceptional value A_0, while in Example 2, $A_0 = 0$, since the function $e^{1/z}$ never vanishes.

Problem 1. Prove that Theorem 11.7 remains valid if z_0 is a limit point of poles.

---

[5] For the proof, see M3, Sec. 51.

Comment. Accordingly, a limit point of poles is sometimes classified as an essential singular point. In any event, a limit point of poles is regarded as a (nonisolated) singular point.

Problem 2. Verify Picard's theorem for each of the functions e^z and $\cos(1/z)$ What are the exceptional values, if any?

58. BEHAVIOR AT INFINITY

Suppose $f(z)$ is analytic at every point of the unbounded domain $|z| > r$, except possibly at the point $z = \infty$. Then, making the preliminary transformation

$$\zeta = \frac{1}{z},$$

which carries the point $z = \infty$ into the point $\zeta = 0$ and every sequence $\{z_n\}$ converging to ∞ into a sequence $\{\zeta_n\} = \{1/z_n\}$ converging to 0, we reduce the study of $f(z)$ to that of the function

$$f^*(\zeta) = f(1/\zeta),$$

which is analytic in a deleted neighborhood of $\zeta = 0$.

DEFINITION. *We say that $z = \infty$ is a regular point, a pole of order k or an essential singular point of $f(z)$, depending on whether $\zeta = 0$ is a regular point, a pole of order k or an essential singular point of $f^*(\zeta)$.*

It is an immediate consequence of the definition that if $z = \infty$ is a regular point of $f(z)$, then

$$f^*(\zeta) = a_0 + a_{-1}\zeta + \cdots + a_{-n}\zeta^n + \cdots,$$
$$f(z) = a_0 + a_{-1}z^{-1} + \cdots + a_{-n}z^{-n} + \cdots.$$

Similarly, if $z = \infty$ is a pole of order k of $f(z)$,

$$f^*(\zeta) = a_k\zeta^{-k} + \cdots + a_1\zeta^{-1} + a_0 + a_{-1}\zeta + \cdots,$$
$$f(z) = a_kz^k + \cdots + a_1z + a_0 + a_{-1}z^{-1} + \cdots,$$

where $a_k \neq 0$, while if $z = \infty$ is an essential singular point of $f(z)$,

$$f^*(\zeta) = \sum_{n=-\infty}^{\infty} a_n\zeta^{-n},$$

$$f(z) = \sum_{n=-\infty}^{\infty} a_nz^n,$$

where infinitely many of the positive powers of z must be nonzero. In other words, the situation is the same as for finite points, except that the roles of positive and negative powers are now interchanged. Correspondingly, the principal part of a Laurent expansion at infinity consists of all terms with positive powers of z, and the regular part consists of all terms with non-positive powers of z.

As we know, it is not actually necessary to study the Laurent expansion of $f^*(\zeta)$ at $\zeta = 0$ in order to determine the behavior of $f^*(\zeta)$ at $\zeta = 0$. In fact, $f^*(\zeta)$ has a regular point at $\zeta = 0$ if $f^*(\zeta)$ is bounded in a deleted neighborhood $\mathscr{N}'(0)$, a pole at $\zeta = 0$ if

$$\lim_{\zeta \to 0} f^*(\zeta) = \infty,$$

and an essential singular point at $\zeta = 0$ if

$$\lim_{\zeta \to 0} f^*(\zeta)$$

fails to exist. Correspondingly, $f(z)$ has a regular point at $z = \infty$ if $f(z)$ is bounded in a deleted neighborhood $\mathscr{N}'(\infty)$, a pole at $z = \infty$ if

$$\lim_{z \to \infty} f(z) = \infty,$$

and an essential singular point at $z = \infty$ if

$$\lim_{z \to \infty} f(z)$$

fails to exist.

Example. Suppose $f(z)$ is an entire function, so that $f(z)$ is analytic at every finite point and has the power series expansion

$$f(z) = a_0 + a_1 z + \cdots + a_n z^n + \cdots \tag{11.28}$$

which converges for all finite z, The series (11.28) converges in a neighborhood $\mathscr{N}'(\infty)$ and can therefore be regarded as the Laurent expansion of $f(z)$ at $z = \infty$. There are now just three possibilities:

1. $f(z)$ is identically constant, i.e.,
$$f(z) \equiv a_0$$
and has a regular point at ∞;
2. $f(z)$ reduces to a polynomial of degree k, i.e.,
$$f(z) = a_0 + a_1 z + \cdots + a_k z^k$$
and has a pole of order k at ∞;
3. $f(z)$ has an essential singular point at ∞.

In the third case, we say that $f(z)$ is an entire *transcendental* function. In other words, an entire transcendental function approaches no limit (finite or infinite) as $z \to \infty$.[6]

---

[6] An entire transcendental function was previously defined as an entire function, like e^z, which is not a polynomial (cf. p. 111). We now see that it is not a coincidence that e^z approaches no limit as $z \to \infty$ (cf. p. 117).

The following result will be needed in Sec. 61:

THEOREM 11.8 *If z_0 is a regular point of $g(z)$ and an A-point or pole of $f(z)$, where $A \neq 0$, then z_0 is a simple pole of the function*

$$g(z)\frac{d}{dz}\operatorname{Ln}[f(z) - A] = g(z)\frac{f'(z)}{f(z) - A}. \tag{11.29}$$

In fact, the Laurent expansion of (11.29) *has the form*

$$\frac{\alpha g(z_0)}{z - z_0} + \cdots$$

at every A-point of order α of $f(z)$, and the form

$$-\frac{\beta g(z_0)}{z - z_0} + \cdots$$

at every pole of order β of $f(z)$.

Proof. To keep things simple, we write only one leading term, using dots to indicate all higher-order terms. Suppose z_0 is an A-point of order α of $f(z)$. Then in a neighborhood of z_0 we have

$$g(z) = g(z_0) + \cdots,$$

$$f(z) - A = a(z - z_0)^\alpha + \cdots,$$

$$f'(z) = a\alpha(z - z_0)^{\alpha-1} + \cdots.$$

Denoting the function (11.29) by $F(z)$, we find that

$$F(z_0) = [g(z_0) + \cdots]\frac{a\alpha(z - z_0)^{\alpha-1} + \cdots}{a(z - z_0)^\alpha + \cdots}$$

$$= \frac{\alpha}{z - z_0}[g(z_0) + \cdots]\frac{1 + \cdots}{1 + \cdots}$$

$$= \frac{\alpha}{z - z_0}[g(z_0) + \cdots] = \frac{\alpha g(z_0)}{z - z_0} + \cdots,$$

as asserted.

On the other hand, suppose z_0 is a pole of order β of $f(z)$. Then z_0 is also a pole of order β of $f(z) - A$, and in a deleted neighborhood of z_0,

$$f(z) - A = b(z - z_0)^{-\beta} + \cdots,$$

$$f'(z) = -\beta b(z - z_0)^{-\beta-1} - \cdots,$$

so that, in particular, z_0 is a pole of order $\beta + 1$ of $f'(z)$. It follows that

$$F(z_0) = [g(z_0) + \cdots]\frac{-\beta b(z - z_0)^{-\beta-1} - \cdots}{b(z - z_0)^{-\beta} + \cdots}$$

$$= \frac{-\beta}{z - z_0}[g(z_0) + \cdots]\frac{1 + \cdots}{1 + \cdots}$$

$$= -\frac{\beta}{z - z_0}[g(z_0) + \cdots] = -\frac{\beta g(z_0)}{z - z_0} + \cdots,$$

and the proof is complete.

Problem 1. Find the singular points and investigate the behavior at infinity of the following functions:

a) $\dfrac{1}{z - z^3}$; b) $\dfrac{z^4}{1 + z^4}$; c) $\dfrac{z^5}{(1 - z)^2}$; d) $\dfrac{1}{z(z^2 + 4)^2}$; e) $\dfrac{e^z}{1 + z^2}$;

f) $\dfrac{z^2 + 1}{e^z}$; g) ze^{-z}; h) $\dfrac{1}{e^z - 1} - \dfrac{1}{z}$; i) $\dfrac{e^z}{z(1 - e^{-z})}$;

j) $\dfrac{1 - e^z}{1 + e^z}$; k) $\tanh z$; l) e^{-1/z^2}; m) $ze^{1/z}$; n) $e^{z/(1-z)}$;

o) $e^{z-(1/z)}$; p) $\dfrac{e^{1/(z-1)}}{e^z - 1}$.

Ans.

a) Simple poles at 0, ± 1; ∞ is a regular point, in fact a zero of order 3;
d) Simple pole at 0 and poles of order 2 at $\pm 2i$; ∞ is a regular point, in fact a zero of order 5;
g) ∞ is an essential singular point;
j) Simple poles at $(2k + 1)\pi i$, where $k = 0, \pm 1, \pm 2, \ldots$; ∞ is a limit point of poles;
m) 0 is an essential singular point; ∞ is a simple pole;
p) Simple poles at $2k\pi i$, where $k = 0, \pm 1, \pm 2, \ldots$; 1 is an essential singular point; ∞ is a limit point of poles.

Problem 2. Find the singular points and investigate the behavior at infinity of the following functions:

a) $\dfrac{\cos z}{z^2}$; b) $\tan z$; c) $\tan^2 z$; d) $\dfrac{\cot z}{z^2}$; e) $\cot z - \dfrac{1}{z}$; f) $\cot z - \dfrac{2}{z}$;

g) $\dfrac{1}{\sin z - \sin a}$; h) $\dfrac{1}{\cos z + \cos a}$; i) $\sin\dfrac{1}{1 - z}$; j) $\cot\dfrac{1}{z}$;

k) $\cot\dfrac{1}{z} - \dfrac{1}{z}$; l) $\sin\dfrac{1}{z} + \dfrac{1}{z^2}$; m) $e^{-z}\cos\dfrac{1}{z}$; n) $e^{\cot(1/z)}$; o) $e^{\tan(1/z)}$;

p) $\sin\left(\dfrac{1}{\sin\dfrac{1}{z}}\right)$; q) $\sin\left(\dfrac{1}{\cos\dfrac{1}{z}}\right)$.

Ans.

b) Simple poles at $(2k + 1)\dfrac{\pi}{2}$, where $k = 0, \pm 1, \pm 2, \ldots$; ∞ is a limit point of poles;
e) Simple poles at $k\pi$, where $k = 0, \pm 1, \pm 2, \ldots$; ∞ is a limit point of poles;
h) Simple poles at $(2k + 1)\pi \pm a$, where $k = 0, \pm 1, \pm 2, \ldots$, if a is not an integral multiple of π, but otherwise poles of order 2 at $2k\pi$ if a is an odd multiple of π or at $(2k + 1)\pi$ if a is an even multiple of π; in every case, ∞ is a limit point of poles;
k) Simple poles at $1/k\pi$, where $k = \pm 1, \pm 2, \ldots$, and a limit point of poles at 0; ∞ is a simple pole;
n) Essential singular points at $1/k\pi$, where $k = \pm 1, \pm 2, \ldots$, and a limit point of essential singular points at 0; ∞ is an essential singular point;
q) Essential singular points at $2/(2k + 1)\pi$, where $k = 0, \pm 1, \pm 2, \ldots$, and a limit point of essential singular points at 0; ∞ is a regular point.

CHAPTER 12

THE RESIDUE THEOREM AND ITS IMPLICATIONS

59. THE RESIDUE THEOREM. RESIDUES AT INFINITY

Given a closed rectifiable Jordan curve L, suppose $f(z)$ is analytic on $\overline{I(L)}$ except for isolated singular points belonging to $I(L)$.[1] Our aim is to evaluate the integral

$$\int_L f(z)\, dz.$$

First we note that $I(L)$ can contain no more than a finite number of singular points of $f(z)$. In fact, if $I(L)$ contained infinitely many singular points, these points would have a limit point $\zeta \in \overline{I(L)}$. But then ζ would be a nonisolated singular point (why?), contrary to hypothesis.

DEFINITION. *Suppose $f(z)$ is analytic in a deleted neighborhood $\mathcal{N}'(z_0)$ of an isolated singular point z_0. Then by the residue of $f(z)$ at z_0, denoted by*

$$\operatorname*{Res}_{z=z_0} f(z),$$

[1] As usual, $I(L)$ denotes the interior of L and $\overline{I(L)}$ denotes the closure of $I(L)$, i.e., the union of $I(L)$ and L.

we mean the coefficient a_{-1} in the Laurent expansion

$$f(z) = \sum_{n=-\infty}^{\infty} a_n(z - z_0)^n, \qquad z \in \mathscr{N}'(z_0).$$

THEOREM 12.1 (*Residue theorem*). *Given a closed rectifiable Jordan curve L, if $f(z)$ is analytic on $\overline{I(L)}$, except for isolated singular points $z_1, \ldots, z_n$ belonging to $I(L)$, then*

$$\int_L f(z)\, dz = 2\pi i \sum_{k=1}^{n} \operatorname*{Res}_{z=z_k} f(z).$$

Proof. Around each point z_k $(k = 1, \ldots, n)$, we draw a circle

$$\gamma_k : |z - z_k| = \rho_k,$$

choosing the radii ρ_k so small that

1. $I(L)$ contains every circle γ_k;
2. For every $k = 1, \ldots, n$, $E(\gamma_k)$ contains γ_j $(j \neq k)$.

Then it follows from Cauchy's integral theorem for a system of contours (Theorem 7.3) that

$$\int_L f(z)\, dz = \int_{\gamma_1} f(z)\, dz + \cdots + \int_{\gamma_n} f(z)\, dz,$$

where all the integrals are taken in the positive (counterclockwise) direction.

Now suppose the Laurent expansion of $f(z)$ at z_k is

$$f(z) = \sum_{m=-\infty}^{\infty} a_m^{(k)}(z - z_k)^m \qquad (k = 1, \ldots, n). \tag{12.1}$$

Then, integrating (12.1) term by term along γ_k (which is permissible because of the uniform convergence of the series on γ_k), we find that

$$\int_{\gamma_k} f(z)\, dz = \int_{\gamma_k} \sum_{m=-\infty}^{\infty} a_m^{(k)}(z - z_k)^m\, dz = \sum_{m=-\infty}^{\infty} a_m^{(k)} \int_{\gamma_k} (z - z_k)^m\, dz$$
$$(k = 1, \ldots, n). \tag{12.2}$$

Since

$$\int_{\gamma_k} (z - z_k)^m\, dz = \int_0^{2\pi} (\rho_k e^{i\theta})^m\, d(\rho_k e^{i\theta})$$
$$= i\rho_k^{m+1} \int_0^{2\pi} e^{i(m+1)\theta}\, d\theta = \begin{cases} 2\pi i & \text{if } m = -1, \\ 0 & \text{otherwise,} \end{cases}$$

(12.2) reduces to

$$\int_{\gamma_k} f(z)\, dz = 2\pi i a_{-1}^{(k)} \qquad (k = 1, \ldots, n). \tag{12.3}$$

Comparing (12.1) and (12.3), we have

$$\int_L f(z)\, dz = \sum_{k=1}^{m} \int_{\gamma_k} f(z)\, dz = 2\pi i \sum_{k=1}^{n} a_{-1}^{(k)} = 2\pi i \sum_{k=1}^{n} \operatorname*{Res}_{z=z_k} f(z),$$

and the theorem is proved.

Before we can apply Theorem 12.1, we must learn how to calculate residues. Suppose z_0 is a simple pole of $f(z)$, so that

$$f(z) = \frac{a_{-1}}{z - z_0} + a_0 + a_1(z - z_0) + \cdots$$

in some deleted neighborhood of z_0. Then

$$(z - z_0)f(z) = a_{-1} + a_0(z - z_0) + a_1(z - z_0)^2 + \cdots,$$

and hence

$$a_{-1} = \operatorname*{Res}_{z=z_0} f(z) = \lim_{z \to z_0} [(z - z_0)f(z)]. \tag{12.4}$$

The calculation is particularly simple if $f(z)$ has the form

$$f(z) = \frac{\varphi(z)}{\psi(z)},$$

where $\varphi(z_0) \neq 0$ and $\psi(z)$ has a simple zero at $z = z_0$, i.e., $\psi(z_0) = 0$, $\psi'(z_0) \neq 0$. Then z_0 is a simple pole of $f(z)$, and according to (12.4),

$$\operatorname*{Res}_{z=z_0} f(z) = \operatorname*{Res}_{z=z_0} \frac{\varphi(z)}{\psi(z)} = \lim_{z \to z_0} \frac{(z - z_0)\varphi(z)}{\psi(z)} = \lim_{z \to z_0} \frac{\varphi(z)}{\dfrac{\psi(z) - \psi(z_0)}{z - z_0}} = \frac{\varphi(z_0)}{\psi'(z_0)}. \tag{12.5}$$

On the other hand, if z_0 is a pole of order $k > 1$ of $f(z)$, the Laurent expansion of $f(z)$ at z_0 is of the form

$$f(z) = \frac{a_{-k}}{(z - z_0)^k} + \cdots + \frac{a_{-1}}{z - z_0} + a_0 + a_1(z - z_0) + \cdots,$$

which implies

$$\begin{aligned}(z - z_0)^k f(z) = a_{-k} + \cdots + a_{-1}(z - z_0)^{k-1} \\ + a_0(z - z_0)^k + a_1(z - z_0)^{k+1} + \cdots.\end{aligned} \tag{12.6}$$

Differentiating (12.6) $k - 1$ times, we obtain

$$\begin{aligned}&\frac{d^{k-1}}{dz^{k-1}} [(z - z_0)^k f(z)] \\ &\qquad = (k - 1)!\, a_{-1} + \frac{k!}{1!} a_0(z - z_0) + \frac{(k + 1)!}{2!} a_1(z - z_0)^2 + \cdots,\end{aligned}$$

which implies

$$(k - 1)!\, a_{-1} = \lim_{z \to z_0} \frac{d^{k-1}}{dz^{k-1}} [(z - z_0)^k f(z)]$$

or

$$a_{-1} = \operatorname*{Res}_{z=z_0} f(z) = \frac{1}{(k - 1)!} \lim_{z \to z_0} \frac{d^{k-1}}{dz^{k-1}} [(z - z_0)^k f(z)]. \tag{12.7}$$

Next let $f(z)$ be analytic in a deleted neighborhood of the point at infinity, so that $f(z)$ has a Laurent expansion of the form

$$f(z) = \cdots + \frac{a_{-m}}{z^m} + \cdots + \frac{a_{-1}}{z} + a_0 + a_1 z + \cdots + a_n z^n + \cdots$$
$$(R < |z| < \infty).$$

Suppose we integrate $f(z)$ along the circle $\gamma_\rho : |z| = \rho$, where $\rho > R$, choosing the following convention for the positive direction of traversing γ_ρ: An observer moving along γ_ρ in the positive direction finds the *exterior* of γ_ρ, and hence the point at infinity, on his left. In other words, the positive direction of traversing a curve surrounding the point at infinity is the *clockwise direction*.[2] It follows that

$$\int_{\gamma_\rho} f(z)\,dz = \int_{\gamma_\rho} \frac{a_{-1}}{z}\,dz = ia_{-1}\int_0^{-2\pi} d\theta = -2\pi i a_{-1}.$$

This suggests defining the quantity $-a_{-1}$ as the *residue of $f(z)$ at infinity*, since then

$$\int_{\gamma_\rho} f(z)\,dz = 2\pi i \operatorname*{Res}_{z=\infty} f(z),$$

in keeping with the residue theorem.

Remark. It should be noted that the residue of $f(z)$ at ∞ is determined by the coefficient of one of the terms of the *regular part* of the Laurent expansion of $f(z)$ at ∞, whereas the residue of $f(z)$ at a finite point z_0 is determined by the coefficient of one of the terms of the *principal part* of the Laurent expansion of $f(z)$ at z_0. (Recall that in defining regular and principal parts at ∞, as opposed to z_0, the roles of positive and negative powers of z are interchanged.) It follows that the residue of $f(z)$ at ∞ can be nonzero even when ∞ is a regular point of $f(z)$. For example, ∞ is a regular point of the function

$$f(z) = \frac{1}{z},$$

in fact, a simple zero, but

$$\operatorname*{Res}_{z=\infty} f(z) = -1 \neq 0.$$

THEOREM 12.2. *If $f(z)$ is analytic except at isolated singular points, then the sum of all the residues of $f(z)$ equals zero.*

[2] Previously, in dealing with a contour L surrounding one or more finite points, we chose the *counterclockwise* direction as positive, since then an observer moving along L finds the *interior* of L, and hence the given finite points, on his left.

Proof. The function $f(z)$ can only have a finite number of singular points, since otherwise the singular points of $f(z)$ would have a limit point ζ (possibly at infinity), and then ζ would be a nonisolated singular point of $f(z)$, contrary to hypothesis. Thus there is a circle $\gamma_\rho : |z| = \rho$ whose radius ρ is so large that all the finite singular points $z_1, \ldots, z_n$ of $f(z)$ lie inside γ_ρ. Then, according to the residue theorem,

$$\int_{\gamma_\rho} f(z)\,dz = 2\pi i \sum_{k=1}^{n} \operatorname*{Res}_{z=z_k} f(z), \tag{12.8}$$

where γ_ρ is traversed in the usual positive direction for finite points, i.e., in the counterclockwise direction. But relative to the point at infinity, this is the negative direction of traversing γ_ρ, and hence the same integral equals

$$\int_{\gamma_\rho} f(z)\,dz = -2\pi i \operatorname*{Res}_{z=\infty} f(z). \tag{12.9}$$

Subtracting (12.9) from (12.8), we find that

$$2\pi i\Big[\operatorname*{Res}_{z=z_1} f(z) + \cdots + \operatorname*{Res}_{z=z_n} f(z) + \operatorname*{Res}_{z=\infty} f(z)\Big] = 0,$$

or

$$\operatorname*{Res}_{z=z_1} f(z) + \cdots + \operatorname*{Res}_{z=z_n} f(z) + \operatorname*{Res}_{z=\infty} f(z) = 0,$$

as asserted.

Example. Theorem 12.2 applies to any rational function, since a rational function has only isolated singular points, namely, its poles.

Problem 1. Find the residues of $f(z)$ at all its isolated singular points and at infinity (if infinity is not a limit point of singular points), where $f(z)$ is given by

a) $\dfrac{1}{z^3 - z^5}$; b) $\dfrac{1}{z(1 - z^2)}$; c) $\dfrac{z^{2n}}{(1 + z)^n}$ (n a positive integer);

d) $\dfrac{z^2}{(z^2 + 1)^2}$; e) $\dfrac{\sin 2z}{(z + 1)^3}$; f) $\dfrac{e^z}{z^2(z^2 + 9)}$;

g) $\cot^2 z$; h) $\cot^3 z$; i) $\cos \dfrac{1}{z - 2}$;

j) $z^3 \cos \dfrac{1}{z - 2}$; k) $e^{z+(1/z)}$; l) $\sin z \sin \dfrac{1}{z}$.

Ans.

a) $\operatorname*{Res}_{z=\pm 1} f(z) = -\frac{1}{2}$, $\operatorname*{Res}_{z=0} f(z) = 1$, $\operatorname*{Res}_{z=\infty} f(z) = 0$;

d) $\operatorname*{Res}_{z=i} f(z) = -\dfrac{i}{4}$, $\operatorname*{Res}_{z=-i} f(z) = \dfrac{i}{4}$, $\operatorname*{Res}_{z=\infty} f(z) = 0$;

g) $\operatorname*{Res}_{z=k\pi} f(z) = 0$ ($k = 0, \pm 1, \pm 2, \ldots$);

j) $\operatorname*{Res}_{z=2} f(z) = -\operatorname*{Res}_{z=\infty} f(z) = -\frac{143}{24}$.

Problem 2. Find the residues of all the single-valued branches of $f(z)$ at the indicated point, where $f(z)$ is the multiple-valued function defined by

a) $\dfrac{\sqrt{z}}{1-z}$ at $z = 1$; b) $\dfrac{1}{\sqrt{2-z}+1}$ at $z = 1$;

c) $\sqrt{(z-a)(z-b)}$ at $z = \infty$; d) $\dfrac{z^a}{1-\sqrt{z}}$ at $z = 1$;

e) $\operatorname{Ln}\dfrac{z-a}{z-b}$ at $z = \infty$; f) $e^z \operatorname{Ln}\dfrac{z-a}{z-b}$ at $z = \infty$.

Ans. a) ± 1; c) $\pm\dfrac{(a-b)^2}{8}$; e) $a - b$ for all branches.

Problem 3. Calculate

$$\operatorname*{Res}_{z=\infty} [f(z)]^2,$$

given that $f(z)$ has a Laurent expansion of the form

$$a_0 + \frac{a_{-1}}{z} + \frac{a_{-2}}{z^2} + \cdots$$

at infinity.

Problem 4. Calculate

$$\operatorname*{Res}_{z=a} f(z)g(z),$$

given that $f(z)$ is analytic at $z = a$, while

a) $g(z)$ has a simple pole with residue A at $z = a$;

b) $g(z)$ has a pole of order k with principal part

$$\frac{a_{-1}}{z-a} + \cdots + \frac{a_{-k}}{(z-a)^k}$$

at $z = a$.

Problem 5. Calculate

$$\operatorname*{Res}_{z=a} \varphi[f(z)],$$

given that

a) $f(z)$ is analytic at $z = a$, with $f'(a) \neq 0$, while $\varphi(\zeta)$ has a simple pole with residue A at the point $\zeta = f(a)$;

b) $f(z)$ has a simple pole with residue A at $z = a$, while $\varphi(\zeta)$ has a simple pole at infinity with principal part $B\zeta$.

Problem 6. Avoiding excessive calculation, evaluate the integral

$$\int_L \frac{dz}{(z-3)(z^5-1)},$$

where L is the circle $|z| = 2$.

Hint. Use Theorem 12.2.

60. JORDAN'S LEMMA. EVALUATION OF DEFINITE INTEGRALS

The residue theorem is a powerful tool for the evaluation of definite integrals of real functions, especially integrals with infinite limits. We now illustrate this technique, known as "the calculus of residues," after first proving the following useful

LEMMA (*Jordan's lemma*). *Given a family of circular arcs*

$$\gamma_R : |z| = R, \qquad \operatorname{Im} z \geqslant -a,$$

where a is a fixed real number, let f(z) be a continuous function defined on every γ_R *such that*

$$\lim_{R\to\infty} M(R) = \lim_{R\to\infty} \max_{z\in\gamma_R} |f(z)| = 0.$$

Then

$$\lim_{R\to\infty} \int_{\gamma_R} f(z)e^{i\lambda z}\, dz = 0$$

for every positive λ.

Proof. If

$$\alpha(R) = \operatorname{arc\,sin} \frac{a}{R},$$

then

$$\lim_{R\to\infty} \alpha(R) = 0,$$

and moreover

$$\lim_{R\to\infty} R\alpha(R) = a$$

(why?). Suppose first that $a > 0$. Then on the arc AB shown in Figure 12.1 we have

$$|e^{i\lambda z}| = e^{-\lambda y} \leqslant e^{\lambda\alpha(R)},$$

and hence

$$\left| \int_{AB} f(z)e^{i\lambda z}\, dz \right| \leqslant M(R)e^{\lambda\alpha(R)}R\alpha(R),$$

where the right-hand side approaches zero as $R \to \infty$. Therefore

$$\lim_{R\to\infty} \int_{AB} f(z)e^{i\lambda z}\, dz = 0,$$

and similarly

$$\lim_{R\to\infty} \int_{CD} f(z)e^{i\lambda z}\, dz = 0.$$

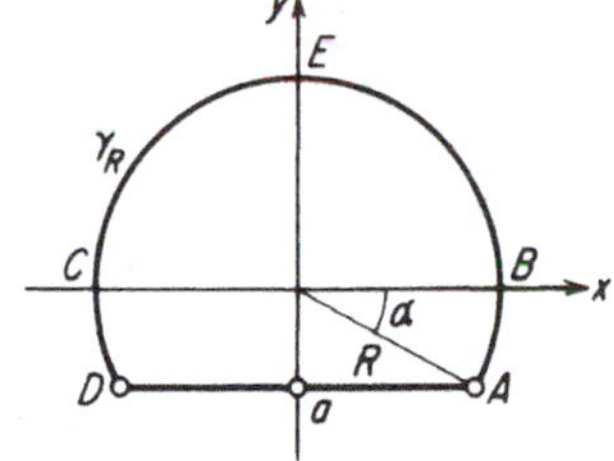

FIGURE 12.1

On the other hand, using the inequality (7.36), we find that

$$|e^{i\lambda z}| = |e^{i\lambda Re^{i\theta}}| = e^{-\lambda R \sin\theta} \leqslant e^{-2\lambda R\theta/\pi}$$

on the arc BE, and hence

$$\left|\int_{BE} f(z)e^{i\lambda z}\,dz\right| \leqslant RM(R)\int_0^{\pi/2} e^{-2\lambda R\theta/\pi}\,d\theta = \frac{\pi M(R)}{2\lambda}(1 - e^{-\lambda R}),$$

which implies

$$\lim_{R\to\infty} \int_{BE} f(z)e^{i\lambda z}\,dz = 0,$$

and similarly

$$\lim_{R\to\infty} \int_{EC} f(z)e^{i\lambda z}\,dz = 0.$$

Therefore

$$\lim_{R\to\infty} \int_{\gamma_R} f(z)e^{i\lambda z}\,dz = \lim_{R\to\infty} \left(\int_{AB} + \int_{BE} + \int_{EC} + \int_{CD}\right) f(z)e^{i\lambda z}\,dz = 0,$$

and the lemma is proved. If $a \leqslant 0$, the proof is even simpler, since then there is no need to estimate the integrals along AB and CD.

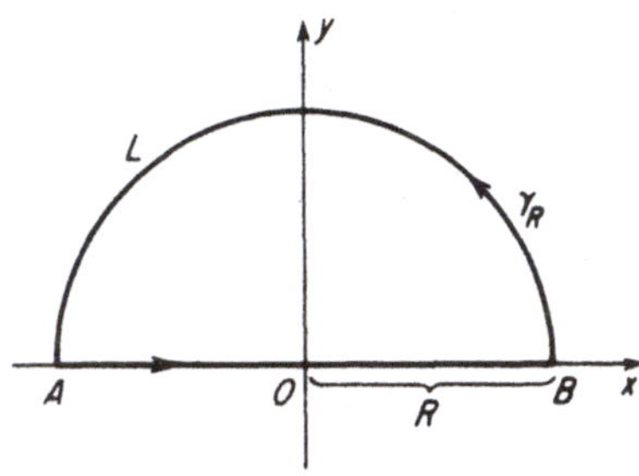

FIGURE 12.2

Example 1. *Evaluate the integral*

$$\mathscr{I} = \int_0^\infty \frac{\cos \lambda x}{x^2 + a^2}\,dx$$

$$(a > 0, \quad \lambda > 0).$$

Consider the function

$$f(z) = \frac{e^{i\lambda z}}{z^2 + a^2}$$

and the contour L shown in Figure 12.2, consisting of the segment $AB\colon -R \leqslant x \leqslant R$ of the real axis and a semicircular arc γ_R of radius R. According to Theorem 12.1,

$$\left(\int_{AB} + \int_{\gamma_R}\right) \frac{e^{i\lambda z}}{z^2 + a^2}\,dz = 2\pi i \operatorname*{Res}_{z=ai} \frac{e^{i\lambda z}}{z^2 + a^2} = \frac{\pi e^{-\lambda a}}{a},$$

provided that $R > a$. But if $z \in \gamma_R$, then

$$\left|\frac{1}{z^2 + a^2}\right| \leqslant \frac{1}{R^2 - a^2} \to 0 \quad \text{as} \quad R \to \infty,$$

and hence the integral along γ_R vanishes as $R \to \infty$. It follows that

$$\lim_{R\to\infty} \int_{AB} \frac{e^{i\lambda z}}{z^2 + a^2}\,dz = \lim_{R\to\infty} \int_{-R}^{R} \frac{e^{i\lambda x}}{x^2 + a^2}\,dx = \frac{\pi e^{-\lambda a}}{a},$$

and hence

$$\int_{-\infty}^{\infty} \frac{e^{i\lambda x}}{x^2 + a^2}\,dx = \frac{\pi e^{-\lambda a}}{a}. \tag{12.10}$$

Taking the real part of (12.10) and using the fact that $\cos \lambda x$ is an even function, we obtain

$$\mathscr{I} = \int_0^\infty \frac{\cos \lambda x}{x^2 + a^2}\, dx = \frac{\pi e^{-\lambda a}}{2a} \qquad (a > 0, \quad \lambda > 0).$$

Example 2. *Evaluate the integral*

$$\mathscr{I} = \int_{-\infty}^{\infty} \frac{e^{ax}}{1 + e^x}\, dx \qquad (0 < a < 1).$$

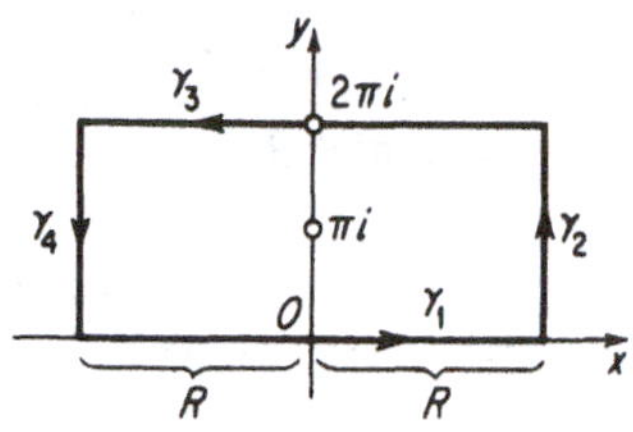

FIGURE 12.3

This time we start from the function

$$f(z) = \frac{e^{az}}{1 + e^z}$$

and the rectangular contour L shown in Figure 12.3 consisting of the line segments $\gamma_1, \gamma_2, \gamma_3, \gamma_4$. According to Theorem 12.1,

$$\left(\int_{\gamma_1} + \int_{\gamma_2} + \int_{\gamma_3} + \int_{\gamma_4}\right) f(z)\, dz = 2\pi i \operatorname*{Res}_{z=\pi i} \frac{e^{az}}{1 + e^z} = -2\pi i e^{a\pi i}$$

[use (12.5)]. Clearly

$$\int_{\gamma_1} f(z)\, dz = \int_{-R}^{R} \frac{e^{ax}}{1 + e^x}\, dx,$$

$$\int_{\gamma_3} f(z)\, dz = \int_{R}^{-R} \frac{e^{a(x+2\pi i)}}{1 + e^{x+2\pi i}}\, dx$$

$$= -e^{2a\pi i} \int_{-R}^{R} \frac{e^{ax}}{1 + x}\, dx,$$

while

$$|f(z)| = \left|\frac{e^{a(R+iy)}}{1 + e^{R+iy}}\right| \leqslant \frac{e^{aR}}{e^R - 1} = \frac{e^{(a-1)R}}{1 - e^{-R}} \qquad (z \in \gamma_2),$$

$$|f(z)| = \left|\frac{e^{a(-R+iy)}}{1 + e^{-R+iy}}\right| \leqslant \frac{e^{-aR}}{1 - e^{-R}} \qquad (z \in \gamma_4).$$

Therefore, as $R \to \infty$, the integrals along γ_2 and γ_4 vanish (recall that $0 < a < 1$). It follows that

$$\lim_{R\to\infty} \left(\int_{\gamma_1} + \int_{\gamma_2} + \int_{\gamma_3} + \int_{\gamma_4}\right) f(z)\, dz = (1 - e^{2a\pi i}) \int_{-\infty}^{\infty} \frac{e^{ax}}{1 + e^x}\, dx = -2\pi i e^{a\pi i},$$

and hence finally

$$\mathscr{I} = \int_{-\infty}^{\infty} \frac{e^{ax}}{1 + e^x}\, dx = \frac{-2\pi i e^{a\pi i}}{1 - e^{2a\pi i}} = \frac{\pi}{\sin a\pi} \qquad (0 < a < 1).$$

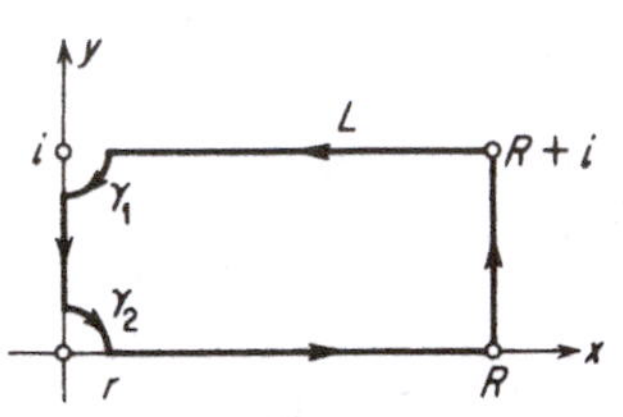

FIGURE 12.4

Example 3. Evaluate the integral

$$\mathscr{I} = \int_0^\infty e^{-\pi x} \frac{\sin ax}{\sinh \pi x}\, dx \qquad (a > 0).$$

Consider the function

$$f(z) = \frac{e^{iaz}}{e^{2\pi z} - 1}$$

and the contour L shown in Figure 12.4, where γ_1 and γ_2 are circular arcs of radius r. Since $f(x + i) = e^{-a}f(x)$, the integrals along the upper and lower horizontal line segments combine to give

$$(1 - e^{-a}) \int_r^R \frac{e^{iax}}{e^{2\pi x} - 1}\, dx,$$

while the integral along the line segment joining the points R and $R + i$ goes to zero as $R \to \infty$ (verify this). Therefore, since $f(z)$ has no singular points inside L, Theorem 12.1 implies

$$(1 - e^{-a}) \int_r^R \frac{e^{iax}}{e^{2\pi x} - 1}\, dx - i \int_r^{1-r} \frac{e^{-ay}}{e^{2\pi iy} - 1}\, dy + \left(\int_{\gamma_1} + \int_{\gamma_2} \right) f(z)\, dz + \delta(R) = 0, \tag{12.11}$$

where $\delta(R) \to 0$ as $R \to \infty$. Near $z = i$ we have

$$f(z) = \frac{e^{iaz}}{e^{2\pi(z-i)} - 1} = \frac{e^{ia} + \alpha(z - i) + \cdots}{2\pi(z - i) + \beta(z - i)^2 + \cdots}$$
$$= \frac{e^{ia}}{2\pi} \frac{1}{z - i} + g(z),$$

where $g(z)$ is regular at $z = i$. Since

$$z = i + re^{i\theta}, \quad dz = ire^{i\theta}\, d\theta \qquad (z \in \gamma_1),$$

we have

$$\int_{\gamma_1} f(z)\, dz = \frac{e^{-a}}{2\pi} \int_0^{-\pi/2} i\, d\theta + \varepsilon(r) = -\frac{ie^{-a}}{4} + \varepsilon(r),$$

where $\varepsilon(r) \to 0$ as $r \to 0$ (why?). Similarly,

$$\int_{\gamma_2} f(z)\, dz = \frac{1}{2\pi} \int_{\pi/2}^0 i\, d\theta + \eta(r),$$

where $\eta(r) \to 0$ as $r \to 0$. Therefore (12.11) implies

$$(1 - e^{-a}) \int_r^R \frac{e^{iax}}{e^{2\pi x} - 1}\, dx = i \int_r^{1-r} \frac{e^{-ay}}{e^{2\pi iy} - 1}\, dy + \frac{i}{4}(1 + e^{-a}) + \delta(R) + \varepsilon(r) + \eta(r),$$

and hence

$$(1 - e^{-a}) \int_0^\infty \frac{e^{iax}}{e^{2\pi x} - 1}\, dx = i \int_0^1 \frac{e^{-ay}}{e^{2\pi i y} - 1}\, dy + \frac{i}{4}(1 + e^{-a}), \qquad (12.12)$$

after passing to the limit as $r \to 0$, $R \to \infty$. Finally, taking the imaginary part of (12.12) and using the fact that

$$\operatorname{Re} \int_r^{1-r} \frac{e^{-ay}}{e^{2\pi i y} - 1}\, dy = - \int_r^{1-r} \frac{e^{-ay}}{2}\, dy = \frac{1}{2a}(e^{-a} - 1) + \omega(r),$$

where $\omega(r) \to 0$ as $r \to 0$, we obtain

$$\mathscr{I} = \int_0^\infty e^{-\pi x} \frac{\sin ax}{\sinh \pi x}\, dx = \frac{1}{2} \frac{1 + e^{-a}}{1 - e^{-a}} - \frac{1}{a} \qquad (a > 0).$$

Our next example involves a multiple-valued function:

Example 4. *Evaluate the integral*

$$\mathscr{I} = \int_0^\infty \frac{\ln x}{(x^2 + 1)^2}\, dx.$$

Consider the function

$$f(z) = \frac{\operatorname{Ln} z}{(z^2 + 1)^2}$$

and the contour L shown in Figure 12.5, where γ_r and γ_R are semicircular

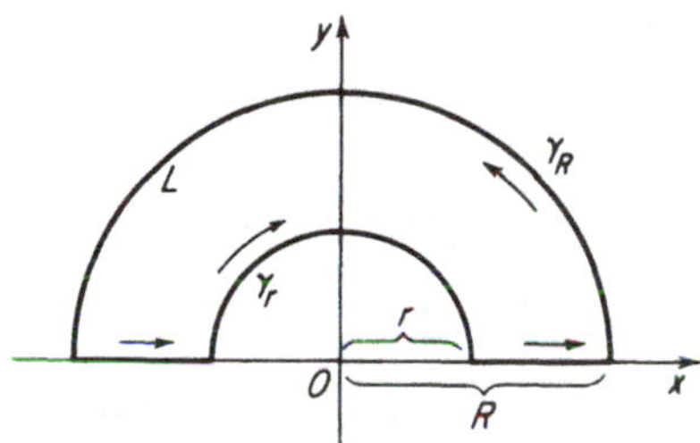

FIGURE 12.5

arcs of radius r and R ($> r$), respectively. Since $I(L)$ does not contain the branch points of Ln z, we can define single-valued branches on $I(L)$. Choosing the principal value $\ln z$ and noting that $f(z)$ has a pole of order 2 at $z = i$, we deduce from Theorem 12.1 that

$$\left(\int_{-R}^{-r} + \int_{\gamma_r} + \int_r^R + \int_{\gamma_R}\right) f(z)\, dz = 2\pi i \operatorname*{Res}_{z=i} f(z) = 2\pi i \left[\frac{d}{dz} \frac{\ln z}{(z + i)^2}\right]_{z=i}$$

$$= 2\pi i\left(\frac{\pi + 2i}{8}\right) = \frac{\pi^2 i}{4} - \frac{\pi}{2} \qquad (12.13)$$

[use (12.7)]. If $z = Re^{i\theta}$, $0 \leqslant \theta \leqslant \pi$, then

$$|\ln z| = \sqrt{\ln^2 R + \theta^2} \leqslant 2 \ln R$$

for sufficiently large R, and hence

$$\left|\int_{\gamma_R} f(z)\,dz\right| < \frac{2\ln R}{(R^2-1)^2}\,\pi R \to 0 \quad \text{as} \quad R \to \infty.$$

Similarly, if $z = re^{i\theta}$, $0 \leqslant \theta \leqslant \pi$, then

$$|\ln z| \leqslant 2 \ln \frac{1}{r}$$

for sufficiently small r, and hence

$$\left|\int_{\gamma_r} f(z)\,dz\right| \leqslant \frac{2\ln \dfrac{1}{r}}{(1-r^2)^2}\,\pi r \to 0 \quad \text{as} \quad r \to 0.$$

Noting that

$$\int_{-R}^{-r} \frac{\ln x}{(x^2+1)^2}\,dx = \int_r^R \frac{\ln x + \pi i}{(x^2+1)^2}\,dx$$

and taking the limit as $r \to 0$, $R \to \infty$, we deduce from (12.13) that

$$2\int_0^\infty \frac{\ln x}{(x^2+1)^2}\,dx + \pi i \int_0^\infty \frac{dx}{(x^2+1)^2} = \frac{\pi^2 i}{4} - \frac{\pi}{2}.$$

Comparing real parts, we finally obtain

$$\mathscr{I} = \int_0^\infty \frac{\ln x}{(x^2+1)^2}\,dx = -\frac{\pi}{4}$$

(how about imaginary parts?).

Our final example concerns an integral with finite limits:

Example 5. *Evaluate the integral*

$$\mathscr{I} = \int_0^{2\pi} \frac{d\varphi}{(a + b\cos\varphi)^2} \qquad (a > b > 0).$$

Setting

$$z = e^{i\varphi}, \quad d\varphi = \frac{dz}{iz}, \quad \cos\varphi = \frac{1}{2}\left(z + \frac{1}{z}\right),$$

we transform $\mathscr{I}$ into an integral along the unit circle:

$$\mathscr{I} = \frac{4}{ib^2}\int_{|z|=1} \frac{z\,dz}{\left(z^2 + \dfrac{2a}{b}z + 1\right)^2}.$$

The integrand has one pole of order 2 inside the unit circle, at the point

$$z_0 = \frac{\sqrt{a^2 - b^2} - a}{b},$$

with residue

$$\left[\frac{d}{dz}\frac{z}{\left(z + \dfrac{\sqrt{a^2 - b^2} + a}{b}\right)^2}\right]_{z=z_0} = \frac{b^2 a}{4(a^2 - b^2)^{3/2}}.$$

Therefore, according to Theorem 12.1,

$$\mathscr{I} = \int_0^{2\pi} \frac{d\varphi}{(a + \cos\varphi)^2} = 2\pi i \frac{4}{ib^2}\frac{b^2 a}{4(a^2 - b^2)^{3/2}} = \frac{2\pi a}{(a^2 - b^2)^{3/2}} \qquad (a > b > 0).$$

Problem 1. Evaluate the following integrals:

a) $\displaystyle\int_0^{2\pi} \frac{d\varphi}{a + \cos\varphi} \quad (a > 1)$; b) $\displaystyle\int_0^{2\pi} \frac{d\varphi}{(a + b\cos^2\varphi)^2} \quad (a > 0, b > 0)$.

Problem 2. Use Jordan's lemma to evaluate

a) $\displaystyle\int_{-\infty}^{\infty} \frac{x\cos x}{x^2 - 2x + 10}\,dx$; b) $\displaystyle\int_{-\infty}^{\infty} \frac{x\sin x}{x^2 + 4x + 20}\,dx$.

Ans. b) $\dfrac{\pi}{2e^4}(2\cos 2 + \sin 2)$.

Problem 3. Evaluate the following integrals:

a) $\displaystyle\int_0^{\infty} \frac{x\cos ax}{\sinh x}\,dx$; b) $\displaystyle\int_0^{\infty} \frac{\cosh ax}{\cosh \pi x}\,dx \quad (-\pi < a < \pi)$;

c) $\displaystyle\int_0^{\infty} \frac{\sin ax}{\sinh x}\,dx$.

Problem 4. Prove that

$$\int_0^{\infty} \frac{\ln x}{(x + a)^2 + b^2}\,dx = \frac{\ln\sqrt{a^2 + b^2}}{b}\arctan\frac{b}{a},$$

where a and b are real.

Problem 5. Prove that

$$\int_0^{\infty} \frac{x^{a-1}}{1 + x}\,dx = \frac{\pi}{\sin a\pi} \qquad (0 < a < 1).$$

Problem 6. Prove that

$$\int_{-1}^{1} \frac{dx}{\sqrt[3]{(1 - x)(1 + x)^2}} = \frac{2\pi}{\sqrt{3}}.$$

61. THE ARGUMENT PRINCIPLE. THE THEOREMS OF ROUCHÉ AND HURWITZ

Our next result is an immediate consequence of Theorems 11.8 and 12.1:

THEOREM 12.3. *Given a closed rectifiable Jordan curve L, suppose $g(z)$ is analytic on $\overline{I(L)}$, while $f(z)$ is analytic on $\overline{I(L)}$ except for poles in $I(L)$ at the points $b_1, \ldots, b_n$. Moreover, suppose $f(z)$ has A-points $a_1, \ldots, a_m$ in $I(L)$, but none on L itself. Then*

$$\frac{1}{2\pi i}\int_L g(z)\frac{f'(z)}{f(z)-A}\,dz = \sum_{k=1}^{m}\alpha_k g(a_k) - \sum_{k=1}^{n}\beta_k g(b_k), \qquad (12.14)$$

where α_k is the order of a_k and β_k the order of b_k.

The first term in the right-hand side of (12.14) is the sum of the values of $g(z)$ at the A-points of $f(z)$, where each value is repeated a number of times equal to the order of the A-point. If we assume that each A-point of $f(z)$ is counted a number of times equal to its order, then the sum

$$\sum_{k=1}^{m}\alpha_k g(a_k)$$

can be regarded as just the sum of the values of $g(z)$ at the A-points of $f(z)$. Similarly, the sum

$$\sum_{k=1}^{n}\beta_k g(b_k)$$

can be regarded as the sum of the values of $g(z)$ at the poles of $f(z)$, if each pole of $f(z)$ is counted a number of times equal to its order. With this convention (which will be in force from now on), we can paraphrase Theorem 12.3 as follows:

THEOREM 12.3′. *Given a closed rectifiable Jordan curve L, suppose $g(z)$ is analytic on $\overline{I(L)}$, while $f(z)$ is analytic on $\overline{I(L)}$ except for poles in $I(L)$ and has no A-points on L. Then the integral*

$$\frac{1}{2\pi i}\int_L g(z)\frac{f'(z)}{f(z)-A}\,dz \qquad (12.15)$$

equals the sum of the values of $g(z)$ at the A-points of $f(z)$ minus the sum of the values of $g(z)$ at the poles of $f(z)$.

Example 1. If $g(z) = z$, then

$$\frac{1}{2\pi i}\int_L \frac{zf'(z)}{f(z)-A}\,dz = \sum_{k=1}^{m}\alpha_k a_k - \sum_{k=1}^{n}\beta_k b_k,$$

i.e., the integral (12.15) is just the sum of the A-points of $f(z)$ inside L minus the sum of the poles of $f(z)$ inside L.

Example 2. If $g(z) = 1$, then

$$\frac{1}{2\pi i}\int_L \frac{f'(z)}{f(z) - A}\,dz = \sum_{k=1}^{m}\alpha_k - \sum_{k=1}^{n}\beta_k, \tag{12.16}$$

where the quantity on the right equals the number of A-points of $f(z)$ inside L minus the number of poles of $f(z)$ inside L.

Now suppose $f(z)$ has N zeros and P poles inside L, where each zero and pole is counted a number of times equal to its order, as already described. (In particular, N or P may vanish.) Then, setting $A = 0$ in (12.16), we have

$$\frac{1}{2\pi i}\int_L \frac{d}{dz}\operatorname{Ln} f(z)\,dz = \frac{1}{2\pi i}\int_L \frac{f'(z)}{f(z)}\,dz = N - P. \tag{12.17}$$

The integral on the left is called the *logarithmic residue* of $f(z)$ relative to the contour L. In other words, *the number of zeros of $f(z)$ inside L minus the number of poles of $f(z)$ inside L equals the logarithmic residue of $f(z)$ relative to L.*

The logarithmic residue of $f(z)$ relative to L has a simple geometric interpretation. Choosing any point $z_0 \in L$ as the initial and final point of the path of integration, we make one circuit around L in the positive (i.e., counterclockwise) direction. Then $\operatorname{Ln} f(z)$ varies continuously, and in general returns to z_0 with a value different from its original value at z_0. In fact, since

$$\operatorname{Ln} f(z) = \ln |f(z)| + i \operatorname{Arg} f(z),$$

the change in $\operatorname{Ln} f(z)$ is entirely due to the change in $\operatorname{Arg} f(z)$. Letting Φ_0 denote the original value of $\operatorname{Arg} f(z_0)$ and Φ_1 its value after the circuit around L, we have

$$\begin{aligned}\frac{1}{2\pi i}\int_L \frac{d}{dz}\operatorname{Ln} f(z)\,dz &= \frac{1}{2\pi i}[\ln |f(z_0)| + i\Phi_1] - \frac{1}{2\pi i}[\ln |f(z_0)| + i\Phi_0]\\ &= \frac{\Phi_1 - \Phi_0}{2\pi}.\end{aligned} \tag{12.18}$$

Comparing (12.17) and (12.18), we find that[3]

$$N - P = \frac{\Phi_1 - \Phi_0}{2\pi} = \frac{1}{2\pi}\Delta_L \operatorname{Arg} f(z). \tag{12.19}$$

In other words, the number of zeros of $f(z)$ inside L minus the number of poles of $f(z)$ inside L equals $1/2\pi$ times the change in $\operatorname{Arg} f(z)$ when the contour L is traversed once in the positive direction.

---

[3] By $\Delta_L h(z)$ is mean the change in $h(z)$ as z traverses the curve L. This quantity is sometimes denoted by $\operatorname*{Var}_{z\in L} h(z)$.

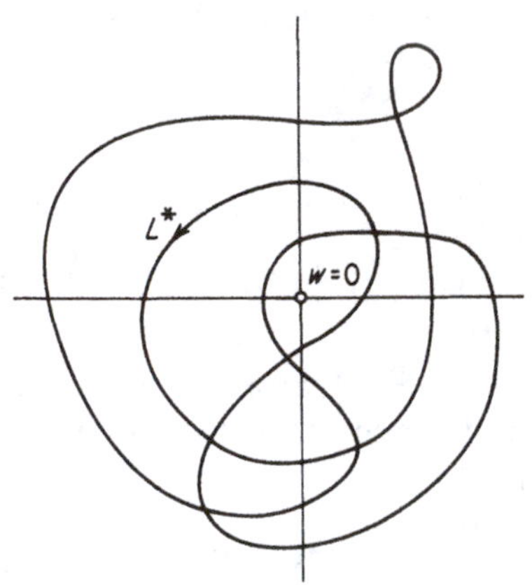

FIGURE 12.6

There is still another way of interpreting this result: As a variable point z describes the closed curve L once in the positive direction, the image point $w = f(z)$ describes a closed curve $L^* = f(L)$ in the w-plane, making some number ν of complete circuits around the origin $w = 0$, where every circuit made in the positive direction is counted as $+1$ and every circuit made in the negative direction is counted as -1. For example, $\nu = 1$ in Figure 12.6. It is easy to see that

$$\Delta_L \operatorname{Arg} f(z) = 2\pi\nu,$$

and therefore the number of zeros of $f(z)$ inside L minus the number of poles of $f(z)$ inside L equals the number of circuits around the origin made by the point $w = f(z)$ as the point z traverses the curve L once in the positive direction.

Making the obvious generalization to the case where $A \neq 0$ in (12.16), we summarize these results in the form of

THEOREM 12.4 (*Argument principle*). *Given a closed rectifiable Jordan curve L, suppose $f(z)$ is analytic on $\overline{I(L)}$ except for poles in $I(L)$ and has no A-points on L. Then the number of A-points of $f(z)$ inside L minus the number of poles of $f(z)$ inside L equals the number of circuits around the point $w = A$ made by the point $w = f(z)$ as the point z traverses the curve L once in the positive direction.*

We now turn to some implications of the argument principle:

THEOREM 12.5 (*Rouché's theorem*). *Given a closed rectifiable Jordan curve L, suppose $f(z)$ and $g(z)$ are analytic on $\overline{I(L)}$, and suppose that*

$$|f(z)| > |g(z)| \tag{12.20}$$

at every point of L. Then $f(z)$ and $f(z) + g(z)$ have the same number of zeros inside L.

Proof. Since $f(z)$ cannot vanish on L, because of (12.20), we have

$$\begin{aligned} \Delta_L \operatorname{Arg}\,[f(z) + g(z)] &= \Delta_L\left\{\operatorname{Arg} f(z)\left[1 + \frac{g(z)}{f(z)}\right]\right\} \\ &= \Delta_L \operatorname{Arg} f(z) + \Delta_L \operatorname{Arg}\left[1 + \frac{g(z)}{f(z)}\right]. \end{aligned} \tag{12.21}$$

Moreover, since

$$\left|\frac{g(z)}{f(z)}\right| < 1$$

for all $z \in L$, the variable point

$$w = 1 + \frac{g(z)}{f(z)}$$

stays in the disk $|w - 1| < 1$ as z describes the curve L. Therefore w cannot wind around the origin, which means that

$$\Delta_L \operatorname{Arg}\left[1 + \frac{g(z)}{f(z)}\right] = 0. \tag{12.22}$$

Combining (12.21) and (12.22), we find that

$$\Delta_L \operatorname{Arg}[f(z) + g(z)] = \Delta_L \operatorname{Arg} f(z),$$

and the theorem is now an immediate consequence of the argument principle.

THEOREM 12.6 (*Hurwitz's theorem*). *Given a closed rectifiable Jordan curve L, suppose $\{f_n(z)\}$ is a sequence of analytic functions on $\overline{I(L)}$ which converges uniformly on $\overline{I(L)}$ to a function $f(z) \not\equiv 0$. If L passes through no zeros of $f(z)$, then there is an integer $N(L) > 0$ such that every function $f_n(z)$ with $n > N(L)$ has the same number of zeros inside L as the function $f(z)$.*

Proof. Let μ be the minimum of $|f(z)|$ on L; by hypothesis, $\mu > 0$. Therefore, since $\{f_n(z)\}$ is uniformly convergent on L, we can find an integer $N(L) > 0$ such that

$$|f_n(z) - f(z)| < \mu \leqslant |f(z)|$$

for all $n > N(L)$ and $z \in L$. But then, according to Rouché's theorem, the functions $f(z)$ and

$$f(z) + [f_n(z) - f(z)] = f_n(z), \qquad n > N(L)$$

have the same number of zeros inside L, as asserted.

Example. *Find the number of roots of the equation*

$$z^8 - 4z^5 + z^2 - 1 = 0 \tag{12.23}$$

of absolute value less than 1.

Writing the left-hand side of (12.23) in the form $f(z) + g(z)$, where

$$f(z) = -4z^5, \qquad g(z) = z^8 + z^2 - 1,$$

we see that $|f(z)| > |g(z)|$ on the circle $|z| = 1$, since

$$|f(z)| = |4z^5| = 4, \qquad |g(z)| = |z^8 + z^2 - 1| \leqslant |z^8| + |z^2| + 1 = 3$$

if $|z| = 1$. Therefore, according to Rouché's theorem, the function

$$f(z) + g(z) = z^8 - 4z^5 + z^2 - 1$$

has the same number of zeros inside the circle $|z| = 1$ as the function $f(z) = -4z^5$. But $f(z)$ obviously has five zeros inside $|z| = 1$, in fact, a zero of order five at the origin (recall our way of counting zeros and poles). In other words, equation (12.33) has five roots inside the unit circle, i.e., five roots of absolute value less than 1.

Problem 1. Find the number of zeros of each of the following polynomials inside the unit circle $|z| = 1$:

a) $z^7 - 5z^4 + z^2 - 2$; b) $z^8 - 5z^5 - 2z + 1$;

c) $z^9 - 2z^6 + z^2 - 8z - 2$; d) $2z^5 - z^3 + 3z^2 - z + 8$.

Problem 2. How many roots has the equation

a) $z^4 - 5z + 1 = 0$ in the annulus $1 < |z| < 2$;

b) $z^4 - 8z + 10 = 0$ in the annulus $1 < |z| < 3$?

Problem 3. How many roots does the equation $e^z = az^n$ have inside the circle $|z| = R$, given that $|a| > e^R/R^n$?

Problem 4. Prove that the equation

$$z + e^{-z} = \lambda \qquad (\lambda > 1)$$

has a unique real root in the right half-plane.

Problem 5. Prove that, given any $\rho > 0$, the function

$$f_n(z) = 1 + \frac{1}{z} + \frac{1}{2!\,z^2} + \cdots + \frac{1}{n!\,z^n}$$

has all its zeros inside the circle $|z| = \rho$ if n is sufficiently large.

Hint. Use Hurwitz's theorem.

Problem 6 (M2, p. 118). Let $\{f_n(z)\}$ be a sequence of one-to-one analytic functions defined on a domain G, and suppose $\{f_n(z)\}$ is uniformly convergent on every compact subset of G. Using Hurwitz's theorem, prove that the limit function

$$f(z) = \lim_{n \to \infty} f_n(z)$$

is also one-to-one (and analytic) on G.

62. LOCAL BEHAVIOR OF ANALYTIC MAPPINGS. THE MAXIMUM MODULUS PRINCIPLE AND SCHWARZ'S LEMMA

Armed with Rouché's theorem, we now investigate the local behavior of the mapping produced by an analytic function $w = f(z)$:

THEOREM 12.7. *Let the function $f(z)$ be analytic at the point $z = z_0$, where it has a simple w_0-point.*[4] *Then there is a neighborhood $\mathcal{N}(z_0)$ [in the z-plane] and a corresponding neighborhood $\mathcal{N}(w_0)$ [in the w-plane] such that every point in $\mathcal{N}(w_0)$ has a unique inverse image in $\mathcal{N}(z_0)$.*

Proof. Let $\gamma: |z - z_0| = \rho$ be a circle such that $f(z)$ has no w_0-points in $\overline{I(\gamma)}$ except at the point z_0 itself. Such a circle must exist, since otherwise z_0 would be a limit point of w_0-points, and then, by the uniqueness theorem for analytic functions (Theorem 10.8), $f(z)$ would be identically equal to w_0, contrary to the hypothesis that z_0 is a *simple* w_0-point [so that $f'(z_0) \neq 0$]. Choose the neighborhood $\mathcal{N}(z_0)$ to be $I(\gamma)$. Let $\Gamma = f(\gamma)$ be the image of the circle γ under the mapping $w = f(z)$, let δ be the distance between w_0 and Γ, and choose $\mathcal{N}(w_0)$ to be the disk $|w - w_0| < \delta$. Then, if w is an arbitrary point of $\mathcal{N}(w_0)$,

$$|f(z) - w_0| \geqslant \delta > |w - w_0|$$

for all $z \in \gamma$, and hence, by Rouché's theorem, the two functions $f(z) - w_0$ and

$$f(z) - w = f(z) - w_0 + (w_0 - w)$$

have the same number of zeros inside γ. But $f(z) - w_0$ has only one zero inside γ, by construction, and therefore the same is true of $f(z) - w$, i.e., w has a unique inverse image in $\mathcal{N}(z_0)$.

COROLLARY. *Let the function $f(z)$ be analytic at the point $z = z_0$, where it has a simple w_0-point. Then the inverse function $f^{-1}(w)$ exists and is single-valued in a neighborhood of the point $w = w_0$.*

Remark. Of course, in general, $\mathcal{N}(z_0)$ contains points whose images do not lie in $\mathcal{N}(w_0)$.

Theorem 12.7 is easily generalized to the case of multiple w_0-points:

THEOREM 12.8. *Let the function $f(z)$ be analytic at the point $z = z_0$, where it has a w_0-point of order $k \geqslant 1$. Then there is a neighborhood $\mathcal{N}(z_0)$ and a corresponding neighborhood $\mathcal{N}(w_0)$ such that every point in $\mathcal{N}(w_0)$ has at least 1 and at most k distinct inverse images in $\mathcal{N}(z_0)$.*

---

[4] I.e., an A-point such that $A = w_0 = f(z_0)$.

Proof. Let γ, $\mathcal{N}(z_0)$, Γ, δ and $\mathcal{N}(w_0)$ be the same as in the proof of Theorem 12.7. Then, since $f(z) - w_0$ has k zeros inside γ (by construction),[5] the same is true of $f(z) - w$. However, $f(z)$ may have multiple w-points at points of $\mathcal{N}(z_0)$ other than z_0 itself.[6] Thus we can only assert that every $w \in \mathcal{N}(w_0)$ has at least 1 but no more than k distinct inverse images in $\mathcal{N}(z_0)$.

Corollary 1. *Let the function $f(z)$ be analytic at the point $z = z_0$, where it has a w_0-point of order $k \geqslant 1$. Then there is a neighborhood $\mathcal{N}(z_0)$ and a corresponding neighborhood $\mathcal{N}(w_0)$ such that every point in $\mathcal{N}(w_0)$ except w_0 itself has precisely k distinct inverse images in $\mathcal{N}(z_0)$.*

Proof. We need only make the circle γ figuring in the proof of Theorem 12.8 so small that *both* $f(z) - w_0$ and $f'(z)$ do not vanish at any point of $\overline{I(\gamma)}$ other than the point z_0 itself.

Corollary 2. *Let the function $f(z)$ be analytic at the point $z = z_0$, where it has a w_0-point of order $k \geqslant 1$. Then the inverse function $f^{-1}(w)$ exists and is k-valued in a neighborhood of the point $w = w_0$.*

Next we prove that the analytic image of a domain is a domain, in the following sense:

Theorem 12.9. *If $f(z) \not\equiv$ const is analytic on a domain G, then $f(G)$ is also a domain.*

Proof. Obviously $f(G)$ is nonempty. The fact that $f(G)$ is connected is almost trivial (see Prob. 5). To prove that $f(G)$ is open, let w_0 be an arbitrary point of $f(G)$ and let $z_0 \in G$ be any inverse image of w_0. Clearly z_0 is a w_0-point of order k of $f(z)$, where k is some positive integer. Therefore, by Theorem 12.8, there are neighborhoods $\mathcal{N}(w_0)$ and $\mathcal{N}(z_0) \subset G$ such that every point in $\mathcal{N}(w_0)$ is the image of a point in $\mathcal{N}(z_0)$. But this implies $\mathcal{N}(w_0) \subset f[\mathcal{N}(z_0)] \subset f(G)$, and hence w_0 is an interior point of $f(G)$. Thus $f(G)$ consists entirely of interior points, and the proof is complete.

Theorem 12.10 has the following important implication:

Theorem 12.11 (*Maximum modulus principle*). *If $f(z) \not\equiv$ const is analytic on a domain G, then $|f(z)|$ cannot have a maximum at any point of G.*[7]

---

[5] Note that now $f^{(k)}(z_0) \neq 0$.

[6] At every such point, the derivative $f'(z)$ vanishes, as it does at z_0.

[7] A real-valued function $h(z)$ defined on a set E is said to have a (*relative*) *maximum* at a point $z_0 \in E$ if there is a neighborhood $\mathcal{N}(z_0)$ such that $h(z_0) \geqslant h(z)$ for all $z \in \mathcal{N}(z_0) \cap E$. A *minimum* is defined similarly, with $\geqslant$ replaced by $\leqslant$.

Proof. Let z_0 be an arbitrary point of G, and let $w_0 = f(z_0)$. The image of any neighborhood $\mathscr{N}(z_0) \subset G$ is a domain containing w_0 and hence contains some neighborhood of w_0. Therefore $\mathscr{N}(z_0)$ contains a point z whose image $w = f(z)$ is further from the origin of the w-plane than the point w_0 itself, i.e., a point z such that $|f(z)| > |f(z_0)|$. In other words, $|f(z)|$ cannot have a maximum at z_0.

Finally we prove a consequence of the maximum modulus principle which will be needed later in studying conformal mapping (see Sec. 71), but is of considerable interest in its own right:

THEOREM 12.12 (*Schwarz's lemma*). *Let $f(z)$ be a function which is analytic on the disk $K: |z| < R$ and vanishes at $z = 0$, and suppose that*

$$|f(z)| \leqslant M < \infty \tag{12.24}$$

for all $z \in K$. Then the inequality

$$|f(z)| \leqslant \frac{M}{R}|z| \tag{12.25}$$

holds for all $z \in K$, and moreover

$$|f'(0)| \leqslant \frac{M}{R} \tag{12.26}$$

Equality is achieved in (12.25) [*for some nonzero $z \in K$*] *or in* (12.26) *if and only if $f(z)$ is a function of the form*

$$f(z) = \frac{M}{R}e^{i\alpha}z, \tag{12.27}$$

where α is a real number.

Proof. From the Taylor series expansion

$$f(z) = f'(0)z + \frac{f''(0)}{2!}z^2 + \cdots \qquad (z \in K)$$

[$f(0) = 0$ by hypothesis], it is apparent that the function

$$\varphi(z) = \frac{f(z)}{z} = f'(0) + \frac{f''(0)}{2!}z + \cdots$$

is analytic on the disk K and takes the value $f'(0)$ at the origin. Let γ_ρ be the circle $|z| = \rho$, where $0 < \rho < R$. Since the inequality

$$|\varphi(z)| = \frac{|f(z)|}{|z|} \leqslant \frac{M}{\rho}$$

holds for all $z \in \gamma_\rho$, it follows from the maximum modulus principle that

$$|\varphi(z)| \leqslant \frac{M}{\rho}$$

for all $z \in I(\gamma_\rho)$. Therefore, letting $\rho \to R$, we find that

$$|\varphi(z)| \leqslant \frac{M}{R} \tag{12.28}$$

for all $z \in K$. If $z \neq 0$, we substitute for $\varphi(z)$ in (12.28), obtaining

$$|f(z)| \leqslant \frac{M}{R}\,|z|.$$

This is the desired inequality (12.25), which is obviously valid for $z = 0$. Moreover $\varphi(0) = f'(0)$, and hence, when $z = 0$, (12.28) takes the form

$$|f'(0)| \leqslant \frac{M}{R},$$

which is just the inequality (12.26). Finally, we note that (12.25) can become an equality for some nonzero $z \in K$ or (12.26) can become an equality if and only if (12.28) becomes an equality for some $z \in K$. But according to the maximum modulus principle, this is only possible if $\varphi(z) \equiv$ const on K. Since this constant must have absolute value M/R, it follows that

$$\varphi(z) = \frac{M}{R}\,e^{i\alpha},$$

or equivalently that

$$f(z) = \frac{M}{R}\,e^{i\alpha}\,z.$$

Problem 1. Give an example of a function $w = f(z)$ with a simple w-point at every point of a domain G which is not one-to-one on G. Reconcile this with Theorem 12.7 and its corollary.

Problem 2 (M2, p. 90). Prove that Theorem 12.8 and its corollaries remain valid if $z_0 = \infty$.

Problem 3 (M2, p. 91). Prove that Theorem 12.8 and its corollaries remain valid if $w_0 = \infty$, i.e., if $f(z)$ has a pole of order $k \geqslant 1$ at z_0 (we now allow $z_0 = \infty$).

Problem 4 (M2, p. 116). Prove the following generalization of Theorem 4.2: Let $f(z)$ be analytic at the point $z = z_0$, where it has a w_0-point of order $k > 1$. Then, under the mapping $w = f(z)$, every angle with its vertex at z_0 is enlarged k times.

Problem 5. Let E be a connected set and let $f(z)$ be continuous on E. Prove that the set $f(E)$ is connected, i.e., that the continuous image of a connected set is connected.

Hint. If w_0, $w' \in f(E)$, let z_0, $z' \in E$ be inverse images of w_0, w'. Then join w_0 to w' by the continuous image of a curve joining z_0 to z'.

Problem 6. Give an example showing that if $f(z) \not\equiv$ const is merely continuous on a domain G, then $f(G)$ need not be a domain.

Comment. However, according to a theorem of plane set topology (see M1, Sec. 26), $f(G)$ is a domain if $f(z)$ is one-to-one on G as well as continuous on G.

Problem 7 (M2, p. 91). Prove that Theorem 12.9 remains valid if $f(z)$ is allowed to have poles in G, and if G and $f(G)$ are domains in the extended plane (see Sec. 12, Prob. 9).

Problem 8 (M1, p. 376). Prove the following *minimum modulus principle*: If $f(z) \not\equiv$ const is analytic on a domain G, then $|f(z)|$ cannot have a minimum at any point of G other than a zero of $f(z)$.

Problem 9 (M1, p. 376). Given a bounded domain G, let $f(z) \not\equiv$ const be analytic on G and continuous on $\bar{G}$, and suppose $|f(z)|$ has the same value at all points of the boundary of G. Prove that $f(z)$ has at least one zero at a point of G.

Problem 10. Let $f(z) \not\equiv$ const be analytic on a domain G. Prove that neither of the functions $\operatorname{Re} f(z)$ and $\operatorname{Im} f(z)$ can have a maximum or a minimum at a point of G.

Problem 11 (M1, p. 298). Given a bounded domain G, let $f(z) \not\equiv$ const be analytic on G and continuous on $\bar{G}$. Prove that there is a point ζ in the boundary of G such that

$$|f(\zeta)| = \max_{z \in \bar{G}} |f(z)|.$$

Problem 12 (M1, p. 298). Let $f(z) \not\equiv$ const be analytic on a domain G containing the circles

$$\gamma_\rho : |z - z_0| = \rho \qquad (0 < \rho < \rho_0)$$

and their interiors. Prove that the maximum modulus function

$$M(\rho) = \max_{z \in \gamma_\rho} |f(z)|$$

is a strictly increasing function of ρ in the interval $(0, \rho_0)$.

Problem 13 (M1, pp. 296, 298, 376). Prove the maximum modulus principle starting from the corollary to Theorem 8.4.

CHAPTER 13

HARMONIC FUNCTIONS

63. LAPLACE'S EQUATION. CONJUGATE HARMONIC FUNCTIONS

A function $u(x, y)$ of two real variables is said to be *harmonic* on a domain G if the second partial derivatives

$$\frac{\partial^2 u}{\partial x^2}, \quad \frac{\partial^2 u}{\partial x\, \partial y}, \quad \frac{\partial^2 u}{\partial y\, \partial x}, \quad \frac{\partial^2 u}{\partial y^2}$$

exist and are continuous on G, and if at every point of G, $u(x, y)$ satisfies the partial differential equation

$$\frac{\partial^2 u}{\partial x^2} + \frac{\partial^2 u}{\partial y^2} = 0, \tag{13.1}$$

known as *Laplace's equation.* Let $u(x, y)$ and $v(x, y)$ be two functions harmonic on a domain G, which satisfy the Cauchy-Riemann equations

$$\frac{\partial u}{\partial x} = \frac{\partial v}{\partial y}, \qquad \frac{\partial u}{\partial y} = -\frac{\partial v}{\partial x} \tag{13.2}$$

at every point of G. Then $u(x, y)$ and $v(x, y)$ are said to be *conjugate harmonic functions* (on G), and each of the functions $u(x, y)$, $v(x, y)$ is said to be the conjugate harmonic function (or simply the *harmonic conjugate*) of the other.

There is an intimate connection between harmonic functions and analytic functions, as shown by

THEOREM 13.1. *A necessary and sufficient condition for a function $f(z) = u(x, y) + iv(x, y)$ to be analytic on a domain G is that its real part $u(x, y)$ and imaginary part $v(x, y)$ be conjugate harmonic functions on G.*

Proof. To prove the necessity, suppose $f(z)$ is analytic on G, and hence differentiable at every point of G. Then, according to Theorem 3.2, the functions $u(x, y)$ and $v(x, y)$ are differentiable and satisfy the Cauchy-Riemann equations (13.2) at every point of G. The function $f'(z)$ is also analytic on G, being the derivative of a function analytic on G (recall Theorem 8.5, Corollary). Since

$$f'(z) = \frac{\partial u}{\partial x} - i\frac{\partial u}{\partial y} = \frac{\partial v}{\partial y} + i\frac{\partial v}{\partial x},$$

we see that

$$\frac{\partial u}{\partial x}, \qquad -\frac{\partial u}{\partial y} \tag{13.3}$$

and

$$\frac{\partial v}{\partial y}, \qquad \frac{\partial v}{\partial x} \tag{13.4}$$

are pairs of differentiable functions on G, satisfying the Cauchy-Riemann equations. Writing the first Cauchy-Riemann equation for (13.3), and the second for (13.4), we find that

$$\frac{\partial}{\partial x}\left(\frac{\partial u}{\partial x}\right) = \frac{\partial}{\partial y}\left(-\frac{\partial u}{\partial y}\right), \qquad \frac{\partial}{\partial y}\left(\frac{\partial v}{\partial y}\right) = -\frac{\partial}{\partial x}\left(\frac{\partial v}{\partial x}\right),$$

or

$$\frac{\partial^2 u}{\partial x^2} + \frac{\partial^2 u}{\partial y^2} = 0, \qquad \frac{\partial^2 v}{\partial x^2} + \frac{\partial^2 v}{\partial y^2} = 0.$$

Thus each of the functions $u(x, y)$ and $v(x, y)$ satisfies Laplace's equation (13.1). It only remains to show that $u(x, y)$ and $v(x, y)$ have *continuous* second derivatives. But this follows at once from the fact that $f''(z)$ is analytic on G (why?) and can be written in any of the forms

$$f''(z) = \frac{\partial^2 u}{\partial x^2} + i\frac{\partial^2 v}{\partial x^2} = -\frac{\partial^2 u}{\partial y^2} - i\frac{\partial^2 v}{\partial y^2}$$

$$= -i\frac{\partial^2 u}{\partial x\,\partial y} + \frac{\partial^2 v}{\partial x\,\partial y} = -i\frac{\partial^2 u}{\partial y\,\partial x} + \frac{\partial^2 v}{\partial y\,\partial x}.$$

To prove the sufficiency, we note that if $u(x, y)$ and $v(x, y)$ are conjugate harmonic functions, then, in particular, they have continuous first derivatives on G, and hence are differentiable on G (recall the remark on p. 51). Since $u(x, y)$ and $v(x, y)$ also satisfy the Cauchy-Riemann equations on G, it follows from Theorem 3.2 that $f(z) = u(x, y) + iv(x, y)$ is analytic on G.

Example 1. The function

$$f(z) = e^z = e^x(\cos y + i \sin y)$$

is analytic in the whole plane, and hence its real and imaginary parts

$$u(x, y) = e^x \cos y, \qquad v(x, y) = e^x \sin y \tag{13.5}$$

are harmonic in the whole plane. It is easily verified that both functions (13.5) satisfy Laplace's equation.

Example 2. Let G be the whole plane cut along the negative real axis (including the origin). Then every single-valued branch of the multiple-valued function

$$f(z) = \operatorname{Ln} z = \ln |z| + i \operatorname{Arg} z$$

is analytic on G. It follows that

$$u(x, y) = \ln |z| = \ln \sqrt{x^2 + y^2}, \qquad v(x, y) = \arg z + 2k\pi$$

is a pair of conjugate harmonic functions on G, for every $k = 0, \pm 1, \pm 2, \ldots$. If $v(x, y)$ is thought of as a multiple-valued function, then it has infinitely many "single-valued harmonic branches."

More generally, let

$$f(z) = \operatorname{Ln} \frac{z - z_1}{z - z_2},$$

and let G be the plane cut along the line segment joining z_1 to z_2. Then our harmonic functions become

$$u(x, y) = \ln \frac{z - z_1}{z - z_2}, \quad v(x, y) = i \arg \frac{z - z_1}{z - z_2} + 2k\pi$$
$$(k = 0, \pm 1, \pm 2, \ldots).$$

Example 3. Let $f(z)$ be single-valued and analytic on an annulus $D: \rho < |z - z_0| < R$. Then $f(z)$ has the Laurent expansion

$$f(z) = \sum_{n=-\infty}^{\infty} a_n(z - z_0)^n \quad . \ (z \in D),$$

or in polar coordinates,

$$f(z) = \sum_{n=-\infty}^{\infty} \rho_n r^n e^{i(n\theta + \varphi_n)} = \sum_{n=-\infty}^{\infty} \rho_n r^n [\cos (n\theta + \varphi_n) + i \sin (n\theta + \varphi_n)],$$

where we write

$$a_n = \rho_n e^{i\varphi_n}, \qquad z - z_0 = re^{i\theta}.$$

This implies

$$u(r, \theta) = \operatorname{Re} f(z) = \sum_{n=-\infty}^{\infty} \rho_n r^n \cos (n\theta + \varphi_n),$$

$$v(r, \theta) = \operatorname{Im} f(z) = \sum_{n=-\infty}^{\infty} \rho_n r^n \sin (n\theta + \varphi_n).$$

It is customary to write $u(x, y)$ and $v(x, y)$ for the expressions obtained from $u(r, \theta)$ and $v(r, \theta)$ after transforming back to rectangular coordinates.[1] Then the nth term of the series for $u(x, y)$ turns out to be a homogeneous polynomial of degree n in x and y, called a *harmonic polynomial* (see Prob. 3). From the standpoint of simplicity, the expressions in polar coordinates are preferable to those in rectangular coordinates.

Example 4. Later on, an important role will be played by the real and imaginary parts of the function

$$f(z) = \frac{\rho e^{i\varphi} + (z - z_0)}{\rho e^{i\varphi} - (z - z_0)},$$

which is analytic on the disk $K: |z - z_0| < \rho$. Writing $z - z_0 = re^{i\theta}$, and denoting the real and imaginary parts of $f(z)$ in polar coordinates by $u(r, \theta)$ and $v(r, \theta)$, as before, we have

$$u(r, \theta) + iv(r, \theta) = \frac{\rho e^{i\varphi} + re^{i\theta}}{\rho e^{i\varphi} - re^{i\theta}} = \frac{(\rho e^{i\varphi} + re^{i\theta})(\rho e^{-i\varphi} - re^{-i\theta})}{(\rho e^{i\varphi} - re^{i\theta})(\rho e^{-i\varphi} - re^{-i\theta})}$$

$$= \frac{\rho^2 - r^2 + 2i\rho r \sin (\theta - \varphi)}{\rho^2 + r^2 - 2\rho r \cos (\theta - \varphi)}.$$

The corresponding (single-valued) conjugate harmonic functions on K are

$$u(r, \theta) = \frac{\rho^2 - r^2}{\rho^2 + r^2 - 2\rho r \cos (\theta - \varphi)}, \qquad v(r, \theta) = \frac{2\rho r \sin (\theta - \varphi)}{\rho^2 + r^2 - 2\rho r \cos (\theta - \varphi)}.$$

Both expressions have the same denominator

$$(\rho e^{i\varphi} - re^{i\theta})(\rho e^{-i\varphi} - re^{-i\theta}) = |\rho e^{i\varphi} - re^{i\theta}|^2 = |\rho e^{i\varphi} - (z - z_0)|^2,$$

equal to the square of the distance between the point z and the point $z_0 + \rho e^{i\varphi}$ on the circle $|z - z_0| = \rho$.

Next we ask whether every function $u(x, y)$ harmonic on a given domain G can be regarded as the real (or imaginary) part of a function $f(z)$ analytic on G. In other words, can we always find another function $v(x, y)$ harmonic

---

[1] This entails the following slight abuse of notation:

$$u(x, y) = u(r, \theta), \qquad v(x, y) = v(r, \theta).$$

on G which is a harmonic conjugate of $u(x, y)$, i.e., which together with $u(x, y)$ satisfies the Cauchy-Riemann equations (13.2) on G? The answer to this question is in the affirmative, as shown by

THEOREM 13.2. *Given a function $u(x, y)$ harmonic on a simply connected domain G, then, to within an arbitrary real constant, the function*

$$v(x, y) = \int_{(x_0, y_0)}^{(x, y)} \left(-\frac{\partial u}{\partial y}\, dx + \frac{\partial u}{\partial x}\, dy \right) \tag{13.6}$$

is the unique harmonic conjugate of $u(x, y)$ on G, where the line integral is evaluated along any rectifiable curve $L \subset G$ joining the fixed point (x_0, y_0) to the variable point (x, y). Similarly, to within an arbitrary purely imaginary constant, the function

$$f(z) = u(x, y) + iv(x, y) = u(x, y) + i \int_{(x_0, y_0)}^{(x, y)} \left(-\frac{\partial u}{\partial y}\, dx + \frac{\partial u}{\partial x}\, dy \right) \tag{13.7}$$

is the unique analytic function on G with $u(x, y)$ as its real part.

Proof. Clearly, the proof reduces to finding all the solutions of the system

$$\frac{\partial v}{\partial x} = P(x, y), \qquad \frac{\partial v}{\partial y} = Q(x, y) \tag{13.8}$$

on G, where

$$P(x, y) = -\frac{\partial u}{\partial y}, \qquad Q(x, y) = \frac{\partial u}{\partial x} \tag{13.9}$$

are given functions with continuous first partial derivatives. In terms of the functions (13.9), Laplace's equation for $u(x, y)$ becomes

$$\frac{\partial Q}{\partial x} = \frac{\partial P}{\partial y}.$$

But this is just the condition for

$$v(x, y) = \int_{(x_0, y_0)}^{(x, y)} \{P(x, y)\, dx + Q(x, y)\, dy\} \tag{13.10}$$

to be a solution of (13.8) on G,[2] and the general solution of (13.8) can only differ from (13.10) by an additive constant (why?). The theorem now follows by combining (13.9) and (13.10).

Remark. Suppose G is multiply connected. Then the integral (13.6) will in general take different values for different paths joining (x_0, y_0) to (x, y),

[2] See e.g., D. V. Widder, *op. cit.*, Theorem 4, p. 227 and Theorem 5, p. 229, where it is assumed that L is a piecewise smooth Jordan curve. Actually, L can be an arbitrary rectifiable curve (see Prob. 4).

giving rise to a multiple-valued function $v(x, y)$, unless the path is confined to a simply connected subdomain of G. Let D be such a subdomain. Then, just as in the proof of Theorem 8.3, (13.6) defines single-valued branches of $v(x, y)$ on D, where the particular branch depends on the path taken from (x_0, y_0) to some fixed intermediate point (x_1, y_1) in D, and any two branches differ only by a real constant [cf. (8.5)]. In other words, if G is multiply connected, Theorem 13.2 remains valid if we replace G by any simply connected subdomain $D \subset G$ and interpret (13.6) as any single-valued branch of $v(x, y)$ on D. Correspondingly, we interpret (13.7) as any single-valued branch of $f(z)$ on D, where $f(z)$ is now multiple-valued. It should be noted, however, that all the branches of $f(z)$ have the same derivative

$$f'(z) = \frac{\partial u}{\partial x} - i\frac{\partial u}{\partial y},$$

i.e., $f'(z)$ is single-valued on G.

Problem 1. If u is harmonic, is the same true of u^2? More generally, if u is harmonic, for which functions f is $f(u)$ also harmonic?

Problem 2. Let $u(x, y)$ and $v(x, y)$ be a pair of conjugate harmonic functions, and let $\chi(u, v)$ be a harmonic function of the variables u and v. Prove that

$$\chi[u(x, y), v(x, y)]$$

is a harmonic function of the variables x and y.

Problem 3. Calculate the harmonic polynomials $p_n(x, y)$ and $q_n(x, y)$ defined by the equation

$$z^n = p_n + iq_n,$$

where $n = 1, 2, 3, 4$.

Problem 4. Suppose the integral (13.10) has a given value when evaluated along a piecewise smooth Jordan curve $L \subset G$ joining the points (x_0, y_0) and (x, y). Prove that it has the same value when evaluated along any rectifiable curve $L \subset G$.

Hint. After proving that $f(z)$ is analytic on G, exploit the fact that $f'(z)$ is also analytic on G. Alternatively, use the key lemma of Sec. 36.

Problem 5. Given a simply connected domain G and a function $v(x, y)$ harmonic on G, prove that to within an arbitrary real constant, there is a unique analytic function on G with $v(x, y)$ as its imaginary part.

Hint. Cf. Theorem 13.2.

Problem 6. Find the conjugate harmonic function $v(x, y)$ corresponding to

a) $u(x, y) = x^2 - y^2 + x$ on the domain $|z| < \infty$;

b) $u(x, y) = \dfrac{x}{x^2 + y^2}$ on the domain $0 < |z| < \infty$.

Why is $v(x, y)$ single-valued in b?

***Problem* 7.** Let G be the complex plane cut along the negative real axis (including the origin). Then the function

$$u(x, y) = \tfrac{1}{2} \ln (x^2 + y^2)$$

is harmonic on G, and in fact equals $\ln |z|$. Verify by direct use of formula (13.6) that to within an arbitrary real constant, the function

$$v(x, y) = \arg z$$

is the harmonic conjugate of $u(x, y)$.

Hint. Let $(x_0, y_0) = (1, 0)$, and use paths of integration consisting of line segments parallel to the real and imaginary axes. Then use the expressions for $\arg z$ in terms of $\arctan (y/x)$ given on p. 4.

64. POISSON'S INTEGRAL. SCHWARZ'S FORMULA

We can use the correspondence between harmonic functions and analytic functions, given by Theorems 13.1 and 13.2, to deduce various properties of harmonic functions from familiar properties of analytic functions. Our first result along these lines is the analogue of the theorem on expansion of an analytic function in power series (Theorem 10.6):

THEOREM 13.3. *Let $u(x, y)$ be a harmonic function on a domain G, with harmonic conjugate $v(x, y)$, let z_0 be an arbitrary (finite) point of G, and let $\Delta = \Delta(z_0)$ be the distance between z_0 and the boundary of G. Then $u(x, y)$ and $v(x, y)$ have expansions of the form*

$$u(x, y) = u(r, \theta) = \alpha_0 + \sum_{n=1}^{\infty} (\alpha_n \cos n\theta - \beta_n \sin n\theta) r^n, \quad (13.11)$$

$$v(x, y) = v(r, \theta) = \beta_0 + \sum_{n=1}^{\infty} (\beta_n \cos n\theta + \alpha_n \sin n\theta) r^n \quad (13.12)$$

on the disk $|z - z_0| < \Delta$,[3] *where $z - z_0 = re^{i\theta}$.*

Proof. Using formula (13.7), we form the function $f(z)$ which is analytic on G and has $u(x, y)$ as its real part. According to Theorem 10.6, $f(z)$ has the power series expansion

$$f(z) = \sum_{n=0}^{\infty} a_n (z - z_0)^n \quad (13.13)$$

on the disk $|z - z_0| < \Delta$. To obtain (13.11), we substitute

$$a_n = \alpha_n + i\beta_n, \qquad z - z_0 = re^{i\theta}$$

[3] If the boundary of G is empty or consists of the single point ∞, we write $\Delta = \infty$. The numbers α_n and β_n are real. Also recall footnote 1, p. 274.

into (13.13) and then take the real part of the resulting equation

$$f(z) = \sum_{n=0}^{\infty}(\alpha_n + i\beta_n)r^n e^{in\theta}. \tag{13.14}$$

To obtain (13.12), we take the imaginary part of (13.14).

Our next result is analogous to the Cauchy-Hadamard theorem (Theorem 10.1):

THEOREM 13.4. *Given two series of the form* (13.11) *and* (13.12), *let*

$$R = \frac{1}{\overline{\lim\limits_{n\to\infty}}\sqrt[n]{|\alpha_n + i\beta_n|}}. \tag{13.15}$$

Then (13.11) *and* (13.12) *converge uniformly on every compact subset of the disk* $|z - z_0| < R$ *to a pair of conjugate harmonic functions. The series do not converge on any larger disk.*

Proof. In addition to Theorem 10.1, use Theorems 10.2 and 10.3.[4]

The following result will be needed below:

LEMMA. *The series*

$$\begin{aligned} \frac{\rho^2 - r^2}{\rho^2 + r^2 - 2\rho r\cos(\theta - \varphi)} &= 1 + 2\sum_{n=1}^{\infty}\left(\frac{r}{\rho}\right)^n \cos n(\theta - \varphi), \\ \frac{2\rho r \sin(\theta - \varphi)}{\rho^2 + r^2 - 2\rho r\cos(\theta - \varphi)} &= 2\sum_{n=1}^{\infty}\left(\frac{r}{\rho}\right)^n \sin n(\theta - \varphi) \end{aligned} \tag{13.16}$$

converge uniformly on every compact subset of the disk $|z - z_0| < \rho$.

Proof. Starting from

$$\begin{aligned} \frac{\rho e^{i\varphi} + (z - z_0)}{\rho e^{i\varphi} - (z - z_0)} &= -1 + \frac{2\rho e^{i\varphi}}{\rho e^{i\varphi} - (z - z_0)} \\ &= -1 + 2\left[1 + \frac{z - z_0}{\rho e^{i\varphi}} + \frac{(z - z_0)^2}{\rho^2 e^{2i\varphi}} + \cdots\right] \\ &= 1 + 2\sum_{n=1}^{\infty}\frac{(z - z_0)^n}{\rho^n}e^{-in\varphi}, \end{aligned}$$

we set $z - z_0 = re^{i\theta}$ $(r < \rho)$ and use Example 4, p. 274 to take real and imaginary parts of the resulting equation.

[4] Obviously, the number Δ figuring in Theorem 13.3 cannot exceed R.

An integral of the form

$$\frac{1}{2\pi}\int_0^{2\pi}\mu(\rho,\varphi)\frac{\rho^2-r^2}{\rho^2+r^2-2\rho r\cos(\theta-\varphi)}\,d\varphi,$$

with the "kernel"

$$\frac{\rho^2-r^2}{\rho^2+r^2-2\rho r\cos(\theta-\varphi)}=\operatorname{Re}\frac{\rho e^{i\varphi}+(z-z_0)}{\rho e^{i\varphi}-(z-z_0)},$$

is called a *Poisson integral.* As we now show, a characteristic feature of harmonic functions is that they can be represented as Poisson integrals:

THEOREM 13.5. *With the same notation as in Theorem* 13.3,

$$u(r,\theta)=\frac{1}{2\pi}\int_0^{2\pi}u(\rho,\varphi)\frac{\rho^2-r^2}{\rho^2+r^2-2\rho r\cos(\theta-\varphi)}\,d\varphi, \quad (13.17)$$

$$v(r,\theta)=\frac{1}{2\pi}\int_0^{2\pi}v(\rho,\varphi)\frac{\rho^2-r^2}{\rho^2+r^2-2\rho r\cos(\theta-\varphi)}\,d\varphi \quad (13.18)$$

for $r<\rho<\Delta$ *and arbitrary* θ. *Moreover*

$$v(r,\theta)=\beta_0+\frac{1}{2\pi}\int_0^{2\pi}u(\rho,\varphi)\frac{2\rho r\sin(\theta-\varphi)}{\rho^2+r^2-2\rho r\cos(\theta-\varphi)}\,d\varphi, \quad (13.19)$$

in terms of $u(r,\theta)$.

Proof. We start from formula (13.11), with r replaced by ρ ($\rho<\Delta$), θ replaced by φ, and n replaced by m:

$$u(\rho,\varphi)=\alpha_0+\sum_{m=1}^{\infty}(\alpha_m\cos m\varphi-\beta_m\sin m\varphi)\rho^m. \quad (13.20)$$

Using the uniform convergence of (13.20) in φ for every $\rho<\Delta$, we multiply (13.20) by $\cos n\varphi$ and integrate term by term with respect to φ between 0 and 2π. The result is

$$\begin{aligned}\alpha_0&=\frac{1}{2\pi}\int_0^{2\pi}u(\rho,\varphi)\,d\varphi,\\ \alpha_n&=\frac{1}{\pi\rho^n}\int_0^{2\pi}u(\rho,\varphi)\cos n\varphi\,d\varphi \qquad (n\geqslant 1).\end{aligned} \quad (13.21)$$

Similarly, multiplying (13.20) by $\sin n\varphi$ and integrating term by term with respect to φ between the same limits, we obtain

$$-\beta_n=\frac{1}{\pi\rho^n}\int_0^{2\pi}u(\rho,\varphi)\sin n\varphi\,d\varphi \qquad (n\geqslant 1). \quad (13.22)$$

Substitution of (13.21) and (13.22) into (13.11) and (13.12) gives

$$u(r, \theta) = \frac{1}{2\pi}\int_0^{2\pi} u(\rho, \varphi)\, d\varphi + \sum_{n=1}^{\infty} \frac{1}{\pi}\int_0^{2\pi} u(\rho, \varphi)\left(\frac{r}{\rho}\right)^n \cos n(\theta - \varphi)\, d\varphi,$$
$$v(r, \theta) = \beta_0 + \sum_{n=1}^{\infty} \frac{1}{\pi}\int_0^{2\pi} u(\rho, \varphi)\left(\frac{r}{\rho}\right)^n \sin n(\theta - \varphi)\, d\varphi. \tag{13.23}$$

Formulas (13.17) and (13.19) now follow at once from the lemma, after multiplying the series (13.16) by

$$\frac{1}{2\pi} u(\rho, \varphi),$$

integrating term by term with respect to φ from 0 to 2π for fixed r and ρ ($r < \rho$), and comparing the results with (13.23).[5] Moreover, since (13.17) holds for an *arbitrary* harmonic function on G, we can replace $u(r, \theta)$ by $v(r, \theta)$, obtaining (13.18).

Another proof of formulas (13.17) and (13.18) exploits the analogy between Poisson's integral and Cauchy's integral. Let γ_ρ be the circle of radius ρ with center z_0, and let $z \in I(\gamma_\rho)$. Inverting the point z in the circle γ_ρ, we obtain the point

$$z^* = z_0 + \frac{\rho^2}{\overline{z - z_0}}$$

[cf. formula (5.26)], where $z^* \in E(\gamma_\rho)$. It follows that

$$f(z) = \frac{1}{2\pi i}\int_{\gamma_\rho} \frac{f(\zeta)}{\zeta - z}\, d\zeta,$$
$$0 = \frac{1}{2\pi i}\int_{\gamma_\rho} \frac{f(\zeta)}{\zeta - z^*}\, d\zeta$$

(cf. Theorem 8.4). Subtracting the second of these equations from the first, we find that

$$f(z) = \frac{1}{2\pi i}\int_{\gamma_\rho} f(\zeta)\left[\frac{1}{\zeta - z} - \frac{1}{\zeta - z^*}\right] d\zeta,$$

which becomes

$$f(z) = \frac{1}{2\pi}\int_0^{2\pi} f(\zeta)\left[\frac{\rho}{\rho - re^{i(\theta-\varphi)}} + \frac{re^{i(\varphi-\theta)}}{\rho - re^{i(\varphi-\theta)}}\right] d\varphi$$
$$= \frac{1}{2\pi}\int_0^{2\pi} f(\zeta)\frac{\rho^2 - r^2}{\rho^2 + r^2 - 2\rho r\cos(\theta - \varphi)}\, d\varphi,$$

[5] In keeping with Theorem 13.2, $v(r, \theta)$ is defined only to within an arbitrary real constant β_0.

if we write

$$\zeta - z = \zeta - z_0 - (z - z_0) = \rho e^{i\varphi} - re^{i\theta},$$

$$\zeta - z^* = \zeta - z_0 - (z^* - z_0) = \rho e^{i\varphi} - \frac{\rho^2}{r} e^{i\theta}.$$

Finally, replacing $f(z)$ by $u(r, \theta) + iv(r, \theta)$ and $f(\zeta)$ by $u(\rho, \varphi) + iv(\rho, \varphi)$, and taking real and imaginary parts, we obtain (13.17) and (13.18).

COROLLARY. *With the same notation as in Theorem* 13.3, *let $f(z)$ be an analytic function on G with $u(r, \theta)$ as its real part. Then*

$$f(z) = i\beta_0 + \frac{1}{2\pi}\int_0^{2\pi} u(\rho, \varphi)\, \frac{\rho e^{i\varphi} + (z - z_0)}{\rho e^{i\varphi} - (z - z_0)}\, d\varphi \tag{13.24}$$

in terms of $u(r, \theta)$, a representation known as Schwarz's formula.[6]

Proof. Multiply (13.19) by i and add the result to (13.17). Then recall Example 4, p. 274.

Although a trivial consequence of Theorem 13.5, the next result is sufficiently important to warrant being called a theorem:

THEOREM 13.6. *With the same notation as in Theorem* 13.3,

$$u(x_0, y_0) = \frac{1}{2\pi}\int_0^{2\pi} u(\rho, \varphi)\, d\varphi, \tag{13.25}$$

where $z_0 = (x_0, y_0)$, $\rho < \Delta(z_0)$, or more explicitly,

$$u(x_0, y_0) = \frac{1}{2\pi}\int_0^{2\pi} u(x_0 + \rho \cos \varphi, y_0 + \rho \sin \varphi)\, d\varphi.$$

In other words, the value of a harmonic function $u(x, y)$ at the point (x_0, y_0) equals the average of its values on the circle $|z - z_0| = \rho$ with center (x_0, y_0).[7]

Proof. Set $r = 0$ in formula (13.17). Alternatively, take the real part of formula (8.11), p. 174.

Remark. Setting $u(r, \theta) \equiv 1$ in (13.17), we obtain the formula

$$\frac{1}{2\pi}\int_0^{2\pi} \frac{\rho^2 - r^2}{\rho^2 + r^2 - 2\rho r \cos(\theta - \varphi)}\, d\varphi = 1, \tag{13.26}$$

which will be needed later.

---

[6] Again in keeping with Theorem 13.2, $f(z)$ is defined only to within an arbitrary purely imaginary constant $i\beta_0 = \operatorname{Im} f(z_0)$.

[7] This result is the analogue of the corollary to Theorem 8.4.

Finally we establish some results on harmonic functions implied by the maximum and minimum modulus principles:

THEOREM 13.7. *Let* $u(x, y) \not\equiv$ const *be harmonic on a domain* G. *Then* $u(x, y)$ *has neither a maximum nor a minimum at any point of* G.

Proof. Let $z_0 = x_0 + iy_0$ be an arbitrary point of G, and consider any neighborhood $\mathcal{N}(z_0) \subset G$. Use Theorem 13.2 to form a function $f(z) = u(x, y) + iv(x, y)$ which is single-valued and analytic on $\mathcal{N}(z_0)$. Then

$$g(z) = e^{f(z)}$$

is analytic on $\mathcal{N}(z_0)$. Moreover, $g(z) \not\equiv$ const [since $u(x, y) \not\equiv$ const], and

$$|g(z)| = e^{u(x,y)}.$$

The function $u(x, y)$ cannot have a maximum at $(x_0. y_0)$, since otherwise $|g(z)|$ would have a maximum at z_0, thereby contradicting the maximum modulus principle (Theorem 12.11). Similarly, $u(x, y)$ cannot have a minimum at (x_0, y_0), since otherwise $|g(z)|$ would have a minimum but not a zero at z_0, thereby contradicting the minimum modulus principle (Sec. 62, Prob. 8).

Alternatively, we can give a proof starting directly from Theorem 12.9. The analytic function $w = u + iv = f(z)$ maps $\mathcal{N}(z_0)$ onto some domain $\mathcal{G}$ in the w-plane, containing the point $w_0 = f(z_0)$. Obviously, $\mathcal{G}$ contains both points with abscissas larger than $u(x_0, y_0)$ and points with abscissas smaller than $u(x_0, y_0)$, i.e., $u(x, y)$ can have neither a maximum nor a minimum at (x_0, y_0). This method of proof has been anticipated in Sec. 62, Prob. 10.

COROLLARY 1. *Let* $u(x, y)$ *be harmonic on a bounded domain* G *and continuous on* $\bar{G}$. *Then* $u(x, y)$ *achieves its maximum and its minimum on the boundary of* G.

Proof. Since $u(x, y)$ is continuous on the compact set $\bar{G}$, it achieves both its maximum and its minimum on $\bar{G}$ (cf. Theorem 2.7). But $u(x, y)$ cannot have a maximum or a minimum at a point of G.

COROLLARY 2. *Let* $u(x, y)$ *be harmonic on a bounded domain* G *and continuous on* $\bar{G}$, *and suppose* $u(x, y) =$ const *on the boundary of* G. *Then* $u(x, y) =$ const *on* $\bar{G}$.

Proof. Apply Corollary 1.

COROLLARY 3. *Let* $u_1(x, y)$ *and* $u_2(x, y)$ *be two functions harmonic on a bounded domain* G *and continuous on* $\bar{G}$, *and suppose* $u_1(x, y) = u_2(x, y)$ *on the boundary of* G. *Then* $u_1(x, y) = u_2(x, y)$ *on* $\bar{G}$.

Proof. The function

$$u(x, y) = u_1(x, y) - u_2(x, y)$$

satisfies the conditions of Corollary 2, where the constant equals zero.

Problem 1. Prove that if $u(x, y)$ is harmonic on a domain G, then $u(x, y)$ is analytic in the variables x and y, i.e., $u(x, y)$ has a "double power series" expansion of the form

$$u(x, y) = \sum_{m,n=0}^{\infty} c_{mn}(x - x_0)^m(y - y_0)^n$$

in a neighborhood of every point $z_0 = x_0 + iy_0 \in G$.

Problem 2. Suppose $f(z) = u(x, y) + iv(x, y)$ is analytic at the point $z_0 = x_0 + iy_0$, where $f(z_0) = c_0$. Prove that

a) $f(z) = 2u\left(\dfrac{z + \bar{z}_0}{2}, \dfrac{z - \bar{z}_0}{2i}\right) - \bar{c}_0$;

b) $f(z) = 2iv\left(\dfrac{z + \bar{z}_0}{2}, \dfrac{z - \bar{z}_0}{2i}\right) + \bar{c}_0$.

Hint. Use the result of the preceding problem.

Comment. If $f(z)$ is analytic at the origin, these formulas can be simplified further by setting $z_0 = 0$.

Problem 3. Find the analytic function $f(z) = u(x, y) + iv(x, y)$, given that

a) $u(x, y) = x^2 - y^2 + 2$; b) $u(x, y) = e^x(x \cos y - y \sin y) - \dfrac{y}{x^2 + y^2}$;

c) $v(x, y) = x + y - 3$; d) $v(x, y) = \cos x \sinh y - \sinh x \sin y$.

Ans. a) $z^2 + 2 + Ci$ (C real); c) $(1 + i)z - 3i + C$.

Problem 4. Derive the following version of Schwarz's formula (13.24):

$$f(z) = \frac{1}{\pi i}\int_{|\zeta|=R} \frac{u(\zeta)}{\zeta - z}\, d\zeta - \overline{f(0)}.$$

Problem 5. Let $u(z)$ be harmonic on the domain $G\colon |z| > R$, and let $f(z)$ be analytic on G. Derive integral representations analogous to Poisson's integral and Schwarz's formula for $u(z)$ and $f(z)$ in terms of their values on the circle $|z| = R$.

Problem 6. Let $u(z)$ be harmonic in the upper half-plane G: $\operatorname{Im} z > 0$, and let $f(z)$ be analytic on G. Derive integral representations analogous to Poisson's integral and Schwarz's formula for $u(z)$ and $f(z)$ in terms of their values on the real axis. State conditions for the convergence of the resulting integrals.

Hint. Map G onto the unit disk.

Ans.

$$u(z) = \int_{-\infty}^{\infty} \frac{yu(t)}{(t - x)^2 + y^2}\, dt, \qquad f(z) = \frac{1}{\pi i}\int_{-\infty}^{\infty} \frac{u(t)}{t - z}\, dt + iC \qquad (C \text{ real}).$$

65. THE DIRICHLET PROBLEM

Let G be a domain with boundary Γ, and let $\mu(z) = \mu(x, y)$ be a continuous real function defined on Γ. Consider the problem of finding a function $u(x, y)$ such that

1. $u(x, y)$ is harmonic on G and continuous on $\bar{G}$;
2. $u(x, y) = \mu(x, y)$ at every point of Γ.

This is the *Dirichlet problem*, of great importance in mathematical physics as well as in function theory.

Remark. To give a physical interpretation of the Dirichlet problem, imagine a rigid wire frame, in the shape of some closed curve, supporting a stretched membrane whose boundary is fastened to the frame.[8] Suppose the projection of the frame onto the xy-plane is a closed Jordan curve Γ, with interior $I(\Gamma)$ [see Figure 13.1], and suppose the membrane, regarded as a surface, has the equation

$$u = u(x, y)$$

where $u(x, y)$ is continuous on $\overline{I(\Gamma)}$. Clearly, the values of $u(x, y)$ on Γ depend on the position of the frame; let these boundary values be denoted by $\mu(x, y)$. Then, as is well known,[9] the equilibrium position of the membrane in the absence of an external load (and of any effect due to the membrane's own weight) is given by the solution of Laplace's equation

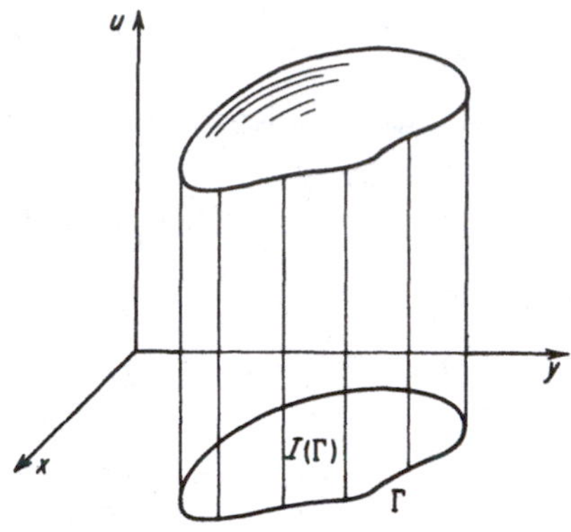

FIGURE 13.1

$$\frac{\partial^2 u}{\partial x^2} + \frac{\partial^2 u}{\partial y^2} = 0$$

which satisfies the boundary condition $u(x, y) = \mu(x, y)$ at every point of Γ. In other words, the problem of determining the equilibrium position of a stretched unloaded membrane reduces to finding a function harmonic on $I(\Gamma)$ and continuous on $\overline{I(\Gamma)}$, which takes given values $\mu(x, y)$ on Γ. But this

---

[8] By a *membrane* we mean a thin sheet whose thickness is negligible compared to its other dimensions, and whose resistance to bending is negligible compared to the tension forces applied to the sheet at its boundary. (These forces are normal to the supporting frame and tangential to the surface of the sheet.) One might think in terms of a thin soap film.

[9] See e.g., R. Courant and D. Hilbert, *Methods of Mathematical Physics, Volume I*, Interscience Publishers, New York (1953), p. 247.

is just the Dirichlet problem for $G = I(\Gamma)$. Conversely, any given Dirichlet problem can be interpreted in terms of finding the equilibrium position of a stretched membrane, supported by a rigid frame corresponding to the given boundary conditions.

We now consider the Dirichlet problem for the simple but important case where the domain G is a disk $|z - z_0| < \rho$ with circular boundary $\Gamma: |z - z_0| = \rho$.[10] Then the function $\mu(z)$ takes the form

$$\mu(z) = \mu(z_0 + \rho e^{i\varphi}) = \mu(\varphi) \qquad (0 \leqslant \varphi \leqslant 2\pi),$$

where $\mu(\varphi)$ must obviously satisfy the condition $\mu(0) = \mu(2\pi)$. If $\mu(\varphi)$ coincides with the values $u(\rho, \varphi)$ taken on Γ by some function $u(r, \theta)$ which is harmonic on a disk $|z - z_0| < \rho'$ of larger radius than G (so that $\rho' > \rho$), then, since

$$u(r, \theta) = \frac{1}{2\pi} \int_0^{2\pi} u(\rho, \varphi) \frac{\rho^2 - r^2}{\rho^2 + r^2 - 2\rho r \cos(\theta - \varphi)} \, d\varphi$$

(see Theorem 13.5), the Dirichlet problem has the obvious solution

$$u(r, \theta) = \frac{1}{2\pi} \int_0^{2\pi} \mu(\varphi) \frac{\rho^2 - r^2}{\rho^2 + r^2 - 2\rho r \cos(\theta - \varphi)} \, d\varphi \qquad (13.27)$$

in terms of Poisson's integral, and moreover this solution is unique, by Theorem 13.7, Corollary 3. However, the true nature of the problem emerges only when the function $\mu(\varphi)$ is specified *a priori*, without reference to a function already known to be harmonic on a larger disk than G. As might be suspected, even in this case, the solution of the Dirichlet problem is still given by (13.27). In fact, we have the following

THEOREM 13.8 (*Dirichlet problem for a disk*). *Let G be the disk $|z - z_0| < \rho$ with boundary $\Gamma: |z - z_0| = \rho$, and let $\mu(\varphi)$ be a continuous real function in the interval* $[0, 2\pi]$ *such that* $\mu(0) = \mu(2\pi)$. *Then the function $u(r, \theta)$ defined by the integral* (13.27) *for any point* (r, θ) *in G, and by*

$$u(\rho, \varphi) = \mu(\varphi)$$

for any point (ρ, φ) *in Γ, solves the Dirichlet problem for the domain G.*

Proof. First we prove that the integral (13.27) defines a harmonic function on G. The reasoning here is essentially the reverse of that used to prove Theorem 13.5. According to the lemma on p. 278, if $r < \rho$

[10] The Dirichlet problem for more general domains is considered in M2, Sec. 27.

we have[11]

$$u(r, \theta) = \frac{1}{2\pi}\int_0^{2\pi} \mu(\varphi) \frac{\rho^2 - r^2}{\rho^2 + r^2 - 2\rho r \cos(\theta - \varphi)}\, d\varphi$$

$$= \frac{1}{2\pi}\int_0^{2\pi} \mu(\varphi)\, d\varphi + \sum_{n=1}^{\infty} \frac{1}{\pi}\int_0^{2\pi} \mu(\varphi)\left(\frac{r}{\rho}\right)^n \cos n(\theta - \varphi)\, d\varphi$$

$$= \frac{1}{2\pi}\int_0^{2\pi} \mu(\varphi)\, d\varphi + \sum_{n=1}^{\infty} \left\{\left[\frac{1}{\pi\rho^n}\int_0^{2\pi} \mu(\varphi) \cos n\varphi\, d\varphi\right] r^n \cos n\theta \right.$$

$$\left. + \left[\frac{1}{\pi\rho^n}\int_0^{2\pi} \mu(\varphi) \sin n\varphi\, d\varphi\right] r^n \sin n\theta\right\}$$

$$= \alpha_0 + \sum_{n=1}^{\infty} (\alpha_n \cos n\theta - \beta_n \sin n\theta) r^n, \tag{13.28}$$

where

$$\alpha_0 = \frac{1}{2\pi}\int_0^{2\pi} \mu(\varphi)\, d\varphi, \quad \alpha_n = \frac{1}{\pi\rho^n}\int_0^{2\pi} \mu(\varphi) \cos n\varphi\, d\varphi \qquad (n \geqslant 1),$$

$$-\beta_n = \frac{1}{\pi\rho^n}\int_0^{2\pi} \mu(\varphi) \sin n\varphi\, d\varphi \qquad (n \geqslant 1) \tag{13.29}$$

[cf. (13.21) and (13.22)]. In particular,

$$|\alpha_n + i\beta_n| = \left|\frac{1}{\pi\rho^n}\int_0^{2\pi} \mu(\varphi) e^{-in\varphi}\, d\varphi\right| \leqslant \frac{1}{\pi\rho^n}\int_0^{2\pi} |\mu(\varphi)|\, d\varphi,$$

and therefore

$$\rho \leqslant \varliminf_{n\to\infty} \frac{\sqrt[n]{\dfrac{1}{\pi}\displaystyle\int_0^{2\pi} |\mu(\varphi)|\, d\varphi}}{\sqrt[n]{|\alpha_n + i\beta_n|}} = \frac{1}{\varlimsup_{n\to\infty} \sqrt[n]{|\alpha_n + i\beta_n|}} = R$$

(recall Sec. 42, Prob. 5), where R is the number given by (13.15). It follows from the representation (13.28) and Theorem 13.4 that $u(r, \theta)$ is harmonic on G.

Next we prove that the function $u(r, \theta)$ defined by the integral (13.27) approaches $\mu(\varphi_0)$ as the point (r, θ) in G approaches any fixed point (ρ, φ_0) in Γ. Using (13.26), we form the difference

$$u(r_n, \theta_n) - \mu(\varphi_0) = \frac{1}{2\pi}\int_0^{2\pi} [\mu(\varphi) - \mu(\varphi_0)] \frac{\rho^2 - r_n^2}{\rho^2 + r_n^2 - 2\rho r_n \cos(\theta_n - \varphi)}\, d\varphi, \tag{13.30}$$

[11] Multiply the first of the series (13.16) by $(1/2\pi)\mu(\varphi)$ and integrate term by term with respect to φ from 0 to 2π.

where $\{(r_n, \theta_n)\}$ is an arbitrary sequence of points in G converging to the boundary point (ρ, φ_0). Since $\mu(\varphi)$ is (uniformly) continuous on Γ, given any $\varepsilon > 0$, there is a number $\delta = \delta(\varepsilon) > 0$ such that

$$|\mu(\varphi) - \mu(\varphi_0)| < \frac{\varepsilon}{2} \tag{13.31}$$

whenever

$$|\varphi - \varphi_0| < 2\delta. \tag{13.32}$$

Assuming that δ is such that (13.32) implies (13.31) and is also so small that $\varphi_0 - 2\delta > 0$ and $\varphi_0 + 2\delta < 2\pi$,[12] we write

$$\begin{aligned} |u(r_n, \theta_n) - \mu(\varphi_0)| &= \left|\frac{1}{2\pi}\int_0^{2\pi} \cdots\right| \\ &\leqslant \left|\frac{1}{2\pi}\int_0^{\varphi_0-2\delta} \cdots\right| + \left|\frac{1}{2\pi}\int_{\varphi_0-2\delta}^{\varphi_0+2\delta} \cdots\right| + \left|\frac{1}{2\pi}\int_{\varphi_0+2\delta}^{2\pi} \cdots\right|, \end{aligned} \tag{13.33}$$

where in each case the dots denote the rest of the integral (13.30). For the middle term in the right-hand side of (13.33), we have the estimate

$$\left|\frac{1}{2\pi}\int_{\varphi_0-2\delta}^{\varphi_0+2\delta} \cdots\right| < \frac{\varepsilon}{2}\,\frac{1}{2\pi}\int_0^{2\pi} \frac{\rho^2 - r_n^2}{\rho^2 + r_n^2 - 2\rho r_n \cos(\theta_n - \varphi)}\,d\varphi = \frac{\varepsilon}{2}, \tag{13.34}$$

where we have used (13.31) and (13.26). To estimate the other two terms, we choose n so large that $|\theta_n - \varphi_0| < \delta$. Then

$$\begin{aligned} &\left|\frac{1}{2\pi}\int_0^{\varphi_0-2\delta} \cdots\right| + \left|\frac{1}{2\pi}\int_{\varphi_0+2\delta}^{2\pi} \cdots\right| \\ &\qquad \leqslant 2M\,\frac{1}{2\pi}\,\frac{\rho^2 - r_n^2}{\rho^2 + r_n^2 - 2\rho r_n \cos\delta}\left(\int_0^{\varphi_0-2\delta} d\varphi + \int_{\varphi_0+2\delta}^{2\pi} d\varphi\right) \\ &\qquad < 2M\,\frac{\rho^2 - r_n^2}{\rho^2 + r_n^2 - 2\rho r_n \cos\delta}, \end{aligned} \tag{13.35}$$

where

$$M = \max_{\varphi\in[0,2\pi]} |\mu(\varphi)|,$$

---

[12] There is no need to consider the case $\varphi_0 = 0$ separately, since the axis of our polar coordinate system (from which the angles θ and φ are measured) can always be rotated to make $\varphi_0 > 0$.

and we have used the fact that

$$\cos(\theta_n - \varphi) < \cos\delta,$$

$$\frac{\rho^2 - r_n^2}{\rho^2 + r_n^2 - 2\rho r_n \cos(\theta_n - \varphi)} < \frac{\rho^2 - r_n^2}{\rho^2 + r_n^2 - 2\rho r \cos\delta}$$

if $|\theta_n - \varphi_0| < \delta$ and φ belongs to either of the intervals $[0, \varphi_0 - 2\delta]$ or $[\varphi_0 + 2\delta, 2\pi]$. The right-hand side of (13.35) obviously approaches zero as $r_n \to \rho$. Therefore, for sufficiently large n,

$$\left|\frac{1}{2\pi}\int_0^{\varphi_0 - 2\delta} \cdots\right| + \left|\frac{1}{2\pi}\int_{\varphi_0+2\delta}^{2\pi} \cdots\right| < \frac{\varepsilon}{2} \tag{13.36}$$

(naturally, we assume that n is also so large that $|\theta_n - \varphi_0| < \delta$). Comparing (13.33), (13.35) and (13.36), we find that

$$|u(r_n, \theta_n) - \mu(\varphi_0)| < \varepsilon$$

for sufficiently large n, i.e.,

$$\lim_{n\to\infty} [u(r_n, \theta_n) - \mu(\varphi_0)] = 0, \tag{13.37}$$

where $\{(r_n, \theta_n)\}$ is an arbitrary sequence of points in G approaching (ρ, φ_0). Equation (13.37) still holds if some or all of the points (r_n, θ_n) lie on Γ, since then we can make direct use of the continuity of $\mu(\varphi)$ [recall that $u(\rho, \varphi) = \mu(\varphi)$, by definition]. In otherwords, $u(r, \theta)$ is continuous on $\bar{G}$, and the proof is complete.

Problem 1 (M2, p. 157). Let $\mu(\varphi)$ be a continuous real function in the interval $[0, 2\pi]$ such that $\mu(0) = \mu(2\pi)$, and let

$$\alpha_0 = \frac{1}{2\pi}\int_0^{2\pi} \mu(\varphi)\, d\varphi, \quad \alpha_n = \int_0^{2\pi} \mu(\varphi) \cos n\varphi\, d\varphi \qquad (n \geqslant 1)$$

$$\beta_n = \int_0^{2\pi} \mu(\varphi) \sin n\varphi\, d\varphi \qquad (n \geqslant 1).$$

Starting from Theorem 13.8, prove that the formal Fourier series

$$\alpha_0 + \sum_{n=1}^{\infty} (\alpha_n \cos n\varphi + \beta_n \sin n\varphi)$$

is uniformly summable by Abel's method to $\mu(\varphi)$, i.e., as $r \to 1-$,[13] the function

$$\mu(r, \varphi) = \alpha_0 + \sum_{n=1}^{\infty} (\alpha_n \cos n\varphi + \beta_n \sin n\varphi) r^n \qquad (0 \leqslant r < 1)$$

converges uniformly to $\mu(\varphi)$ in $[0, 2\pi]$.

---

[13] We write $r \to a-$ if r approaches a from the left (and $r \to a+$ if r approaches a from the right).

Problem 2 (M2, p. 158). Using the result of the preceding problem, prove that if $\mu(\varphi)$ is a continuous real function in the interval $[0, 2\pi]$ such that $\mu(0) = \mu(2\pi)$, then, given any $\varepsilon > 0$, there is a trigonometric polynomial

$$\tau(\varphi) = \sum_{n=0}^{N} (a_n \cos n\varphi + b_n \sin n\varphi)$$

such that

$$|\tau(\varphi) - \mu(\varphi)| < \varepsilon \qquad (0 \leqslant \varphi \leqslant 2\pi).$$

CHAPTER 14

INFINITE PRODUCT AND PARTIAL FRACTION EXPANSIONS

66. PRELIMINARY RESULTS. INFINITE PRODUCTS

The factorization (4.1) of a polynomial suggests the possibility of finding an analogous representation for an arbitrary entire function $f(z)$. This is in fact possible, as will be shown in Sec. 68. First we dispose of the trivial case where $f(z)$ has no zeros or only finitely many zeros:

THEOREM 14.1. *If an entire function $f(z)$ has no zeros, then $f(z)$ is of the form*

$$f(z) = e^{g(z)},$$

where $g(z)$ is an entire function.

Proof. Since $f(z)$ does not vanish, the function

$$h(z) = \frac{f'(z)}{f(z)} = \frac{d}{dz} \operatorname{Ln} f(z)$$

is entire, and hence, by Theorem 8.1, so is

$$\int_0^z h(\zeta)\,d\zeta = \Big[\operatorname{Ln} f(\zeta)\Big]_{\zeta=0}^{\zeta=z} = \operatorname{Ln} f(z) - \operatorname{Ln} f(0).$$

Therefore

$$f(z) = e^{g(z)},$$

where

$$g(z) = \int_0^z h(\zeta)\,d\zeta + \operatorname{Ln} f(0)$$

is entire.

Example. If $g(z)$ is a constant, then so is $f(z)$. Otherwise, $f(z)$ is an entire transcendental function, as defined on p. 243.

THEOREM 14.2. *Let $\zeta_1, \ldots, \zeta_m$ be the distinct zeros of an entire function $f(z)$, where ζ_j is of order k_j $(j = 1, \ldots, m)$. Then $f(z)$ is of the form*

$$f(z) = (z - \zeta_1)^{k_1} \cdots (z - \zeta_m)^{k_m} e^{g(z)}, \tag{14.1}$$

where $g(z)$ is an entire function.

Proof. By hypothesis, the function

$$F(z) = \frac{f(z)}{(z - \zeta_1)^{k_1} \cdots (z - \zeta_m)^{k_m}}$$

is entire and has no zeros. Therefore, according to Theorem 14.1,

$$F(z) = e^{g(z)},$$

where $g(z)$ is entire. But this is just another way of writing (14.1).

Example. If $g(z)$ is a constant, then $f(z)$ is a polynomial. Otherwise, $f(z)$ is an entire transcendental function.

We must now digress to develop the theory of infinite products, since it turns out that such products are the natural tool for representing $f(z)$ in the case where $f(z)$ has infinitely many zeros. By an *infinite product* we mean an expression of the form

$$\prod_{n=1}^{\infty} u_n \tag{14.2}$$

with complex nonzero factors u_n. Let

$$\{p_n\} = \{u_1 u_2 \cdots u_n\} \tag{14.3}$$

be the sequence of *partial products* of (14.2). Then the infinite product (14.2) is said to *converge* if the sequence (14.3) converges to a nonzero limit, and the limit of the sequence is called the *value* of the infinite product.

Moreover, if

$$\lim_{n\to\infty} p_n = u \neq 0,$$

we write

$$\prod_{n=1}^{\infty} u_n = u.$$

An infinite product which does not converge to a nonzero (finite) limit is said to be *divergent.*

Obviously, a necessary condition for the infinite product (14.2) to converge is that

$$\lim_{n\to\infty} u_n = 1,$$

since if

$$\lim_{n\to\infty} \prod_{k=1}^{n} u_k = u \neq 0,$$

then

$$\lim_{n\to\infty} u_n = \lim_{n\to\infty} \frac{\prod_{k=1}^{n} u_k}{\prod_{k=1}^{n-1} u_k} = 1.$$

In view of this, it is convenient to write the general term of the infinite product (14.2) in the form

$$u_n = 1 + v_n,$$

where now a necessary condition for convergence is that

$$\lim_{n\to\infty} v_n = 0.$$

THEOREM 14.3. *The infinite product*

$$\prod_{n=1}^{\infty} (1 + v_n) \tag{14.4}$$

converges if and only if the infinite series

$$\sum_{n=1}^{\infty} \ln(1 + v_n) \tag{14.5}$$

converges.

Proof. Convergence of (14.5) implies convergence of the two series

$$\sum_{n=1}^{\infty} \ln|1 + v_n|, \qquad \sum_{n=1}^{\infty} \arg(1 + v_n),$$

i.e., of the two sequences $\{s_n\}$, $\{\sigma_n\}$ with terms

$$s_n = \sum_{k=1}^{n} \ln|1 + v_k| = \ln\left|\prod_{k=1}^{n} (1 + v_k)\right|$$

and

$$\sigma_n = \sum_{k=1}^{n} \arg(1 + v_k) = \operatorname{Arg}_n \left[\prod_{k=1}^{n} (1 + v_k) \right], \tag{14.6}$$

where Arg_n denotes that value of the argument for which the equality (14.6) holds. Therefore the sequence

$$\left\{ \prod_{k=1}^{n} (1 + v_k) \right\} = \{ e^{s_n} (\cos \sigma_n + i \sin \sigma_n) \} \tag{14.7}$$

converges to a nonzero limit, i.e., the infinite product (14.4) converges.

Conversely, if

$$\prod_{n=1}^{\infty} (1 + v_n) = u \neq 0,$$

then the limit

$$\lim_{n \to \infty} |1 + v_1| \cdots |1 + v_n| = |u| \neq 0$$

exists (why?), and hence the series

$$\sum_{n=1}^{\infty} \ln |1 + v_n| \tag{14.8}$$

converges. Moreover, according to Sec. 6, Prob. 3, there exists a sequence of values

$$\operatorname{Arg} \left[\prod_{k=1}^{n} (1 + v_k) \right] = \sum_{k=1}^{n} \arg(1 + v_k) + 2\pi\mu_n = \Phi_n$$

(with integral μ_n), converging to any given value Φ of $\operatorname{Arg} u$. Since

$$\begin{aligned} |\Phi_{n+1} - \Phi_n| &= |\arg(1 + v_{n+1}) + 2\pi(\mu_{n+1} - \mu_n)| \\ &\geqslant 2\pi\, |\mu_{n+1} - \mu_n| - \pi, \end{aligned}$$

and since

$$\lim_{n \to \infty} |\Phi_{n+1} - \Phi_n| = 0,$$

it follows that all the nonnegative integers $|\mu_{n+1} - \mu_n|$ vanish if n exceeds a certain integer $N > 0$, i.e.,

$$\mu_{N+1} = \mu_{N+2} = \cdots = \mu.$$

But then the sequence $\{\Phi_n - 2\pi\mu_n\}$ converges, as well as the sequence $\{\Phi_n\}$ itself, and therefore the series

$$\sum_{n=1}^{\infty} \arg(1 + v_n)$$

converges. This fact, together with the convergence of (14.8), implies the convergence of (14.5).

Remark. Suppose the infinite product (14.4) converges. Then, according to Theorem 14.3, the infinite series (14.5) also converges, and hence, using the continuity of the exponential, we can take the limit as $\nu \to \infty$ of the obvious identity

$$\prod_{n=1}^{\nu} (1 + v_n) = \exp\left[\sum_{n=1}^{\nu} \ln(1 + v_n)\right],$$

obtaining

$$\prod_{n=1}^{\infty} (1 + v_n) = \exp\left[\sum_{n=1}^{\infty} \ln(1 + v_n)\right]. \tag{14.9}$$

The infinite product (14.4) is said to be *absolutely convergent* if the infinite series (14.5) is absolutely convergent. Since we can arbitrarily rearrange the terms of an absolutely convergent series without destroying the convergence of the series or the value of its sum, it follows from (14.9) that we can arbitrarily rearrange the factors of an absolutely convergent product without destroying the convergence of the product or changing the value of the product.

THEOREM 14.4. *The infinite product*

$$\prod_{n=1}^{\infty} (1 + v_n)$$

is absolutely convergent if and only if the infinite series

$$\sum_{n=1}^{\infty} v_n \tag{14.10}$$

is absolutely convergent.

Proof. It follows from the expansion

$$\ln(1 + v_n) = v_n - \frac{v_n^2}{2} + \frac{v_n^3}{3} - \cdots = v_n\left(1 - \frac{v_n}{2} + \frac{v_n^2}{3} - \cdots\right),$$

proved in Example 2, pp. 206–207, that

$$\frac{1}{2}|v_n| = |v_n|\left(1 - \frac{1}{2^2} - \frac{1}{2^3} - \cdots\right) < |\ln(1 + v_n)|$$

$$< |v_n|\left(1 + \frac{1}{2^2} + \frac{1}{2^3} + \cdots\right) = \frac{3}{2}|v_n|$$

if $|v_n| < \frac{1}{2}$ (why is this no restriction?). Therefore, as asserted,

$$\sum_{n=1}^{\infty} \ln(1 + v_n)$$

is absolutely convergent if and only if (14.10) is absolutely convergent.

Finally we consider infinite products whose factors are functions. An infinite product of the form

$$\prod_{n=1}^{\infty} [1 + v_n(z)], \tag{14.11}$$

where every $v_n(z)$ is a function defined on a given set E, is said to be *uniformly convergent on E* if the sequence of partial products

$$\{p_n(z)\} = \{[1 + v_1(z)] \cdots [1 + v_n(z)]\}$$

is uniformly convergent on E.

THEOREM 14.5. *Suppose every term of the sequence $\{v_n(z)\}$ is analytic on a domain G, and suppose the infinite series*

$$\sum_{n=1}^{\infty} \ln [1 + v_n(z)] \tag{14.12}$$

is uniformly convergent on every compact subset of G.[1] *Then the infinite product* (14.11) *converges uniformly to a nonvanishing analytic function $f(z)$ on every compact subset of G.*

Proof. The convergence of (14.11) to a nonvanishing function $f(z)$ on G follows from Theorem 14.3. The fact that $f(z)$ is analytic on G follows from the analyticity of (14.12) [see Theorem 9.4] and the representation

$$f(z) = \exp\left\{\sum_{n=1}^{\infty} \ln [1 + v_n(z)]\right\}$$

[see (14.9)]. Finally, if F is a compact subset of G, let

$$M = \max_{z \in F} |f(z)|$$

(see Theorem 2.7), and let ε be an arbitrary positive number less than M. Since (14.12) is uniformly convergent on F, there is an integer $N(\varepsilon) > 0$ such that

$$\left|\sum_{k=n+1}^{\infty} \ln [1 + v_k(z)]\right| < \frac{\varepsilon}{2M} < \frac{1}{2}$$

for all $n > N(\varepsilon)$ and $z \in F$. Therefore

---

[1] In particular, none of the terms $v_n(z)$ can take the value -1 at any point of G.

$$\left|f(z)-\prod_{k=1}^{n}[1+v_k(z)]\right|$$
$$=\left|\exp\left\{\sum_{k=1}^{\infty}\ln[1+v_k(z)]\right\}-\exp\left\{\sum_{k=1}^{n}\ln[1+v_k(z)]\right\}\right|$$
$$=\left|\exp\left\{\sum_{k=1}^{\infty}\ln[1+v_k(z)]\right\}\right|\left|1-\exp\left\{-\sum_{k=n+1}^{\infty}\ln[1+v_k(z)]\right\}\right|$$
$$< M\left[\frac{\varepsilon}{2M}+\frac{1}{2!}\left(\frac{\varepsilon}{2M}\right)^2+\frac{1}{3!}\left(\frac{\varepsilon}{2M}\right)^3+\cdots\right]$$
$$=\frac{\varepsilon}{2}\left[1+\frac{1}{2!}\frac{\varepsilon}{2M}+\frac{1}{3!}\left(\frac{\varepsilon}{2M}\right)^2+\cdots\right]<\frac{\varepsilon}{2}\left(1+\frac{1}{2}+\frac{1}{2^2}+\cdots\right)=\varepsilon$$

for all $n > N(\varepsilon)$ and $z \in F$, i.e., the infinite product (14.11) is uniformly convergent on F.

Problem 1. Prove the following generalization of Theorem 14.1: If an entire function $f(z)$ has no A_0-points ($A_0 \neq \infty$), then $f(z)$ is of the form

$$f(z) = A_0 + e^{g(z)},$$

where $g(z)$ is an entire function.

Problem 2. Prove the following generalization of Theorem 14.2: Let $\zeta_1, \ldots, \zeta_m$ be the distinct A_0-points ($A_0 \neq \infty$) of an entire function $f(z)$, where ζ_j is of order k_j ($j = 1, \ldots, m$). Then $f(z)$ is of the form

$$f(z) = A_0 + (z-\zeta_1)^{k_1}\cdots(z-\zeta_m)^{k_m}e^{g(z)},$$

where $g(z)$ is an entire function.

Problem 3. Evaluate the following infinite products:

a) $\displaystyle\prod_{n=2}^{\infty}\left(1-\frac{1}{n^2}\right)$; b) $\displaystyle\prod_{n=1}^{\infty}\left(1+\frac{1}{n(n+2)}\right)$; c) $\displaystyle\prod_{n=3}^{\infty}\frac{n^2-4}{n^2-1}$;

d) $\displaystyle\prod_{n=2}^{\infty}\left(1-\frac{2}{n(n+1)}\right)$; e) $\displaystyle\prod_{n=2}^{\infty}\frac{n^3-1}{n^3+1}$; f) $\displaystyle\prod_{n=1}^{\infty}\left(1+\frac{(-1)^{n+1}}{n}\right)$.

Ans. b) 2; d) $\frac{1}{3}$; f) 1.

Problem 4. Give an example showing that $\lim_{n\to\infty} u_n = 1$ is not a sufficient condition for the infinite product

$$\prod_{n=1}^{\infty} u_n \tag{14.13}$$

to converge.

Problem 5. Prove that the infinite product (14.13) converges if and only if given any $\varepsilon > 0$, there exists an integer $N(\varepsilon) > 0$ such that

$$|u_{n+1}u_{n+2}\cdots u_{n+p}-1| < \varepsilon$$

for all $n > N(\varepsilon)$ and $p > 0$.

Hint. Cf. Theorem 1.2.

Problem 6. Prove that the infinite product

$$\prod_{n=1}^{\infty} (1 + v_n) \tag{14.14}$$

is absolutely convergent (as defined on p. 295) if and only if the infinite product

$$\prod_{n=1}^{\infty} (1 + |v_n|) \tag{14.15}$$

is convergent.

Hint. Using the inequality $1 + x \leqslant e^x$ for $x \geqslant 0$, prove that (14.15) is convergent if and only if

$$\sum_{n=1}^{\infty} |v_n|$$

is convergent. Then use Theorem 14.4.

Comment. Thus, we could just as well have defined an absolutely convergent infinite product (14.14) as one for which (14.15) converges.

Problem 7. Prove that the infinite product

$$\prod_{n=1}^{\infty} (1 + z^{2^n})$$

is uniformly convergent on every compact subset of the unit disk $K: |z| < 1$, and hence represents an analytic function on K. What is this function?

Ans. $\dfrac{1}{1 - z^2}$.

67. WEIERSTRASS' THEOREM

We are now in a position to find representations of entire (transcendental) functions with infinitely many zeros:

THEOREM 14.6 (*Weierstrass' theorem*). *Let λ be a nonnegative integer, and let $\{\zeta_n\}$ be a sequence of complex numbers converging to infinity such that*[2]

$$|\zeta_1| \leqslant |\zeta_2| \leqslant \cdots \leqslant |\zeta_n| \leqslant \cdots.$$

Then there exists an entire function $f(z)$ whose zeros coincide with the points

$$\underbrace{0, \ldots, 0}_{\lambda \text{ times}}, \zeta_1, \ldots, \zeta_n, \ldots. \tag{14.16}$$

[2] Such a sequence $\{\zeta_n\}$ will be called an *increasing sequence.*

Proof. Consider the sequence of entire functions

$$f_m(z) = z^\lambda \prod_{n=1}^{m} \left(1 - \frac{z}{\zeta_n}\right) e^{P_n(z)} \qquad (m = 1, 2, \ldots),$$

where the $P_n(z)$ are polynomials, to be suitably chosen later. Obviously, the zeros of $f_m(z)$ coincide with the first $m + \lambda$ points of the sequence (14.16). In general, $f_m(z)$ has multiple zeros, since (14.16) can contain the same point several times (this possibility is explicitly indicated for the point $z = 0$). The idea of the proof is to choose the polynomials $P_n(z)$ in such a way that the sequence $\{f_m(z)\}$ is uniformly convergent on every compact subset, since then we can invoke the corollary to Theorem 9.4 to deduce that the limit function

$$f(z) = \lim_{m\to\infty} f_m(z) \tag{14.17}$$

is entire.

With this in mind, let K_R denote the disk $|z| < R$, and let $N(R)$ be the smallest integer such that $|\zeta_n| > 2R$ for all $n > N(R)$. Then, if $z \in K_R$ and $m > N(R)$, we can write $f_m(z)$ in the form

$$\begin{aligned} f_m(z) &= f_{N(R)}(z) \prod_{n=N(R)+1}^{m} \left(1 - \frac{z}{\zeta_n}\right) e^{P_n(z)} \\ &= f_{N(R)}(z) \exp\left\{\sum_{n=N(R)+1}^{n} \left[\ln\left(1 - \frac{z}{\zeta_n}\right) + P_n(z)\right]\right\}, \end{aligned} \tag{14.18}$$

where every logarithmic term can be expanded as a power series

$$\ln\left(1 - \frac{z}{\zeta_n}\right) = -\frac{z}{\zeta_n} - \cdots - \frac{z^n}{n\zeta_n^n} - \frac{z^{n+1}}{(n+1)\zeta_n^{n+1}} - \cdots,$$

since $|z/\zeta_n| < \frac{1}{2}$ for all $z \in K_R$ and $n > N(R)$. Choosing $P_n(z)$ so as to cancel the first n terms of this series, i.e.,

$$P_n(z) = \frac{z}{\zeta_n} + \cdots + \frac{z^n}{n\zeta_n^n}, \tag{14.19}$$

we have

$$\ln\left(1 - \frac{z}{\zeta_n}\right) + P_n(z) = -\frac{z^{n+1}}{(n+1)\zeta_n^{n+1}} - \cdots, \tag{14.20}$$

which implies

$$\begin{aligned} \left|\ln\left(1 - \frac{z}{\zeta_n}\right) + P_n(z)\right| &\leqslant \frac{1}{n+1}\left|\frac{z}{\zeta_n}\right|^{n+1} + \frac{1}{n+2}\left|\frac{z}{\zeta_n}\right|^{n+2} + \cdots \\ &< \frac{1}{2^{n+1}} + \frac{1}{2^{n+2}} + \cdots = \frac{1}{2^n}. \end{aligned} \tag{14.21}$$

Then the series

$$\sum_{n=N(R)+1}^{\infty}\left[\ln\left(1-\frac{z}{\zeta_n}\right)+P_n(z)\right] \tag{14.22}$$

is uniformly convergent on K_R, since

$$\sum_{n=N(R)+1}^{\infty}\left|\ln\left(1-\frac{z}{\zeta_n}\right)+P_n(z)\right|<\sum_{n=1}^{\infty}\frac{1}{2^n}=1<\infty.$$

Therefore (14.22) represents an analytic function $\chi_R(z)$ on K_R (cf. Theorem 9.4).

Comparing (14.17) and (14.18), and using the continuity of the exponential function, we find that

$$f(z)=f_{N(R)}(z)e^{\chi_R(z)} \qquad (z\in K_R). \tag{14.23}$$

This shows that $f(z)$ is analytic on K_R, a fact which also follows from the uniform convergence of the sequence $\{f_m(z)\}$ on K_R (implied by Theorem 14.5). Since the disk K_R can have arbitrarily large radius, $\{f_m(z)\}$ is uniformly convergent on every compact set, and hence the function $f(z)$ is analytic in the whole plane, i.e., $f(z)$ is entire. [This can also be seen directly from (14.23).] The fact that the zeros of $f(z)$ coincide with the points (14.16) is almost obvious, and follows from the representation (14.23) and the arbitrariness of R, since $\exp\,[\chi_R(z)]$ is nonvanishing, while, by construction, the zeros of $f_{N(R)}(z)$ in K_R are precisely those points of the sequence (14.16) which lie in K_R. Finally, recalling the definition of $f_m(z)$, we note that

$$f(z)=\lim_{m\to\infty} z^{\lambda}\prod_{n=1}^{m}\left(1-\frac{z}{\zeta_n}\right)e^{P_n(z)}. \tag{14.24}$$

COROLLARY 1. *The infinite product*

$$f(z)=z^{\lambda}\prod_{n=1}^{\infty}\left(1-\frac{z}{\zeta_n}\right)\exp\left(\frac{z}{\zeta_n}+\cdots+\frac{z^n}{n\zeta_n^n}\right) \tag{14.25}$$

is an entire function satisfying the requirements of Theorem 14.6.

Proof. Equation (14.25) is another way of writing (14.24).

COROLLARY 2. *Let $f(z)$ be an entire function with zeros given by the increasing sequence*

$$\underbrace{0,\ldots,0}_{\lambda\ \text{times}},\zeta_1,\ldots,\zeta_n,\ldots$$

Then $f(z)$ can be represented in the form

$$f(z)=e^{g(z)}z^{\lambda}\prod_{n=1}^{\infty}\left(1-\frac{z}{\zeta_n}\right)\exp\left(\frac{z}{\zeta_n}+\cdots+\frac{z^n}{n\zeta_n^n}\right), \tag{14.26}$$

where $g(z)$ is an entire function.

Proof. The function

$$\varphi(z) = z^\lambda \prod_{n=1}^{\infty} \left(1 - \frac{z}{\zeta_n}\right) \exp\left(\frac{z}{\zeta_n} + \cdots + \frac{z^n}{n\zeta_n^n}\right)$$

is entire, with the same zeros as $f(z)$, and hence the quotient $f(z)/\varphi(z)$ is entire and nonvanishing. Therefore, according to Theorem 14.1,

$$\frac{f(z)}{\varphi(z)} = e^{g(z)},$$

where $g(z)$ is entire.

In Theorem 14.6 we assumed no detailed information about the sequence $\{\zeta_n\}$. However, if appropriate information is available, it can be used to draw stronger conclusions:

THEOREM 14.7. *Let* λ *and* $\{\zeta_n\}$ *be the same as in Theorem* 14.6, *and let* $\varkappa$ *be the largest nonnegative integer* k *for which the series*

$$\sum_{n=1}^{\infty} \frac{1}{|\zeta_n|^k}$$

diverges. Then the expression

$$f(z) = z^\lambda \prod_{n=1}^{\infty} \left(1 - \frac{z}{\zeta_n}\right) \exp\left(\frac{z}{\zeta_n} + \cdots + \frac{z^\varkappa}{\varkappa\zeta_n^\varkappa}\right),$$

known as a canonical product, where the exponential factors disappear if $\varkappa = 0$, *represents an entire function whose zeros coincide with the points*

$$\underbrace{0, \ldots, 0}_{\lambda \text{ times}}, \zeta_1, \ldots, \zeta_n, \ldots$$

Proof. The proof parallels that of Theorem 14.6, with the important difference that we now have

$$P_n(z) = \begin{cases} \dfrac{z}{\zeta_n} + \cdots + \dfrac{z^\varkappa}{\varkappa\zeta_n^\varkappa} & \text{if } \varkappa \geqslant 1, \\ 0 & \text{if } \varkappa = 0 \end{cases}$$

and

$$\ln\left(1 - \frac{z}{\zeta_n}\right) + P_n(z) = -\frac{z^{\varkappa+1}}{(\varkappa + 1)\zeta_n^{\varkappa+1}} - \cdots$$

instead of (14.19) and (14.20), and

$$\left|\ln\left(1 - \frac{z}{\zeta_n}\right) + P_n(z)\right| = \left|-\frac{z^{\varkappa+1}}{\zeta_n^{\varkappa+1}} \sum_{p=1}^{\infty} \frac{z^{p-1}}{(\varkappa + p)\zeta_n^{p-1}}\right|$$

$$< \frac{R^{\varkappa+1}}{|\zeta_n|^{\varkappa+1}} \sum_{p=1}^{\infty} \frac{1}{2^{p-1}} = \frac{2R^{\varkappa+1}}{|\zeta_n|^{\varkappa+1}}$$

instead of (14.21). Therefore (14.22) is again uniformly convergent on $K_R: |z| < R$, but this time because of the convergence of the series

$$\sum_{n=1}^{\infty} \frac{1}{|\zeta_n|^{\varkappa+1}}$$

instead of the series

$$\sum_{n=1}^{\infty} \frac{1}{2^n}$$

The rest of the proof is identical with that of Theorem 14.6.

COROLLARY. *Let $f(z)$ be an entire function with zeros given by the increasing sequence*[3]

$$\underbrace{0, \ldots, 0}_{\lambda \text{ times}}, \zeta_1, \ldots, \zeta_n, \ldots,$$

and let $\varkappa$ be the largest nonnegative integer k for which the series

$$\sum_{n=1}^{\infty} \frac{1}{|\zeta_n|^k}$$

diverges. Then $f(z)$ can be represented in the form

$$f(z) = e^{g(z)} z^{\lambda} \prod_{n=1}^{\infty} \left(1 - \frac{z}{\zeta_n}\right) \exp\left(\frac{z}{\zeta_n} + \cdots + \frac{z^{\varkappa}}{\varkappa \zeta_n^{\varkappa}}\right), \tag{14.27}$$

where $g(z)$ is an entire function and the exponential factors disappear if $\varkappa = 0$.

Remark. Obviously Theorem 14.7 and its corollary remain valid if we replace $\varkappa$ by any larger integer. The great advantage of formula (14.27) over the general formula (14.26) is that polynomials of the same fixed degree $\varkappa$ can be used in all the exponential factors, instead of polynomials whose degree becomes arbitrarily large (with n). In the special case where $\varkappa = 0$, the exponential factors in (14.27) disappear, and we have the particularly simple infinite product expansion

$$f(z) = e^{g(z)} z^{\lambda} \prod_{n=1}^{\infty} \left(1 - \frac{z}{\zeta_n}\right).$$

It will be recalled from p. 79 that by a meromorphic function $f(z)$ we mean a function which can be written as a quotient

$$f(z) = \frac{g(z)}{h(z)}$$

[3] Note that the sequence $\{\zeta_n\}$ necessarily converges to infinity, since otherwise $f(z) \equiv 0$ (why?).

of two entire functions $g(z)$ and $h(z)$, where $h(z) \not\equiv 0$. We now use Weierstrass' theorem to establish an equivalent characterization of such functions:

THEOREM 14.8. *A function $f(z)$ is meromorphic if and only if its only singular points in the finite plane are poles.*

Proof. If $f(z)$ has a representation of the form $g(z)/h(z)$ where $g(z)$ and $h(z)$ are entire, then $f(z)$ is analytic at every point where $h(z) \neq 0$. Suppose $h(z)$ has a zero of order $l \geqslant 1$ at a finite point z_0, and suppose z_0 is a zero of order $k \geqslant 0$ of $g(z)$. [If z_0 is not a zero of $g(z)$, we set $k = 0$.] Then $f(z)$ has a pole of order $l - k$ at z_0 if $l > k$ and a regular point at z_0 if $l \leqslant k$, where the regular point is a zero of order $k - l$ if $l < k$ (see Sec. 56, Prob. 2). Moreover, since the zeros of $h(z)$ cannot have a finite limit point [otherwise $h(z)$ would vanish identically], neither can the poles of $f(z)$. Thus the only singular points of $f(z)$ in the finite plane are poles, of which there are countably many (by Theorem 10.10).

Conversely, let $f(z)$ be a function whose only singular points in the finite plane are poles. If $f(z)$ has no poles, then $f(z)$ is entire and hence (trivially) meromorphic. Suppose $f(z)$ has a finite number of poles $\zeta_1, \ldots, \zeta_n$, of orders $\beta_1, \ldots, \beta_n$, respectively. Then, forming the polynomial

$$P(z) = (z - \zeta_1)^{\beta_1} \cdots (z - \zeta_n)^{\beta_n},$$

we see that $g(z) = f(z)P(z)$ is entire. In other words, $f(z)$ can be written as a quotient $g(z)/P(z)$ of two entire functions, as required. Finally, suppose $f(z)$ has an infinite number of poles. Then no compact set F can contain more than a finite number of poles, since otherwise F would contain a limit point of poles, contrary to the assumption that $f(z)$ has no singular points other than poles. Therefore, by the same argument as used to prove Theorem 10.10, $f(z)$ has countably many poles, which we write as an increasing sequence

$$\underbrace{0, \ldots, 0}_{\lambda \text{ times}}, \zeta_1, \ldots, \zeta_n, \ldots,$$

where $\zeta_n \to \infty$ as $n \to \infty$ (why?), and each pole is repeated a number of times equal to its order.[4] According to Weierstrass' theorem, there is an entire function $h(z)$ whose zeros coincide with the poles of $f(z)$ and have the same orders. But then $g(z) = f(z)h(z)$ is entire, i.e., we can write $f(z)$ as a quotient $g(z)/h(z)$ of two entire functions, and the proof is complete.

---

[4] This way of writing the poles of $f(z)$ has been chosen to match the statement of Weierstrass' theorem. If $f(z)$ has no poles at the origin, then $\lambda = 0$.

Problem 1. Find the "simplest" entire function $f(z)$ with zeros $\zeta_n = n$ $(n = 1, 2, \ldots)$.

Ans. $f(z) = \prod_{n=1}^{\infty}\left(1 - \frac{z}{n}\right)e^{z/n}.$

Problem 2. Prove that

$$\sin z = e^{g(z)}z\prod_{n=1}^{\infty}\left(1 - \frac{z^2}{n^2\pi^2}\right).$$

Comment. It will be shown on p. 320 that $e^{g(z)} \equiv 1$.

Problem 3. Suppose it is known that the function $g(z)$ in the preceding problem is of the form $c_0 + c_1z$. Prove that then $g(z)$ is an integral multiple of $2\pi i$.

Hint. Use the fact that $e^{g(z)}$ is even.

Comment. The fact that $g(z)$ is of the form $c_0 + c_1z$ can be deduced from a factorization theorem due to Hadamard which lies beyond the scope of this book (see M2, p. 289).

Problem 4 (M2, p. 298). Prove that a meromorphic function $f(z)$ is a rational function if and only if the point at infinity is a regular point or a pole of $f(z)$.

68. MITTAG-LEFFLER'S THEOREM

Weierstrass' theorem asserts that there exists an entire function with arbitrarily preassigned zeros. Similarly, as we now prove, there exists a meromorphic function with arbitrarily preassigned poles and principal parts:

THEOREM 14.9 (*Mittag-Leffler's theorem*). *Let*

$$\zeta_0 = 0, \zeta_1, \ldots, \zeta_n, \ldots \tag{14.28}$$

be an increasing sequence of distinct complex numbers converging to infinity, and let

$$G_0(z), G_1(z), \ldots, G_n(z), \ldots$$

be a sequence of rational functions of the form[5]

$$G_n(z) = \frac{a_{-\beta_n}^{(n)}}{(z-\zeta_n)^{\beta_n}} + \cdots + \frac{a_{-1}^{(n)}}{z-\zeta_n} \qquad (n = 0, 1, 2, \ldots),$$

[5] Thus $G_n(z)$ has a pole of order β_n at the point ζ_n, but no other poles.

where $\beta_n \neq 0$, $a_{-\beta_n}^{(n)} \neq 0$ *if* $n = 0$.[6] *Then there exists a meromorphic function* $f(z)$ *whose poles coincide with the points* (14.28), *and whose principal part at the pole* ζ_n *equals* $G_n(z)$ *for each* $n = 0, 1, 2, \ldots$

Proof. It is not surprising that the proof bears a strong resemblance to the proof of Weierstrass' theorem (Theorem 14.6). We start from the Taylor series expansion

$$G_n(z) = a_0^{(n)} + a_1^{(n)}z + \cdots + a_k^{(n)}z^k + \cdots \qquad (n = 1, 2, \ldots),$$

which is convergent on the disk $|z| < |\zeta_n|$ and uniformly convergent on every smaller disk, in particular on $D_n: |z| < \frac{1}{2}|\zeta_n|$. Let $\{\varepsilon_n\}$ be any sequence of positive numbers such that

$$\sum_{n=1}^{\infty} \varepsilon_n < \infty. \tag{14.29}$$

Then, choosing integers $k_1, k_2, \ldots$ such that

$$|G_n(z) - [a_0^{(n)} + a_1^{(n)}z + \cdots + a_{k_n}^{(n)}z^{k_n}]| < \varepsilon_n \qquad (n = 1, 2, \ldots) \tag{14.30}$$

for all $z \in D_n$, we introduce the polynomials

$$P_n(z) = -a_0^{(n)} - a_1^{(n)}z - \cdots - a_{k_n}^{(n)}z^{k_n} \qquad (n = 1, 2, \ldots) \tag{14.31}$$

and an arbitrary polynomial $P_0(z)$. Given any disk $K_R: |z| < R$, let $N(R)$ be the smallest integer such that $|\zeta_n| > 2R$ for all $n > N(R)$. Consider the series

$$\sum_{n=N(R)+1}^{\infty} [G_n(z) + P_n(z)], \tag{14.32}$$

noting that $K_R \subset D_n$ for all $n > N(R)$, while K_R contains none of the points $\zeta_{N(R)+1}, \zeta_{N(R)+2}, \ldots$. It follows from (14.30) and (14.31) that

$$|G_n(z) + P_n(z)| < \varepsilon_n$$

for all $n > N(R)$ and $z \in K_R$. Therefore, because of (14.29) and Weierstrass' M-test (Theorem 9.1), the series (14.32) is uniformly convergent on K_R, and hence represents an analytic function $\chi_R(z)$ on K_R (see Theorem 9.4).

Thus, if

$$f(z) = \sum_{n=0}^{\infty} [G_n(z) + P_n(z)],$$

[6] We reserve the symbol ζ_0 for the point $z = 0$, i.e., for a possible pole of $f(z)$ at the origin, with principal part $G_0(z)$. The case where $f(z)$ has no pole at the origin is included by allowing $G_0(z)$ to vanish identically; formally this corresponds to setting $\beta_0 = 0$, and regarding a regular point as a pole of order zero.

we have the representation[7]

$$f(z) = f_{N(R)}(z) + \chi_R(z) \qquad (z \in K_R), \tag{14.33}$$

where $\chi_R(z)$ is analytic on K_R, and the partial sum

$$f_{N(R)}(z) = \sum_{n=0}^{N(R)} [G_n(z) + P_n(z)]$$

is a rational function whose poles in K_R are precisely those points of the sequence (14.28) which lie in K_R. Moreover, the principal part of $f_{N(R)}(z)$, and hence of $f(z)$, at any point $\zeta_n \in K_R$ is just $G_n(z)$. The theorem now follows at once from the observation that K_R can have arbitrarily large radius.

COROLLARY. *Let $f(z)$ be a meromorphic function whose poles are given by an increasing sequence of distinct complex numbers*[8]

$$\zeta_0 = 0, \zeta_1, \ldots, \zeta_n, \ldots,$$

with corresponding principal parts

$$G_0(z), G_1(z), \ldots, G_n(z), \ldots.$$

Then $f(z)$ can be represented in the form

$$f(z) = g(z) + \sum_{n=0}^{\infty} [G_n(z) + P_n(z)],$$

called a partial fraction expansion,[9] *where $g(z)$ is an entire function and the $P_n(z)$ are polynomials.*

Proof. Use Mittag-Leffler's theorem to construct a function

$$\varphi(z) = \sum_{n=0}^{\infty} [G_n(z) + P_n(z)]$$

with the same poles and principal parts as $f(z)$. Then $f(z) - \varphi(z)$ is analytic in the whole plane, and hence equals an entire function, which we denote by $g(z)$.

Example. *Given an increasing sequence $\{\zeta_n\}$ of distinct nonzero complex numbers converging to infinity, and an arbitrary complex sequence $\{A_n\}$, find an entire function $f(z)$ such that*

$$f(\zeta_n) = A_n \qquad (n = 1, 2, \ldots). \tag{14.34}$$

---

[7] Note the analogy between formulas (14.33) and (14.23).

[8] Note that the sequence $\{\zeta_n\}$ necessarily converges to infinity, since $\{\zeta_n\}$ cannot have a finite limit point.

[9] By analogy with the familiar case where $f(z)$ is a rational function (see e.g., G. Birkhoff and S. MacLane, *op. cit.*, Chap. 3, Sec. 11).

We begin by using Theorem 14.6, Corollary 1 to construct an entire function $g(z)$ with simple zeros at the points $\zeta_1, \zeta_2, \ldots$, i.e.,

$$g(z) = \prod_{n=1}^{\infty} \left(1 - \frac{z}{\zeta_n}\right) \exp\left(\frac{z}{\zeta_n} + \cdots + \frac{z^n}{n\zeta_n^n}\right).$$

Then we calculate the derivative $g'(z)$ at every point ζ_n, obtaining a sequence of nonzero complex numbers $\{g'(\zeta_n)\}$. Next we use Mittag-Leffler's theorem to find a meromorphic function $\varphi(z)$, with simple poles at the points $\zeta_1, \zeta_2, \ldots$ and corresponding principal parts

$$\frac{A_n/g'(\zeta_n)}{z - \zeta_n} \qquad (n = 1, 2, \ldots).$$

Thus

$$\varphi(z) = \sum_{n=1}^{\infty} \left[\frac{A_n/g'(\zeta_n)}{z - \zeta_n} + P_n(z)\right],$$

where the $P_n(z)$ are suitably chosen polynomials. Then the function

$$f(z) = g(z)\varphi(z)$$

is obviously entire, and moreover satisfies the condition (14.34), since

$$f(\zeta_n) = \lim_{z \to \zeta_n} g(z)\varphi(z) = \lim_{z \to \zeta_n} \left[\frac{g(z) - g(\zeta_n)}{z - \zeta_n} \varphi(z)(z - \zeta_n)\right]$$

$$= \frac{g'(\zeta_n)A_n}{g'(\zeta_n)} \qquad (n = 1, 2, \ldots).$$

Problem 1 (M2, pp. 302–304). Given an increasing sequence $\{\zeta_n\}$ of distinct nonzero complex numbers converging to infinity, find a meromorphic function $f(z)$ with

a) Simple poles at the points $\zeta_1, \zeta_2, \ldots$ and corresponding principal parts

$$\frac{1}{z - \zeta_n} \qquad (n = 1, 2, \ldots);$$

b) Poles of order 2 at the points $\zeta_1, \zeta_2, \ldots$ and corresponding principal parts

$$\frac{1}{(z - \zeta_n)^2} \qquad (n = 1, 2, \ldots).$$

Problem 2. Deduce Weierstrass' theorem from Mittag-Leffler's theorem.

Problem 3. Generalize Mittag-Leffler's theorem to the case of functions with essential singular points as well as poles.

69. THE GAMMA FUNCTION

The *gamma function* $\Gamma(z)$ is the simplest and most important of an infinite set of meromorphic functions which extend the concept of the factorial function $n!$ (originally defined only for positive integral n) to the case of arbitrary complex z. For historical reasons, the relation between $\Gamma(z)$ and $n!$ is given by the formula

$$\Gamma(n+1) = n!,$$

rather than the more natural formula $\Gamma(n) = n!$, which would allow $\Gamma(z)$ to be thought of as $z!$ Thus the relation

$$n(n-1)! = n!,$$

characterizing the factorial function, takes the form

$$n\Gamma(n) = \Gamma(n+1) \tag{14.35}$$

in terms of the gamma function. In particular, substituting $n = 1$ into (14.35), we find that[10]

$$\Gamma(1) = \Gamma(2) = 1! = 1.$$

These considerations suggest that we study the solutions of the functional equation

$$zf(z) = f(z+1) \qquad [f(1) = 1] \tag{14.36}$$

for complex z. The solution of (14.36) must be a meromorphic function, as can be seen by successively applying (14.36) to the values $z, z + 1, \ldots, z + n - 1$ (where n is a positive integer), and then combining the results to obtain the formula

$$z(z+1)\cdots(z+n-1)f(z) = f(z+n). \tag{14.37}$$

In fact, letting z approach $-(n - 1) = -m$ $(m = 0, 1, \ldots)$ in (14.37), we find that

$$\lim_{z\to -m} (z+m)f(z) = \frac{f(1)}{(-1)^m m!} = \frac{(-1)^m}{m!},$$

i.e., $f(z)$ has simple poles at the points

$$z = -m \qquad (m = 0, 1, \ldots), \tag{14.38}$$

with corresponding residues

$$\frac{(-1)^m}{m!}.$$

[10] Since $\Gamma(1)$ is formally equal to $0!$, this justifies the familiar formula $0! = 1$.

By itself, equation (14.36) cannot be used to define the gamma function, since the solution of (14.36) is not unique. To see this, let $\varphi(z) \not\equiv 1$ be any meromorphic function such that[11]

$$\varphi(z+1) = \varphi(z) \qquad [\varphi(1) = 1]. \tag{14.39}$$

Then the function

$$f^*(z) = \varphi(z)f(z) \tag{14.40}$$

also satisfies (14.36). By imposing the extra requirement that $f(z)$ have no poles other than the points (14.38) and no zeros at all, we can narrow down the class of functions satisfying (14.36). For example, if $f(z)$ satisfies (14.36), so does the function

$$f^*(z) = \frac{\tan(i + 2\pi z)}{\tan i} f(z),$$

but $f^*(z)$ has infinitely many (imaginary) zeros introduced by the factor $\tan(i + 2\pi z)$. However, even this additional requirement is not enough to make the solution of (14.36) unique. In fact, this time let $\varphi(z) \not\equiv 1$ be any nonvanishing entire function satisfying (14.39).[12] Then, if $f(z)$ is a solution of (14.36), so is the function (14.40). The following theorem shows the extent to which the solution of (14.36) remains arbitrary:

THEOREM 14.10. *Let $f(z)$ be a meromorphic function satisfying* (14.36) *with no zeros, and no poles other than the points $z = -m$ $(m = 0, 1, \ldots)$. Then $f(z)$ can be represented in the form*

$$f(z) = e^{-g(z)} \frac{1}{z \prod_{m=1}^{\infty} \left(1 + \frac{z}{m}\right) e^{-z/m}}, \tag{14.41}$$

in terms of an entire function $g(z)$ such that

$$g(z+1) - g(z) = C + 2k\pi i \qquad [g(1) = C + 2l\pi i], \tag{14.42}$$

where k and l are integers, and C is Euler's constant, defined by[13]

$$C = \lim_{n\to\infty} \left(\sum_{m=1}^{n} \frac{1}{m} - \ln n \right) = 0.5772\ldots \tag{14.43}$$

Proof. Since $f(z)$ has no zeros, and no poles except at the points $z = -m$ $(m = 0, 1, \ldots)$, the function

$$F(z) = \frac{1}{f(z)}$$

[11] Thus $\varphi(z)$ is periodic, with period 1.

[12] Such a function is $\varphi(z) = e^{2\pi i z}$.

[13] For a proof that the limit (14.43) exists, see e.g., D. V. Widder, *op. cit.*, p. 387.

is entire, with its only zeros at the same points. Since 1 is the largest nonnegative integer k for which

$$\sum_{m=1}^{\infty} \frac{1}{m^k}$$

diverges, it follows from Theorem 14.7 that $F(z)$ has the representation

$$F(z) = e^{g(z)} z \prod_{m=1}^{\infty} \left(1 + \frac{z}{m}\right) e^{-z/m},$$

where $g(z)$ is an entire function. Recalling the meaning of $F(z)$, we obtain (14.41). To see what (14.36) implies about $g(z)$, we write (14.41) in the form

$$f(z) = \lim_{n\to\infty} f_n(z), \tag{14.44}$$

where

$$f_n(z) = \frac{e^{-g(z)}}{z \prod_{m=1}^{n} \left(1 + \frac{z}{m}\right) e^{-z/m}} = \frac{n! \exp\left[-g(z) + \sum_{m=1}^{n} \frac{z}{m}\right]}{z(z+1)\cdots(z+n)}. \tag{14.45}$$

Then

$$\begin{aligned} \frac{zf(z)}{f(z+1)} &= \lim_{n\to\infty} \frac{zf_n(z)}{f_n(z+1)} \\ &= \lim_{n\to\infty} (z+n+1) \exp\left[-g(z) + g(z+1) - \sum_{m=1}^{n} \frac{1}{m}\right] \\ &= \lim_{n\to\infty} \left(1 + \frac{z+1}{n}\right) \exp\left[-g(z) + g(z+1) - \left(\sum_{m=1}^{n} \frac{1}{m} - \ln n\right)\right] \\ &= \exp\left[-g(z) + g(z+1) - C\right] \end{aligned}$$

for every z, where C is given by (14.43), and moreover

$$\begin{aligned} f(1) = \lim_{n\to\infty} f_n(1) &= \lim_{n\to\infty} \frac{\exp\left[-g(1) + \sum_{m=1}^{n} \left(\frac{1}{m} - \ln n\right)\right]}{1 + \frac{1}{n}} \\ &= \exp\left[-g(1) + C\right]. \end{aligned}$$

Therefore (14.36) implies

$$\exp\left[-g(z) + g(z+1) - C\right] = 1, \qquad \exp\left[-g(1) + C\right] = 1,$$

or equivalently (14.42) and the theorem is proved.

The simplest entire function $g(z)$ satisfying (14.42) is obviously the linear function $g(z) = Cz$, and we complete our definition of the gamma function by making precisely this choice. In other words, by the gamma function $\Gamma(z)$ we mean the meromorphic function

$$\Gamma(z) = e^{-Cz}\frac{1}{z\prod_{m=1}^{\infty}\left(1+\frac{z}{m}\right)e^{-z/m}}, \tag{14.46}$$

whose reciprocal is the entire function

$$\frac{1}{\Gamma(z)} = e^{Cz}z\prod_{m=1}^{\infty}\left(1+\frac{z}{m}\right)e^{-z/m}. \tag{14.47}$$

It follows from (14.47) and Sec. 68, Prob. 2 that

$$\frac{1}{\Gamma(z)\Gamma(-z)} = -z^2\prod_{m=1}^{\infty}\left(1-\frac{z^2}{m^2}\right) = -\frac{z}{\pi}\pi z\prod_{m=1}^{\infty}\left(1-\frac{\pi^2z^2}{m^2\pi^2}\right) = -\frac{z\sin \pi z}{\pi}$$

or

$$\frac{1}{\Gamma(z)[-z\Gamma(-z)]} = \frac{\sin \pi z}{\pi}. \tag{14.48}$$

But according to (14.36), with $f(z)$ replaced by $\Gamma(z)$,

$$z\Gamma(z) = \Gamma(z+1), \tag{14.49}$$

and hence

$$-z\Gamma(-z) = \Gamma(1-z).$$

Therefore (14.48) becomes

$$\frac{1}{\Gamma(z)\Gamma(1-z)} = \frac{\sin \pi z}{\pi}$$

or

$$\Gamma(z)\Gamma(1-z) = \frac{\pi}{\sin \pi z}, \tag{14.50}$$

an important formula in the theory of the gamma function. In particular, (14.50) implies

$$[\Gamma(\tfrac{1}{2})]^2 = \pi$$

or

$$\Gamma(\tfrac{1}{2}) = \sqrt{\pi}, \tag{14.51}$$

since $\Gamma(\frac{1}{2}) > 0$, according to (14.46). Together (14.49) and (14.51) imply

$$\Gamma\left(n+\frac{1}{2}\right) = \frac{1\cdot 3\cdots(2n-1)}{2^n}\sqrt{\pi} \qquad (n = 1, 2, \ldots).$$

Finally, replacing $g(z)$ by Cz and $f(z)$ by $\Gamma(z)$ in (14.44) and (14.45), we find that

$$\Gamma(z) = \lim_{n\to\infty} \frac{n!\exp\left\{\left(\sum_{m=1}^{n}\frac{1}{m} - C\right)z\right\}}{z(z+1)\cdots(z+n)}$$
$$= \lim_{n\to\infty} \frac{n!\exp\left\{\left[\left(\sum_{m=1}^{n}\frac{1}{m} - \ln n - C\right) + \ln n\right]z\right\}}{z(z+1)\cdots(z+n)}. \tag{14.52}$$

But according to (14.43),

$$\lim_{n\to\infty}\left(\sum_{m=1}^{n}\frac{1}{m} - \ln n - C\right) = 0,$$

while obviously

$$\exp(z\ln n) = n^z.$$

Therefore (14.52) reduces to the following basic representation of the gamma function:

$$\Gamma(z) = \lim_{n\to\infty} \frac{n!\,n^z}{z(z+1)\cdots(z+n)}.$$

Problem 1. Let $\psi(z)$ be the logarithmic derivative of the gamma function, i.e.,

$$\psi(z) = \frac{\Gamma'(z)}{\Gamma(z)}.$$

Prove that

$$\text{a) } \psi(z+1) - \psi(z) = \frac{1}{z}; \quad \text{b) } \psi(1-z) - \psi(z) = \pi\cot\pi z.$$

Problem 2. Prove that the gamma function satisfies the following relation, known as the *duplication formula*:

$$\sqrt{\pi}\,\Gamma(2z) = 2^{2z-1}\Gamma(z)\Gamma(z+\tfrac{1}{2}).$$

Hint. Starting from (14.46), show that

$$\psi'(z) + \psi'(z+\tfrac{1}{2}) = 2\psi'(2z), \tag{14.53}$$

where $\psi(z)$ is the same as in the preceding problem. Then integrate (14.53).

70. CAUCHY'S THEOREM ON PARTIAL FRACTION EXPANSIONS

The existence of a partial fraction expansion of a meromorphic function $f(z)$ is guaranteed by the corollary to Mittag-Leffler's theorem. We now use the residue theorem to explicitly find this partial fraction expansion, after first proving the following

LEMMA. *Let $f(z)$ be a meromorphic function, and let L be any closed rectifiable Jordan curve such that*

1. *The origin lies inside L;*
2. *L passes through no poles of $f(z)$.*

Let $b_0 = 0, b_1, \ldots, b_n$ be the poles of $f(z)$ inside L, with corresponding principal parts

$$G_k(z) = \frac{a_{-\beta_k}^{(k)}}{(z - b_k)^{\beta_k}} + \cdots + \frac{a_{-1}^{(k)}}{z - b_k} \qquad (k = 0, 1, \ldots, n),$$

where $G_0(z) \equiv 0$ if $b_0 = 0$ is a regular point of $f(z)$. Then

$$f(z) = \sum_{k=0}^{n} G_k(z) + \frac{1}{2\pi i}\int_L \frac{f(\zeta)}{\zeta - z}\, d\zeta, \tag{14.54}$$

where z is an arbitrary regular point of $f(z)$ inside L.

Proof. The only poles inside L of the function

$$\varphi(\zeta) = \frac{f(\zeta)}{\zeta - z} \tag{14.55}$$

are at the points $b_0, b_1, \ldots, b_n$ and at the point $\zeta = z$. In a deleted neighborhood of the point b_k $(k = 0, 1, \ldots, n)$, we have

$$f(\zeta) = \frac{a_{-\beta_k}^{(k)}}{(\zeta - b_k)^{\beta_k}} + \cdots + \frac{a_{-1}^{(k)}}{\zeta - b_k} + a_0^{(k)} + a_1^{(k)}(\zeta - b_k) + \cdots, \tag{14.56}$$

while

$$\begin{aligned} \frac{1}{\zeta - z} &= -\frac{1}{(z - b_k) - (\zeta - b_k)} = -\frac{1}{z - b_k}\,\frac{1}{1 - \dfrac{\zeta - b_k}{z - b_k}} \\ &= -\frac{1}{z - b_k} - \frac{\zeta - b_k}{(z - b_k)^2} - \cdots - \frac{(\zeta - b_k)^{\beta_k - 1}}{(z - b_k)^{\beta_k}} - \cdots \end{aligned} \tag{14.57}$$

if $|\zeta - b_k| < |z - b_k|$. Multiplying (14.56) by (14.57), we find that the term proportional to $(\zeta - b_k)^{-1}$ in the Laurent expansion of the function (14.55) at the point $\zeta = b_k$ equals

$$-\left[\frac{a_{-\beta_k}^{(k)}}{(z - b_k)^{\beta_k}} + \cdots + \frac{a_{-1}^{(k)}}{z - b_k}\right]\frac{1}{\zeta - b_k} = -\frac{G_k(z)}{\zeta - b_k}.$$

In other words,

$$\operatorname*{Res}_{\zeta = b_k} \varphi(\zeta) = -G_k(z) \qquad (k = 0, 1, \ldots, n).$$

Since obviously

$$\operatorname*{Res}_{\zeta=z} \varphi(\zeta) = f(z),$$

it follows from the residue theorem (Theorem 12.1) that

$$\frac{1}{2\pi i}\int_L \frac{f(\zeta)}{\zeta - z}\,d\zeta = \frac{1}{2\pi i}\int_L \varphi(\zeta)\,d\zeta = f(z) - \sum_{k=0}^{n} G_k(z),$$

which is equivalent to (14.54).

Theorem 14.11 (*Cauchy's theorem on partial fraction expansions*). *Let $f(z)$ be a meromorphic function, and let $\{L_n\}$ be a sequence of closed rectifiable Jordan curves with the following properties:*

1. *The origin lies inside each curve L_n $(n = 1, 2, \ldots)$;*
2. *None of the curves passes through poles of $f(z)$;*
3. *Each curve in the sequence is contained in the next, i.e., $L_n \subset I(L_{n+1})$;*
4. *If r_n is the distance from the origin to L_n, then*

$$\lim_{n\to\infty} r_n = \infty. \tag{14.58}$$

Let the poles of $f(z)$ be ordered in such a way that L_n contains the first $m_n + 1$ poles $b_0 = 0, b_1, \ldots, b_{m_n}$, with principal parts $G_0(z), G_1(z), \ldots, G_{m_n}(z)$, so that $m_n \leqslant m_{n+1}$ $(n = 1, 2, \ldots)$.[14] *Moreover, suppose that*

$$\overline{\lim_{n\to\infty}} \int_{L_n} \frac{|f(\zeta)|}{|\zeta|^{p+1}}\,ds = M < \infty \tag{14.59}$$

for some integer $p \geqslant -1$, where $ds = |d\zeta|$, and let z be an arbitrary regular point of $f(z)$. Then, if $p = -1$,

$$f(z) = \lim_{n\to\infty} \sum_{k=0}^{m_n} G_k(z), \tag{14.60}$$

while, if $p > -1$,

$$f(z) = \lim_{n\to\infty} \sum_{k=0}^{m_n} [G_k(z) + P_k(z)], \tag{14.61}$$

where the $P_k(z)$ are polynomials of degree no higher than p, and the convergence in both (14.60) and (14.61) is uniform on every compact set containing no poles of $f(z)$.[15]

Proof. First suppose that $p = -1$, so that the condition (14.59) becomes simply

$$\overline{\lim_{n\to\infty}} \int_{L_n} |f(\zeta)|\,ds = M < \infty.$$

[14] As in the lemma, $b_0 = 0$ may not be a pole of $f(z)$, in which case $G_0(z) \equiv 0$.

[15] We wish to exclude the possibility of any of the terms in (14.60) or (14.61) becoming infinite.

Then, if $|z| \leqslant R < r_n$, where r_n is the distance from the origin to L_n, for some $M' < \infty$ we have

$$\begin{aligned} \left| \frac{1}{2\pi i} \int_{L_n} \frac{f(\zeta)}{\zeta - z} d\zeta \right| &\leqslant \frac{1}{2\pi(r_n - R)} \int_{L_n} |f(\zeta)|\, ds \\ &\leqslant \frac{M'}{2\pi(r_n - R)} \to 0 \qquad \text{as} \quad n \to \infty \end{aligned} \tag{14.62}$$

[cf. (14.58) and Sec. 34, Prob. 10]. Therefore, replacing L by L_n and n by m_n in formula (14.54) [the lemma obviously applies here], and letting $n \to \infty$, we obtain the expansion (14.60). Moreover, according to (14.62), the remainder term approaches zero uniformly on the disk $|z| \leqslant R$, and hence the convergence in (14.60) is uniform on every compact set containing no poles of $f(z)$.

Next, assuming that $p > -1$, we replace $(\zeta - z)^{-1}$ in formula (14.54) by the expansion

$$\frac{1}{\zeta - z} = \frac{1}{\zeta} \frac{1}{1 - \dfrac{z}{\zeta}} = \frac{1}{\zeta} + \frac{z}{\zeta^2} + \cdots + \frac{z^p}{\zeta^{p+1}} + \frac{1}{\zeta - z} \frac{z^{p+1}}{\zeta^{p+1}}.$$

Then (14.54) becomes

$$f(z) = \sum_{k=0}^{n} G_k(z) + \sum_{j=0}^{p} \frac{z^j}{2\pi i} \int_L \frac{f(\zeta)}{\zeta^{j+1}} d\zeta + \frac{1}{2\pi i} \int_L \frac{f(\zeta)}{\zeta - z} \frac{z^{p+1}}{\zeta^{p+1}} d\zeta. \tag{14.63}$$

Observing that the poles of the function $f(\zeta)/\zeta^{j+1}$ inside L are the points $b_0, b_1, \ldots, b_n$, we write

$$\operatorname*{Res}_{\zeta=b_k} \frac{f(\zeta)}{\zeta^{j+1}} = A_k^{(j)},$$

where $j = 0, 1, \ldots, p$ and $k = 0, 1, \ldots, n$. It follows that

$$\sum_{j=0}^{p} \frac{z^j}{2\pi i} \int_L \frac{f(\zeta)}{\zeta^{j+1}} d\zeta = \sum_{j=0}^{p} (A_0^{(j)} + A_1^{(j)} + \cdots + A_n^{(j)}) z^j = \sum_{k=0}^{n} P_k(z),$$

where the polynomials

$$P_k(z) = A_k^{(0)} + A_k^{(1)} z + \cdots + A_k^{(p)} z^p \tag{14.64}$$

have degree no higher than p. Therefore we can write (14.63) in the form

$$f(z) = \sum_{k=0}^{n} [G_k(z) + P_k(z)] + \frac{1}{2\pi i} \int_L \frac{f(\zeta)}{\zeta - z} \frac{z^{p+1}}{\zeta^{p+1}} d\zeta. \tag{14.65}$$

We now replace L by L_n and n by m_n in (14.65), and examine the remainder term

$$\frac{1}{2\pi i} \int_{L_n} \frac{f(\zeta)}{\zeta - z} \frac{z^{p+1}}{\zeta^{p+1}} d\zeta.$$

If $|z| \leqslant R < r_n$, the analogue of the estimate (14.62) is

$$\left|\frac{1}{2\pi i}\int_{L_n}\frac{f(\zeta)}{\zeta - z}\frac{z^{p+1}}{\zeta^{p+1}}\,d\zeta\right| \leqslant \frac{R^{p+1}}{2\pi(r_n - R)}\int_{L_n}\frac{|f(\zeta)|}{|\zeta|^{p+1}}\,ds$$
$$\leqslant \frac{MR^{p+1}}{2\pi(r_n - R)} \to 0 \qquad \text{as} \quad n \to \infty. \tag{14.66}$$

Therefore, taking the limit of (14.65) as $n \to \infty$ (with L_n and m_n instead of L and n), we obtain the expansion (14.61). Moreover, according to (14.66), the remainder term approaches zero uniformly on the disk $|z| \leqslant R$, and hence the convergence in (14.61) is uniform on every compact set containing no poles of $f(z)$.

Remark. Suppose that besides the properties of the sequence $\{L_n\}$ listed in the statement of Theorem 14.11, there exists a constant $\lambda > 0$ such that

$$\frac{l_n}{r_n} \leqslant \lambda \qquad (n = 1, 2, \ldots),$$

where l_n is the length of L_n.[15] Then, since

$$\int_{L_n}\frac{|f(\zeta)|}{|\zeta|^{p+1}}\,ds \leqslant \max_{\zeta\in L_n}|f(\zeta)|\,\frac{l_n}{r_n^{p+1}} \leqslant \lambda\,\frac{\max_{\zeta\in L_n}|f(\zeta)|}{r_n^p},$$

(14.59) is implied by the simpler condition

$$\varlimsup_{n\to\infty}\frac{\max_{\zeta\in L_n}|f(\zeta)|}{r_n^p} < \infty, \tag{14.67}$$

which permits

$$\max_{\zeta\in L_n}|f(\zeta)|$$

to approach infinity as $n \to \infty$, but no faster than some integral power of r_n.

Example 1. *Expand the function* sec z *in partial fractions.*

For the contours L_n figuring in Theorem 14.11, we choose squares with centers at the origin and sides of length $2n\pi$ parallel to the coordinate axes. On the sides parallel to the imaginary axis,

$$z = \pm n\pi + iy,$$

and hence

$$|\sec z| = \frac{1}{|\cos(\pm n\pi + iy)|} = \frac{1}{|\cos iy|} = \frac{1}{\cosh y},$$

[15] For example, this is the case if the L_n are homothetic figures with ray center $z = 0$ (see p. 60).

while on the sides parallel to the real axis,

$$z = x \pm in\pi,$$

and hence

$$|\sec z| = \frac{1}{|\cos(x \pm in\pi)|} \leqslant \frac{1}{\sinh n\pi}$$

[see (6.33)]. It follows that

$$\int_{L_n} |\sec \zeta|\, ds \leqslant 2\int_{-n\pi}^{n\pi} \frac{dy}{\cosh y} + 4n\pi \frac{1}{\sinh n\pi}.$$

Since the integral

$$\int_{-\infty}^{\infty} \frac{dy}{\cosh y}$$

converges, and since

$$\frac{4n\pi}{\sinh n\pi} \to 0 \qquad \text{as} \quad n \to \infty,$$

the condition (14.59) is satisfied for $p = -1$, and we can use the expansion (14.60).

Inside L_n, the function

$$\sec z = \frac{1}{\cos z}$$

has poles at the points

$$z = (k - \tfrac{1}{2})\pi \qquad (-n + 1 \leqslant k \leqslant n),$$

where the poles are all simple, since the zeros of $\cos z$ are simple. According to (12.5),

$$\operatorname*{Res}_{z=(k-\frac{1}{2})\pi} \sec z = -\frac{1}{\sin(k - \frac{1}{2})\pi} = (-1)^k,$$

and therefore the principal part of $\sec z$ at the pole $z = (k - \frac{1}{2})\pi$ is

$$G_k(z) = \frac{(-1)^k}{z - (k - \frac{1}{2})\pi}.$$

Moreover, we have $G_0(z) \equiv 0$, since $z = 0$ is not a pole of $\sec z$. Therefore formula (14.60) gives

$$\sec z = \lim_{n\to\infty} \sum_{k=-n+1}^{n} \frac{(-1)^k}{z - (k - \frac{1}{2})\pi}$$

$$= \lim_{n\to\infty} \left[\sum_{k=1}^{n} \frac{(-1)^k}{z - (k - \frac{1}{2})\pi} + \sum_{k=-n+1}^{0} \frac{(-1)^k}{z - (k - \frac{1}{2})\pi} \right].$$

Replacing k by $1 - j$ in the second sum on the right, we find that

$$\sum_{k=-n+1}^{0} \frac{(-1)^k}{z - (k - \frac{1}{2})\pi} = \sum_{j=1}^{n} \frac{(-1)^{j+1}}{z + (j - \frac{1}{2})\pi},$$

which implies

$$\sec z = \lim_{n\to\infty}\left[\sum_{k=1}^{n} \frac{(-1)^k}{z-(k-\frac{1}{2})\pi} + \sum_{k=1}^{n} \frac{(-1)^{k+1}}{z+(k-\frac{1}{2})\pi}\right]$$

$$= \lim_{n\to\infty} \sum_{k=1}^{n}(-1)^k \frac{(2k-1)\pi}{z^2-(k-\frac{1}{2})^2\pi^2},$$

after changing from j to k. Thus, finally, we have the expansion

$$\sec z = \sum_{k=1}^{\infty}(-1)^k \frac{(2k-1)\pi}{z^2-(k-\frac{1}{2})^2\pi^2},$$

which converges uniformly on every compact set containing none of the points

$$z = (k-\tfrac{1}{2})\pi \qquad (k = 0, \pm 1, \pm 2, \ldots).$$

Example 2. *Expand the function* cot z *in partial fractions.*

This time we choose the contours L_n to be squares with centers at the origin and sides of length $(2n+1)\pi$ parallel to the coordinate axes. On the sides parallel to the imaginary axis,

$$z = \pm(n+\tfrac{1}{2})\pi + iy,$$

and hence

$$|\cot z| = \left|\frac{\cos[\pm(n+\frac{1}{2})\pi + iy]}{\sin[\pm(n+\frac{1}{2})\pi + iy]}\right| = \frac{|\sin(iy)|}{|\cos(iy)|} = \left|\frac{e^y-e^{-y}}{e^y+e^{-y}}\right| < 1,$$

while on the sides parallel to the real axis,

$$z = x \pm i(n+\tfrac{1}{2})\pi,$$

and hence

$$|\cot z| = \left|\frac{\cos[x \pm i(n+\frac{1}{2})\pi]}{\sin[x \pm i(n+\frac{1}{2})\pi]}\right| \leqslant \frac{\cosh(n+\frac{1}{2})\pi}{\sinh(n+\frac{1}{2})\pi}$$

$$= \frac{1+e^{-(2n+1)\pi}}{1-e^{-(2n+1)\pi}} \leqslant \frac{1+e^{-\pi}}{1-e^{-\pi}} = \frac{e^{\pi}+1}{e^{\pi}-1}$$

[see (6.33)]. Therefore

$$|\cot z| \leqslant \frac{e^{\pi}+1}{e^{\pi}-1}$$

on every square L_n. This means that the condition (14.67), and consequently the condition (14.59) is satisfied for $p = 0$, which allows us to use the expansion (14.61).

Inside L_n, the function

$$\cot z = \frac{\cos z}{\sin z}$$

has poles at the points

$$z = k\pi \qquad (-n \leqslant k \leqslant n).$$

These poles are all simple, since the zeros of $\sin z$ are simple. Clearly

$$\operatorname*{Res}_{z=k\pi} \cot z = \frac{\cos k\pi}{\cos k\pi} = 1,$$

and therefore the principal part of $\cot z$ at the pole $z = k\pi$ is

$$G_k(z) = \frac{1}{z - k\pi}.$$

In the present case, the polynomials $P_k(z)$ defined by (14.64) reduce to constants:

$$P_k(z) = A_k^{(0)} = \operatorname*{Res}_{\zeta=k\pi} \frac{\cot \zeta}{\zeta}.$$

The function $\cot \zeta/\zeta$ is obviously even, and hence its Laurent expansion at the origin contains only even powers of ζ, which implies

$$\operatorname*{Res}_{\zeta=0} \frac{\cot \zeta}{\zeta} = 0.$$

Moreover, the points $\zeta = k\pi$ $(k \neq 0)$ are simple poles of $\cot \zeta/\zeta$, and hence

$$\operatorname*{Res}_{\zeta=k\pi} \frac{\cot \zeta}{\zeta} = \frac{\cos k\pi}{k\pi \cos k\pi} = \frac{1}{k\pi} \qquad (k \neq 0).$$

Thus

$$P_0(z) = 0, \quad P_k(z) = \frac{1}{k\pi} \qquad (k \neq 0),$$

and formula (14.61) gives

$$\begin{aligned} \cot z &= \lim_{n\to\infty} \left[\frac{1}{z} + \sum_{k=1}^{n}\left(\frac{1}{z - k\pi} + \frac{1}{k\pi}\right) + \sum_{k=1}^{n}\left(\frac{1}{z + k\pi} - \frac{1}{k\pi}\right)\right] \\ &= \lim_{n\to\infty} \left[\frac{1}{z} + \sum_{k=1}^{n}\left(\frac{1}{z - k\pi} + \frac{1}{z + k\pi}\right)\right] \qquad (14.68) \\ &= \frac{1}{z} + \sum_{k=1}^{\infty}\left(\frac{1}{z - k\pi} + \frac{1}{z + k\pi}\right) = \frac{1}{z} + \sum_{k=1}^{\infty} \frac{2z}{z^2 - k^2\pi^2}, \end{aligned}$$

where the convergence is uniform on every compact set containing none of the points

$$z = k\pi \qquad (k = 0, \pm 1, \pm 2, \ldots).$$

Integrating (14.68) term by term along an arbitrary rectifiable curve γ which starts at the origin and passes through none of the points $z = k\pi$ $(k \neq 0)$, we obtain

$$\int_0^z \left(\cot \zeta - \frac{1}{\zeta}\right) d\zeta = \sum_{k=1}^{\infty} \operatorname{Ln}\left(\frac{k\pi - z}{k\pi}\frac{k\pi + z}{k\pi}\right) = \sum_{k=1}^{\infty} \operatorname{Ln}\left(1 - \frac{z^2}{k^2\pi^2}\right), \quad (14.69)$$

where the logarithms on the right have perfectly definite values, i.e., those corresponding to the values of the integral

$$\int_\gamma \left(\frac{1}{z - k\pi} + \frac{1}{z + k\pi}\right) dz$$

(cf. Sec. 40, Prob. 2). The integral on the left in (14.69) equals one of the values of the function

$$\operatorname{Ln} \frac{\sin z}{z}.$$

Therefore

$$\operatorname{Ln} \frac{\sin z}{z} = \sum_{k=1}^{\infty} \operatorname{Ln} \left(1 - \frac{z^2}{k^2\pi^2}\right),$$

which implies

$$\frac{\sin z}{z} = \prod_{k=1}^{\infty} \left(1 - \frac{z^2}{k^2\pi^2}\right),$$

or

$$\sin z = z \prod_{k=1}^{\infty} \left(1 - \frac{z^2}{k^2\pi^2}\right) \tag{14.70}$$

(cf. Sec. 68, Prob. 2).

Problem 1. Verify the following partial fraction expansions:

a) $\csc z = \dfrac{1}{z} + \displaystyle\sum_{k=1}^{\infty} (-1)^k \frac{2z}{z^2 - k^2\pi^2}$;

b) $\tan z = \displaystyle\sum_{k=1}^{\infty} \frac{2z}{(k - \frac{1}{2})^2\pi^2 - z^2}$;

c) $\dfrac{z}{e^z - 1} = 1 - \dfrac{z}{2} + \displaystyle\sum_{k=1}^{\infty} \frac{2z^2}{z^2 + 4k^2\pi^2}$.

Problem 2. Derive partial fraction expansions of the functions

a) $\tanh z$; b) $\operatorname{csch} z$.

Problem 3. Let $f(z)$ be a function whose only singular points in the finite plane are poles at the points $a_1, a_2, \ldots, a_m$, none of which equals $0, \pm 1, \pm 2, \ldots$. Prove that if there is a sequence of contours L_n with the same properties as in Theorem 14.11 such that

$$\lim_{n \to \infty} \int_{L_n} f(z) \cot \pi z \, dz = 0, \tag{14.71}$$

then

$$\sum_{n=-\infty}^{\infty} f(n) = -\pi \sum_{k=1}^{m} \operatorname*{Res}_{z=a_k} \{f(z) \cot \pi z\}. \tag{14.72}$$

Problem 4. Let $f(z)$, the points a_k and the contours L_n be the same as in the preceding problem, but suppose (14.71) is replaced by the condition

$$\lim_{n\to\infty} \int_{L_n} f(z) \csc z \, dz = 0.$$

Prove that (14.72) is then replaced by

$$\sum_{n=-\infty}^{\infty} (-1)^n f(n) = -\pi \sum_{k=1}^{m} \operatorname*{Res}_{z=a_k} \{f(z) \csc z\}.$$

Problem 5. Sum the following series, where in the first four cases a is such that none of the denominators vanishes:

$$\text{a) } \sum_{n=-\infty}^{\infty} \frac{1}{(n+a)^2}; \quad \text{b) } \sum_{n=-\infty}^{\infty} \frac{(-1)^n}{(n+a)^2}; \quad \text{c) } \sum_{n=0}^{\infty} \frac{1}{n^2+a^2};$$

$$\text{d) } \sum_{n=0}^{\infty} \frac{(-1)^n}{n^2+a^2}; \quad \text{e) } \sum_{n=0}^{\infty} \frac{1}{(2n+1)^2}; \quad \text{f) } \sum_{n=0}^{\infty} \frac{(-1)^n}{(2n+1)^3}.$$

Ans. c) $\dfrac{1}{2a^2}(1 + \pi a \coth \pi a)$; f) $\dfrac{\pi^3}{32}$.

Problem 6. Derive *Wallis' product*

$$\frac{\pi}{2} = \frac{2}{1}\frac{2}{3}\frac{4}{3}\frac{4}{5}\frac{6}{5}\frac{6}{7} \cdots \frac{2n}{2n-1}\frac{2n}{2n+1} \cdots.$$

Problem 7. Show that

$$\frac{\sin(a-z)}{\sin a} = e^{-z \cot a} \prod_{n=-\infty}^{\infty} \left(1 - \frac{z}{a+n\pi}\right) e^{z/(a+n\pi)},$$

if a is not a multiple of π.

Hint. Recall (14.68).

Problem 8. Deduce the periodicity of $\sin z$ (see p. 121) from its infinite product expansion (14.71).

Hint. Note that

$$\frac{\sin \pi z}{\pi} = \lim_{n\to\infty} \varphi_n(z),$$

where

$$\varphi_n(z) = \left(1 - \frac{z}{n}\right) \cdots (1-z) z (1+z) \cdots \left(1 + \frac{z}{n}\right).$$

Then show that

$$\varphi_n(z+1) = -\frac{n+z+1}{n-z} \varphi_n(z).$$

CHAPTER 15

CONFORMAL MAPPING

71. GENERAL PRINCIPLES OF CONFORMAL MAPPING

It will be recalled that a single-valued function $w = f(z)$ analytic on a domain G is conformal at every point $z_0 \in G$ such that $f'(z_0) \neq 0$ (see Theorem 3.4). Moreover, according to Theorem 12.9, the image $\mathscr{G} = f(G)$ of the domain G under the mapping $w = f(z)$ is itself a domain. Thus, given a function $f(z)$ analytic on G, where $f'(z)$ is nonvanishing on G, one might be tempted to describe the mapping of G onto $\mathscr{G} \doteq f(G)$ as a conformal mapping. However, in any theory involving "mappings," it seems reasonable to insist that the roles of the domains G and $\mathscr{G}$ be interchangeable. This leads to the additional requirement that the inverse function $z = f^{-1}(w)$ be single-valued, or equivalently that the original function $w = f(z)$ be one-to-one:

DEFINITION. *Let G and $\mathscr{G}$ be two domains in the finite plane,*[1] *and let $w = f(z)$ be a one-to-one analytic function mapping G onto $\mathscr{G}$. Then $w = f(z)$ is called a conformal mapping of G onto $\mathscr{G}$, and $w = f(z)$ is said to map G conformally onto $\mathscr{G}$. Moreover, $\mathscr{G}$ is called a conformal image of G.*

---

[1] The case of domains in the extended plane (see Sec. 12, Prob. 9) is considered in Probs. 2 and 3.

Remark. Note that if $f(z)$ is one-to-one on G, then $f'(z)$ is automatically nonvanishing on G (see Theorem 12.8 and its corollaries). Hence, any conformal mapping of G is conformal at every point of G.

Example 1. Let $w = f(z) = e^z$, and let G be any domain containing two points z_1, z_2 differing by an integral multiple of $2\pi i$. Then $f'(z)$ is nonvanishing on G, and hence $f(z)$ is conformal at every point of G and one-to-one in a sufficiently small neighborhood of every point of G (recall Theorem 12.7). However, $f(z)$ is not one-to-one on G itself, since it takes the same value at the points z_1 and z_2. Therefore $w = f(z)$ is not a conformal mapping of G onto $\mathcal{G} = f(G)$.

Example 2. Let $w = f(z) = e^z$ as in Example 1, but this time let G be a domain containing no pair of points differing by an integral multiple of $2\pi i$.[2] Then $w = f(z)$ is a conformal mapping of G onto $\mathcal{G} = f(G)$.

THEOREM 15.1. *If $\mathcal{G}$ is a conformal image of a domain G, then G is a conformal image of $\mathcal{G}$.*

Proof. Apply Rule 5, p. 48 after noting that the inverse of a one-to-one analytic function is continuous (why?).

THEOREM 15.2. *If $\mathcal{G}$ is a conformal image of a domain G and if $\mathcal{G}^*$ is a conformal image of $\mathcal{G}$, then $\mathcal{G}^*$ is a conformal image of G.*

Proof. Apply Rule 4, p. 46.

It is now a natural problem to look for classes of domains which are conformal images of each other. For multiply connected domains, this is a complicated matter, but for simply connected domains the problem has a definitive solution given by

THEOREM 15.3 (*Riemann's mapping theorem*).[3] *Any simply connected domain G in the finite plane other than the whole plane is a conformal image of the open unit disk.*

COROLLARY. *Let G and $\mathcal{G}$ be any two simply connected domains in the finite plane other than the whole plane. Then G and $\mathcal{G}$ are conformal images of each other.*

Proof. Use Theorems 15.1 and 15.2.

---

[2] Thus G is a domain of univalence for $w = e^z$ as defined on p. 73. (A one-to-one analytic function is often called a *univalent* function.)

[3] The proof of Theorem 15.3 is beyond the scope of this book, but can be found in the more advanced literature on complex analysis (see M3, Sec. 2).

Remark 1. It is clear why the case where G is the whole plane must be excluded in Theorem 15.3. In fact, the existence of a one-to-one analytic function $w = f(z)$ mapping the whole plane onto the unit disk implies $|f(z)| < 1$ for all z, i.e., $f(z)$ is a bounded entire function. But then by Liouville's theorem (Theorem 10.7), $f(z) \equiv$ const, which is absurd.

Remark 2. The function $w = f(z)$ mapping G conformally onto $\mathscr{G} = f(G)$ is not unique. To see this, map G conformally onto the unit disk, then map the unit disk conformally onto itself, and finally map the unit disk conformally onto $\mathscr{G}$. The result is a conformal mapping of G onto $\mathscr{G}$. However, the unit disk can be mapped conformally onto itself in infinitely many ways (see Example 2, p. 109), and hence G can be mapped conformally onto $\mathscr{G}$ in infinitely many ways.

The freedom of choice in the conformal mapping of G onto $\mathscr{G}$ is eliminated in precisely the same way as for the conformal mapping of a disk onto itself (see Sec. 28, Prob. 4):

THEOREM 15.4 (*Uniqueness theorem for conformal mappings*). *Let G be the same as in Theorem* 15.3, *and let z_0 be an arbitrary point of G. Then there exists a unique function $w = f(z)$ which maps G conformally onto the disk $|w| < 1$ and satisfies the conditions*

$$f(z_0) = 0, \qquad f'(z_0) > 0.$$

Proof. Suppose $w = g(z)$ is another function satisfying the same conditions as $w = f(z)$. Then the function

$$\varphi(w) = f[g^{-1}(w)]$$

is analytic on the unit disk, satisfies the conditions

$$\varphi(0) = f[g^{-1}(0)] = f(z_0) = 0, \qquad \varphi'(0) = \frac{1}{g'(z_0)} f'(z_0) > 0,$$

and maps the unit disk conformally onto itself. Therefore, according to Schwarz's lemma (Theorem 12.12),

$$|\varphi(w)| \leqslant |w|,$$

i.e.,

$$|f(z)| \leqslant |g(z)| \tag{15.1}$$

for all $z \in G$. But then, interchanging the roles of $f(z)$ and $g(z)$, we obtain

$$|g(z)| \leqslant |f(z)|,$$

which together with (15.1) implies

$$|f(z)| = |g(z)|,$$

or equivalently

$$|\varphi(w)| = |w|. \tag{15.2}$$

Applying Schwarz's lemma again to (15.2), we conclude that

$$\varphi(w) = e^{i\alpha}w.$$

But $\varphi'(0) > 0$ and hence $e^{i\alpha} = 1$, i.e.,

$$\varphi(w) = w.$$

It follows that

$$f(z) = g(z)$$

for all $z \in G$, as asserted.

It is often possible to infer that $w = f(z)$ is a conformal mapping of a domain G onto $\mathscr{G} = f(G)$ from a knowledge of the behavior of $f(z)$ on the boundary of G. The following is a typical result of this kind:

THEOREM 15.5. *Given a closed rectifiable Jordan curve γ, suppose $f(z)$ is analytic on $\overline{I(\gamma)}$ and one-to-one on γ. Then $f(z)$ is analytic on $I(\gamma)$, which it maps conformally onto $I(\Gamma)$ where $\Gamma = f(\gamma)$.*

Proof. First we note that Γ is itself a closed rectifiable Jordan curve (see Prob. 7), while $f[I(\gamma)]$ is a domain (see Theorem 12.9). Let w_1 be an arbitrary point of $I(\Gamma)$. According to the argument principle (Theorem 12.4), the number of w_1-points of $f(z)$ inside γ minus the number of poles of $f(z)$ inside γ equals ν, the number of circuits around the point w_1 made by the point $w = f(z)$ as the point z traverses γ once in the positive direction. *A priori*, ν can only equal -1 or $+1$, since Γ is a closed Jordan curve.[4] But $f(z)$ has no poles inside γ, and hence ν must equal $+1$, i.e., $f(z)$ takes every value $w_1 \in I(\Gamma)$ once, and we have incidentally proved that $w = f(z)$ traverses Γ in the same direction as z traverses γ. It follows that

$$f[I(\gamma)] \supset I(\Gamma). \tag{15.3}$$

Next let w_2 be an arbitrary point of $E(\Gamma)$. Then, as z traverses the curve γ, the point $w = f(z)$ traverses the curve Γ and obviously cannot wind around w_2. Again applying the argument principle, and the fact that $f(z)$ has no poles inside γ, we find that $f(z)$ has no w_2-points inside γ, i.e., $E(\Gamma) \cap f[I(\gamma)] = 0$. We also have $\Gamma \cap f[I(\gamma)] = 0$, since $w \in \Gamma$, $w \in f[I(\gamma)]$ implies that some neighborhood $\mathscr{N}(w)$ is contained in $f[I(\gamma)]$. But this contradicts what was just proved, since every neighborhood of a point of Γ must contain points of $E(\Gamma)$. Therefore

$$f[I(\gamma)] \subset I(\Gamma), \tag{15.4}$$

[4] This fact, which is easily verified in the simplest cases (e.g., where γ is a circle or a polygon), can be proved in complete generality. See e.g., P. S. Aleksandrov, *op. cit.*, Chap. 2.

and, together, (15.3) and (15.4) imply the desired result $f[I(\gamma)] = I(\Gamma)$. Moreover, as already noted, every point of $I(\Gamma)$ has precisely one inverse image in $I(\gamma)$. It follows that $f(z)$ is one-to-one on $I(\gamma)$, which it maps conformally onto $I(\Gamma)$.

Theorem 15.5 can be generalized in a number of ways. One of these is particularly important in the applications and will be needed in the next two sections:

THEOREM 15.6.[5] *Let γ be the real axis and D the open upper half-plane. Suppose $f(z)$ is continuous on $\bar{D}$, analytic on $\bar{D}$ except at a finite number of points of γ, and one-to-one on γ which it maps into a closed rectifiable Jordan curve Γ. Then $f(z)$ is one-to-one on D, which it maps conformally onto $I(\Gamma)$.*

Problem 1 (M1, p. 382). Interpret Schwarz's lemma geometrically.

Problem 2. Generalize Riemann's mapping theorem to the case where G is a simply connected domain in the extended plane and simple connectedness has the same meaning as in Sec. 12, Prob. 9.

Hint. Use the preliminary transformation $\zeta = 1/z$ to map G conformally onto a domain in the finite plane.

Problem 3. Generalize the corollary to Riemann's mapping theorem to the case where G and $\mathcal{G}$ are both simply connected domains in the extended plane.

Hint. The function mapping G onto $\mathcal{G}$ is now allowed to have one pole.

Problem 4. Can the finite plane with the origin deleted be mapped conformally onto the open unit disk with the origin deleted?

Problem 5 (M2, p. 116). Prove that if $f(z)$ is one-to-one and analytic in the whole plane except for one pole, then $f(z)$ is a Möbius transformation.

Problem 6 (M2, p. 118). Prove that if a sequence $\{f_n(z)\}$ is uniformly convergent on every compact subset of a domain G and if every term $f_n(z)$ is one-to-one and analytic on G, then the limit function

$$f(z) = \lim_{n\to\infty} f_n(z)$$

is also one-to-one and analytic on G, unless $f(z) \equiv \text{const.}$

Hint. Use Hurwitz's theorem (Theorem 12.6).

Problem 7 (M1, p. 305). Prove that the analytic image of a rectifiable curve is rectifiable.

---

[5] We omit the proof of Theorem 15.6 which is a straightforward but rather tedious deduction from the argument principle. The interested reader can prove it by specializing the considerations given in M2, pp. 120–123.

Problem 8 (M2, p. 119). Given a closed rectifiable Jordan curve γ, suppose $f(z)$ is analytic on $\overline{E(\gamma)}$ and one-to-one on γ. Prove that $f(z)$ is one-to-one on $E(\gamma)$, which it maps conformally onto $I(\Gamma)$ where $\Gamma = f(\gamma)$.

72. MAPPING OF THE UPPER HALF-PLANE ONTO A RECTANGLE

To illustrate the application of the results of the preceding section, consider the function

$$w = f(z) = \int_0^z \frac{dt}{\sqrt{(1-t^2)(1-k^2t^2)}} \qquad (0 < k < 1), \tag{15.5}$$

known as the *elliptic integral of the first kind* (*in Legendre's form*), where the path of integration is any rectifiable curve joining the points 0 and z. The function

$$\sqrt{(1-t^2)(1-k^2t^2)} \tag{15.6}$$

is double-valued, and has first-order branch points at $t = \pm 1$ and $t = \pm 1/k$ (cf. Remark 3, p. 78). Since these points all lie on the real axis, we can define two single-valued analytic branches of (15.6) in the z-plane cut along the segments $-1/k \leqslant x \leqslant -1$ and $1 \leqslant x \leqslant 1/k$ of the real axis (see Prob. 1), in particular, in the upper half-plane D: $\operatorname{Im} z > 0$. Of these two branches which take values with opposite signs at every point, we choose the one which takes positive values when t belongs to the interval $(0, 1)$.[6] In a neighborhood $|z| > 1/k$ of the point at infinity, the integrand in (15.5) has the expansion

$$\begin{aligned}(1-t^2)^{-1/2}(1-k^2t^2)^{-1/2} &= k^{-1}t^{-2}(1-t^{-2})^{-1/2}(1-k^{-2}t^{-2})^{-1/2} \\ &= \pm k^{-1}t^{-2}(1+\tfrac{1}{2}t^{-2}+\cdots)\left(1+\frac{1}{2k^2}t^{-2}+\cdots\right) \\ &= \pm[k^{-1}t^{-2}+\tfrac{1}{2}(k^{-1}+k^{-3})t^{-4}+\cdots],\end{aligned}$$

which implies

$$f(z) = C \mp \left[k^{-1}z^{-1} + \frac{1}{2\cdot 3}(k^{-1}+k^{-3})z^{-3}+\cdots\right],$$

where C is a constant. Therefore it is clear that both branches of $f(z)$ are analytic at infinity. Moreover, both branches are continuous on $\bar{D}$.

With a view to eventually applying Theorem 15.6, we now consider the image of γ under the mapping (15.5). In the interval $0 < x < 1$, we have

$$w = f(x) = \int_0^x \frac{dt}{\sqrt{(1-t^2)(1-k^2t^2)}},$$

[6] This is in keeping with what we mean by $\sqrt{a}$ when $a > 0$, i.e., the *positive* square root.

which is positive and increases from 0 to

$$K = \int_0^1 \frac{dt}{\sqrt{(1-t^2)(1-k^2t^2)}}$$

as x goes from 0 to 1. In the interval $1 < x < 1/k$, the integrand in (15.5) becomes

$$\frac{1}{\pm i\sqrt{(t^2-1)(1-k^2t^2)}}. \tag{15.7}$$

Here the sign cannot be chosen arbitrarily, but must agree with our previous choice of the branch of (15.6), in order to guarantee continuity of this branch on $\bar{D}$. Writing $(1-t^2)(1-k^2t^2)$ in the form

$$k^2(t-1)[t-(-1)]\left(t-\frac{1}{k}\right)\left[t-\left(-\frac{1}{k}\right)\right] = g(t), \tag{15.8}$$

we observe that in going from the interval $0 < x < 1$ to the interval $1 < x < 1/k$, the change in $\operatorname{Arg} g(t)$ equals $-\pi$.[7] In fact, the change in $\operatorname{Arg} g(t)$ equals the sum of the changes in arguments of all the separate factors in (15.8), but none of these arguments changes except $\operatorname{Arg}(t-1)$, which changes by $-\pi$. Therefore the change in $\operatorname{Arg}\sqrt{g(t)}$ is $-\pi/2$, so that $\operatorname{Arg}\sqrt{g(t)}$ becomes

$$-\frac{\pi}{2} + 2k\pi, \tag{15.9}$$

since $\operatorname{Arg}\sqrt{g(t)} = 2k\pi$ for $0 < t < 1$. Thus we see that of the two values $\pm i\sqrt{(t^2-1)(1-k^2t^2)}$ in (15.7), the correct choice is $-i\sqrt{(t^2-1)(1-k^2t^2)}$. It follows that

$$\begin{aligned} w = f(x) &= \int_0^x \frac{dt}{\sqrt{(1-t^2)(1-k^2t^2)}} \\ &= \int_0^1 \frac{dt}{\sqrt{(1-t^2)(1-k^2t^2)}} + i\int_1^x \frac{dt}{\sqrt{(t^2-1)(1-k^2t^2)}} \\ &= K + i\int_1^x \frac{dt}{\sqrt{(t^2-1)(1-k^2t^2)}} \end{aligned}$$

for $1 < x < 1/k$. Therefore, as the point $z = x$ traverses the interval $1 < x < 1/k$, the image point $w = f(x)$ traverses the line segment parallel

[7] This is easily seen by drawing a small semicircle in the upper half-plane with center $t = 1$, and examining the behavior of $\operatorname{Arg} g(t)$ as t traverses the semicircle going from left to right.

to the imaginary axis going from the point K to the point $K + iK'$, where

$$K' = \int_1^{1/k} \frac{dt}{\sqrt{(t^2 - 1)(1 - k^2t^2)}}.$$

By making the substitution

$$t = \frac{1}{\sqrt{1 - k'^2\tau^2}} \qquad (k'^2 = 1 - k^2),$$

and then changing τ back to t, we can write the integral for K' in the same form as the integral for K:

$$K' = \int_0^1 \frac{dt}{\sqrt{(1 - t^2)(1 - k'^2t^2)}} \qquad (k'^2 = 1 - k^2).$$

Next we consider the interval $1/k < x < +\infty$. In this interval, the integrand in (15.5) becomes

$$\frac{1}{\pm\sqrt{(t^2 - 1)(k^2t^2 - 1)}}, \tag{15.10}$$

where, as before, the sign cannot be chosen arbitrarily. In fact, writing $(t^2 - 1)(1 - k^2t^2)$ in the form

$$-k^2(t - 1)(t + 1)\left(t - \frac{1}{k}\right)\left(t + \frac{1}{k}\right) = h(t),$$

we observe that in going from the interval $1 < x < 1/k$ to the interval $1/k < x < +\infty$, the change in $\operatorname{Arg} h(t)$ equals $-\pi$, i.e., the change in $\operatorname{Arg}[t - (1/k)]$. Therefore the change in $\operatorname{Arg}\sqrt{h(t)}$ is $-\pi/2$, so that $\operatorname{Arg}\sqrt{h(t)}$ becomes

$$-\frac{\pi}{2} + \operatorname{Arg}\sqrt{g(t)} = -\pi + 2k\pi$$

[cf. (15.9)]. Thus we see that of the two values $\pm\sqrt{(t^2 - 1)(k^2t^2 - 1)}$ in (15.10), the correct choice is $-\sqrt{(t^2 - 1)(k^2t^2 - 1)}$. It follows that

$$\begin{aligned} w = f(x) &= \int_0^x \frac{dt}{\sqrt{(1 - t^2)(1 - k^2t^2)}} \\ &= \int_0^1 \frac{dt}{\sqrt{(1 - t^2)(1 - k^2t^2)}} + i\int_1^{1/k} \frac{dt}{\sqrt{(t^2 - 1)(1 - k^2t^2)}} \\ &\quad - \int_{1/k}^x \frac{dt}{\sqrt{(t^2 - 1)(k^2t^2 - 1)}} \\ &= K + iK' - \int_{1/k}^x \frac{dt}{\sqrt{(t^2 - 1)(k^2t^2 - 1)}}. \end{aligned}$$

As x increases from $1/k$ to $+\infty$, the last integral on the right becomes

$$\int_{1/k}^{\infty} \frac{dt}{\sqrt{(t^2-1)(k^2t^2-1)}} = \int_0^1 \frac{d\tau}{\sqrt{(1-\tau^2)(1-k^2\tau^2)}} = K,$$

where we make the change of variables $t = 1/k\tau$. Correspondingly, the image point $w = f(x)$ describes the line segment parallel to the real axis going from the point $K + iK'$ to the point iK'

In just the same way, we see that as x moves along the real axis first from 0 to -1, then from -1 to $-1/k$ and finally from $-1/k$ to $-\infty$, the point $w = f(x)$ first describes the segment from 0 to $-K$, then the segment from $-K$ to $-K + iK'$, and finally the segment from $-K + iK'$ to iK'. Thus $w = f(x)$ is a one-to-one mapping of the real axis onto the rectangular contour with vertices $-K$, K, $K + iK'$ and $-K + iK'$, where

$$K = \int_0^1 \frac{dt}{\sqrt{(1-t^2)(1-k^2t^2)}},$$

$$K' = \int_0^1 \frac{dt}{\sqrt{(1-t^2)(1-k'^2t^2)}} \qquad (k'^2 = 1 - k^2).$$

It follows from Theorem 15.6 that $w = f(z)$ is one-to-one in the upper half-plane, which it maps conformally onto the indicated rectangle.

Now consider the expression

$$\lambda(k) = \frac{K'}{2K} = \frac{\displaystyle\int_0^1 \frac{dt}{\sqrt{(1-t^2)(1-k'^2t^2)}}}{\displaystyle 2\int_0^1 \frac{dt}{\sqrt{(1-t^2)(1-k^2t^2)}}},$$

which is the ratio of the altitude of the rectangle to its base. As the parameter k $(0 < k < 1)$ increases from 0 to 1, the denominator increases from

$$2\int_0^1 \frac{dt}{\sqrt{1-t^2}} = \pi$$

to ∞. At the same time, $k' = \sqrt{1-k^2}$ decreases from 1 to 0, so that the numerator decreases from ∞ to $\pi/2$. Therefore, as k increases from 0 to 1, the ratio $\lambda(k) = K'/2K$ decreases continuously from ∞ to 0. It follows that given any rectangle with base $2a$ and altitude b, we can find a unique value of k $(0 < k < 1)$ such that

$$\frac{b}{2a} = \frac{\displaystyle\int_0^1 \frac{dt}{\sqrt{(1-t^2)(1-k'^2t^2)}}}{\displaystyle\int_0^1 \frac{dt}{\sqrt{(1-t^2)(1-k^2t^2)}}} = \frac{K'}{2K}.$$

Then the function (15.5) corresponding to this value of k maps the upper half-plane conformally onto a rectangle similar to the given rectangle. To obtain a mapping onto the given rectangle, we need only introduce an extra factor $\mu = a/K = b/K'$, writing

$$w = f(z) = \mu \int_0^z \frac{dt}{\sqrt{(1-t^2)(1-k^2t^2)}} \tag{15.11}$$

instead of (15.5). Then (15.15) maps the upper half-plane onto the rectangle with vertices

$$-\mu K = -a, \quad \mu K = a, \quad \mu K + i\mu K' = a + ib, \quad -\mu K + i\mu K' = -a + ib,$$

and hence onto any given rectangle, provided the parameters k and μ are suitably chosen.

Problem 1 (M1, Sec. 55). Let

$$f(z) = \sqrt[n]{P(z)},$$

where $P(z)$ is an arbitrary polynomial of degree N, with zeros $a_1, \ldots, a_m$ of orders $\alpha_1, \ldots, \alpha_m$, respectively, where $\alpha_1 + \cdots + \alpha_m = N$. Prove that

a) No point of the finite plane other than a zero of $P(z)$ can be a branch point of $f(z)$;

b) Every zero a_k of $P(z)$ whose order is not a multiple of n is a branch point of $f(z)$ [of what order?];

c) If γ is a closed Jordan curve whose interior contains a set of zeros $a_{k_1}, \ldots, a_{k_q}$ $(q < m)$ of $P(z)$ such that $\alpha_{k_1} + \cdots + \alpha_{k_q}$ is a multiple of n and whose exterior contains all the other zeros, then a circuit around γ does not change the value of $f(z)$.

Problem 2. Specialize formula (15.11) to the case of a square.

73. THE SCHWARZ-CHRISTOFFEL TRANSFORMATION

The mapping (15.5) studied in the preceding section is a special case of a more general mapping known as the *Schwarz-Christoffel transformation*, defined by the integral

$$w = f(z) = C \int_0^z (t - a_1)^{\alpha_1 - 1}(t - a_2)^{\alpha_2 - 1} \cdots (t - a_n)^{\alpha_n - 1}\, dt. \tag{15.12}$$

Here C is a positive constant, $a_1, \ldots, a_n$ are distinct real numbers (assumed to be arranged in increasing order $a_1 < \cdots < a_n$), and the exponents $\alpha_1 - 1, \ldots, \alpha_n - 1$ are also real but not necessarily distinct. For example, (15.12) reduces to the elliptic integral (15.5) if we make the following choice of constants

$$C = \frac{1}{k}, \quad a_1 = -\frac{1}{k}, \quad a_2 = -1, \quad a_3 = 1, \quad a_4 = \frac{1}{k},$$

$$\alpha_1 = \alpha_2 = \alpha_3 = \alpha_4 = \frac{1}{2}.$$

To guarantee convergence of the integral (15.12), we require that

$$\alpha_k - 1 > -1 \quad \text{or} \quad \alpha_k > 0 \qquad (k = 1, \ldots, n), \tag{15.13}$$

and

$$\alpha_1 + \cdots + \alpha_n - n < -1, \tag{15.14}$$

where (15.14) stems from the fact that the integrand

$$\beta(t) = (t - a_1)^{\alpha_1 - 1} \cdots (t - a_n)^{\alpha_n - 1} \tag{15.15}$$

behaves like $t^{\alpha_1 + \cdots + \alpha_n - n}$ for large t.

Of the various single-valued analytic branches of the function $\beta(t)$ which can be defined in the upper half-plane, we now choose the branch such that the argument of every factor $(t - a_1)^{\alpha_1 - 1}, \ldots, (t - a_n)^{\alpha_n - 1}$ equals π when all the differences $t - a_1, \ldots, t - a_n$ are negative, i.e., when t belongs to the interval $(-\infty, a_1)$. This corresponds to choosing the value

$$(\alpha_1 - 1)\pi + \cdots + (\alpha_n - 1)\pi \tag{15.16}$$

for the argument of $\beta(t)$ when $t \in (-\infty, a_1)$.[8] Then in the other intervals $(a_1, a_2), \ldots, (a_{n-1}, a_n), (a_n, +\infty)$, we assign $\operatorname{Arg} \beta(t)$ values which are consistent with this choice, noting that the only factor in (15.15) whose argument changes when we go from the interval (a_{k-1}, a_k) to the interval (a_k, a_{k+1}) is $t - a_k$, giving rise to a change of $-\pi$ in $\operatorname{Arg}(t - a_k)$ and a change of $-(\alpha_k - 1)\pi$ in $\operatorname{Arg} \beta(t)$ itself (cf. footnote 7, p. 329). Since $\operatorname{Arg} \beta(t)$ has the value (15.16) in the interval $(-\infty, a_1)$, it has the value $(\alpha_2 - 1)\pi + \cdots + (\alpha_n - 1)\pi$ in the interval (a_1, a_2), the value $(\alpha_3 - 1)\pi + \cdots + (\alpha_n - 1)\pi$ in the interval (a_2, a_3), and so on. In general, we have

$$\operatorname{Arg} \beta(t) = (\alpha_k - 1)\pi + \cdots + (\alpha_n - 1)\pi \quad \text{if} \quad t \in (a_{k-1}, a_k), \tag{15.17}$$

where $k = 2, \ldots, n$. Finally, to the right of all the points $a_1, \ldots, a_n$, i.e., in the interval $(a_n, +\infty)$, $\operatorname{Arg} \beta(t)$ has the value 0.[9]

Setting

$$w_k = C \int_0^{a_k} (t - a_1)^{\alpha_1 - 1} \cdots (t - a_n)^{\alpha_n - 1} \, dt \qquad (k = 0, 1, \ldots, n + 1), \tag{15.18}$$

[8] In (15.16), (15.17), etc., an extra term $+2k\pi$ $(k = 0, \pm 1, \pm 2, \ldots)$ is always understood, since $\operatorname{Arg} z$ is only defined to within an integral multiple of 2π.

[9] Formula (15.17) also holds in the intervals $(-\infty, a_1)$ and $(a_n, +\infty)$, if we introduce extra constants $a_0 = -\infty$, $a_{n+1} = +\infty$ and let k range from 1 to $n + 1$, regarding the right-hand side as zero if $k = n + 1$. Similarly, in (15.18), we allow the upper limit of integration to be any of the points $-\infty = a_0, a_1, \ldots, a_n, a_{n+1} = +\infty$. It should be noted that $w_0 = w_{n+1}$ (why?).

we find that (15.12) can be written in the form

$$w = f(z) = w_{k-1} + C\int_{a_{k-1}}^{z} \beta(t)\,dt$$
$$= w_{k-1} + C\exp\,[(\alpha_k - 1)\pi i + \cdots + (\alpha_n - 1)\pi i]\int_{a_{k-1}}^{z} |\beta(t)|\,dt$$
$$\text{if} \quad z \in (a_{k-1}, a_k).$$

The last term on the right has the same argument for all $z \in (a_{k-1}, a_k)$, but its absolute value increases continuously from 0 to

$$C\int_{a_{k-1}}^{a_k} |\beta(t)|\,dt = l_k \tag{15.19}$$

as $z = x$ increases from a_{k-1} to a_k. Therefore, as z traverses the interval (a_{k-1}, a_k) from left to right, the image point $w = f(z)$ traverses the line segment Δ_{k-1} of length l_k, with initial point w_{k-1} and final point w_k, making an angle of $(\alpha_k - 1)\pi + \cdots + (\alpha_n - 1)\pi$ radians with the positive real axis. This applies equally well to the line segment Δ_0, with initial point

$$w_0 = C\int_0^{\infty} (t - a_1)^{\alpha_1 - 1} \cdots (t - a_n)^{\alpha_n - 1}\,dt$$

and final point w_1, and the line segment Δ_n, with initial point w_n and final point $w_{n+1} = w_0$. It follows that the function (15.12) maps the real axis onto a closed polygonal curve Λ with segments $\Delta_0, \Delta_1, \ldots, \Delta_n$ and vertices $w_0, w_1, \ldots, w_n, w_{n+1} = w_0$. It cannot be asserted that this mapping is one-to-one, since the polygonal curve may have self-intersections, i.e., the segments Δ_k may have points in common other than the common end points of consecutive segments. However, if we make the additional assumption that there are no self-intersections [as, for example, in the case of the elliptic integral (15.5)], then Λ will be a closed rectifiable Jordan curve, bounding a polygon with vertices $w_0, w_1, \ldots, w_n$. Therefore, in this case, we can apply Theorem 15.6, concluding that the Schwarz-Christoffel transformation (15.12) is one-to-one in the upper half-plane, which it maps conformally onto the indicated polygon.

According to the construction shown in Figure 15.1, the interior angle of the polygon at the vertex w_k $(0 < k < n + 1)$ equals $\alpha_k\pi$. Although all the angles in the figure are only defined to within integral multiples of 2π, it is easy to see that $\alpha_k\pi$ gives the interior angle without any correction term, i.e., that the inequality $\alpha_k < 2$ holds as well as the inequality $\alpha_k > 0$ assumed in (15.13). In fact, we can write

$$\beta(t) = [(t - a_1)^{\alpha_1 - 1} \cdots (t - a_{k-1})^{\alpha_{k-1} - 1}(t - a_{k+1})^{\alpha_{k+1} - 1} \cdots (t - a_n)^{\alpha_n - 1}]$$
$$\times (t - a_k)^{\alpha_k - 1},$$

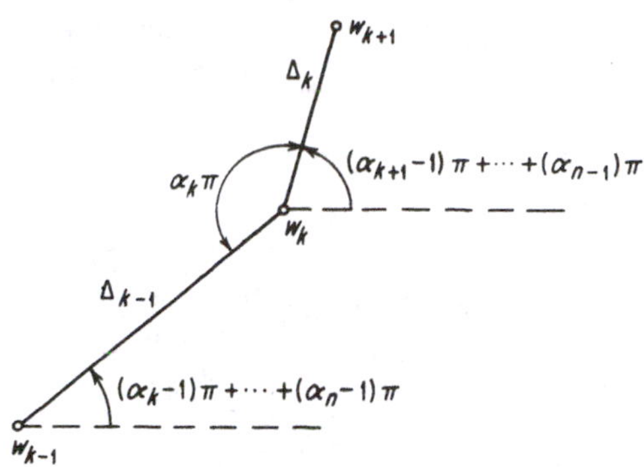

FIGURE 15.1

where the expression in brackets is analytic and nonvanishing in a neighborhood of the point a_k, and hence has a power series expansion of the form

$$A_0^{(k)} + A_1^{(k)}(t - a_k) + \cdots \qquad (A_0^{(k)} \neq 0)$$

which converges uniformly in a neighborhood of a_k. But then

$$\begin{aligned} w &= w_k + C\int_{a_k}^{z} \beta(t)\,dt \\ &= w_k + C\int_{a_k}^{z} [A_0^{(k)} + A_1^{(k)}(t - a_k) + \cdots](t - a_k)^{\alpha_k - 1}dt \\ &= w_k + C\frac{A_0^{(k)}}{\alpha_k}(z - a_k)^{\alpha_k} + C\frac{A_1^{(k)}}{\alpha_k + 1}(z - a_k)^{\alpha_k+1} + \cdots, \end{aligned}$$

which implies

$$w - w_k = C\frac{A_0^{(k)}}{\alpha_k}(z - a_k)^{\alpha_k}\left[1 + \frac{\alpha_k}{\alpha_k + 1}\frac{A_1^{(k)}}{A_0^{(k)}}(z - a_k) + \cdots\right] \qquad (A_0^{(k)} \neq 0).$$

Therefore

$$\lim_{w \to w_k} \operatorname{Arg}(w - w_k) = \operatorname{Arg} A_0^{(k)} + \alpha_k\theta,$$

where we use the fact that $C > 0$, $\alpha_k > 0$. Thus, as θ varies from 0 to π along an "arbitrarily small" semicircle with center a_k, the image point $w = f(z)$ varies continuously along some Jordan curve joining a point on the segment Δ_k to a point on the segment Δ_{k-1}, always staying inside the interior angle with vertex at w_k, while Arg w increases from Arg $A_0^{(k)}$ to Arg $A_0^{(k)} + \alpha_k\pi$, without winding around the point $w = 0$ (why?). Therefore $\alpha_k\pi$ is just the size of some interior angle of the polygon, and hence must be less than 2π. In other words, instead of the condition (15.13), we have the stronger condition

$$0 < \alpha_k < 2\pi \qquad (k = 1, \ldots, n). \tag{15.20}$$

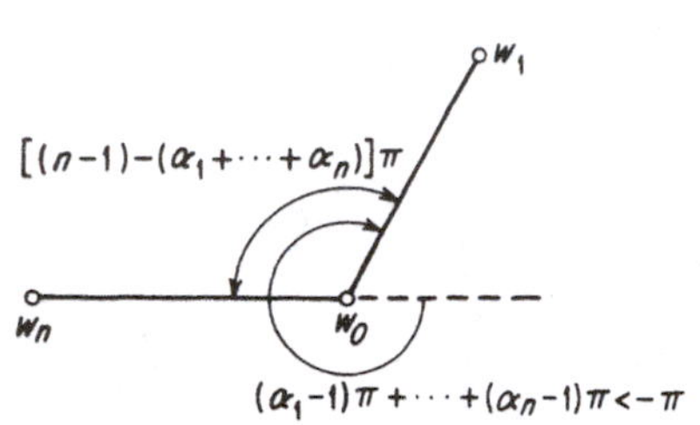

FIGURE 15.2

At the vertex $w_0 = w_{n+1}$, the interior angle of the polygon equals

$$[(n - 1) - (\alpha_1 + \cdots + \alpha_n)]\pi, \tag{15.21}$$

as shown by Figure 15.2. From (15.14) we can only deduce that this quantity is positive. However, our polygon has a total of $n + 1$ interior angles, whose sum must equal $[(n + 1) - 2]\pi = (n - 1)\pi$. Therefore (15.21) is precisely the size of the $(n + 1)$st interior angle, and hence

$$[(n - 1) - (\alpha_1 + \cdots + \alpha_n)]\pi < 2\pi,$$

or

$$\alpha_1 + \cdots + \alpha_n > n - 3.$$

In other words, instead of the condition (15.14), we have the stronger condition

$$n - 3 < \alpha_1 + \cdots + \alpha_n < n - 1. \tag{15.22}$$

The conditions (15.20) and (15.22) are necessary, but not sufficient, for the Schwarz-Christoffel transformation (15.12) to be a conformal mapping of the upper half-plane onto a bounded polygon, with vertices $w_0, w_1, \ldots, w_n$, and interior angles

$$[(n - 1) - (\alpha_1 + \cdots + \alpha_n)]\pi, \alpha_1\pi, \ldots, \alpha_n\pi.$$

In the special case where

$$\alpha_1 + \cdots + \alpha_n = n - 2,$$

we find that the angle with vertex w_0 equals π. Geometrically, this means that w_0 is not really a vertex, but just an interior point of the side with vertices w_n and w_1 (cf. Figure 15.2). Then (15.12) maps the upper half-plane onto an n-gon rather than an $(n + 1)$-gon. This occurs in the case of the elliptic integral (15.5), where

$$n = 4, \qquad \alpha_1 + \alpha_2 + \alpha_3 + \alpha_4 = 4 \cdot \tfrac{1}{2} = n - 2,$$

and in the following example:

Example 1. We now use the Schwarz-Christoffel transformation to map the upper half-plane onto a triangle with angles $\alpha_1\pi$, $\alpha_2\pi$, $\alpha_3\pi$ ($\alpha_k > 0$, $\alpha_1 + \alpha_2 + \alpha_3 = 1$), where the side opposite the angle $\alpha_1\pi$ has length l. This problem can be solved in two ways. First we use formula (15.12) with $n = 3$, choosing three arbitrary real points a_1, a_2, a_3, which are mapped into the vertices of the triangle (in this approach, $\alpha_1 + \alpha_2 + \alpha_3 = n - 2$). Thus, for example, suppose that

$$a_1 = -1, \quad a_2 = 0, \quad a_3 = 1.$$

Then (15.12) gives

$$w = C \int_0^z (t+1)^{\alpha_1 - 1} t^{\alpha_2 - 1} (t-1)^{\alpha_3 - 1}\, dt, \tag{15.23}$$

and we need only determine the factor $C > 0$. But since l must equal

$$C \int_0^1 |(t+1)^{\alpha_1 - 1} t^{\alpha_2 - 1} (t-1)^{\alpha_3 - 1}|\, dt = C \int_0^1 (t+1)^{\alpha_1 - 1} t^{\alpha_2 - 1} (1-t)^{\alpha_3 - 1}\, dt$$

[cf. (15.19)], we have

$$C = \frac{l}{\displaystyle\int_0^1 (t+1)^{\alpha_1 - 1} t^{\alpha_2 - 1} (1-t)^{\alpha_3 - 1}\, dt}.$$

Therefore (15.23) becomes

$$w = l\, \frac{\displaystyle\int_0^z (t+1)^{\alpha_1 - 1} t^{\alpha_2 - 1} (t-1)^{\alpha_3 - 1}\, dt}{\displaystyle\int_0^1 (t+1)^{\alpha_1 - 1} t^{\alpha_2 - 1} (1-t)^{\alpha_3 - 1}\, dt}. \tag{15.24}$$

This function actually solves the problem, since it is one-to-one on the real axis, which it maps onto a triangle with the given vertex angles and side length (note that a triangle can have no self-intersections). In particular, (15.24) reduces to

$$w = l\, \frac{\displaystyle\int_0^z \frac{dt}{\sqrt[3]{t^2(t^2-1)^2}}}{\displaystyle\int_0^1 \frac{dt}{\sqrt[3]{t^2(1-t\,)^2}}}$$

if $\alpha_1 = \alpha_2 = \alpha_3 = \frac{1}{3}$ (corresponding to an equilateral triangle with side length l), and to

$$w = l\, \frac{\displaystyle\int_0^z \frac{dt}{\sqrt{t}\sqrt[4]{(t^2-1)^3}}}{\displaystyle\int_0^1 \frac{dt}{\sqrt{t}\sqrt[4]{(1-t^2)^3}}}$$

if $\alpha_2 = \frac{1}{2}$, $\alpha_1 = \alpha_3 = \frac{1}{4}$ (corresponding to an isosceles right triangle with hypotenuse l).

Example 2. Another approach to the problem of mapping the upper half-plane onto a triangle is based on the use of formula (15.12) with $n = 2$. This time we choose *two* arbitrary real points a_1 and a_2, which are to be mapped into two vertices of the triangle, with angles $\alpha_1\pi$ and $\alpha_2\pi$. Since

$\alpha_1 + \alpha_2$ cannot equal $n - 2 = 0$, the point at infinity must go into the third vertex of the triangle, with angle

$$[(2 - 1) - (\alpha_1 + \alpha_2)]\pi = \pi - \alpha_1\pi - \alpha_2\pi = \alpha_3\pi.$$

Thus, for example, suppose that

$$a_1 = 0, \qquad a_2 = -1.$$

Then (15.12) gives

$$w = C \int_0^z t^{\alpha_1-1}(t - 1)^{\alpha_2-1}\, dt,$$

and to determine C we have the equation

$$C \int_1^\infty |t^{\alpha_1-1}(t - 1)^{\alpha_2-1}|\, dt = l,$$

which implies

$$C = \frac{l}{\displaystyle\int_1^\infty t^{\alpha_1-1}(t - 1)^{\alpha_2-1}\, dt}.$$

It follows that

$$w = l\,\frac{\displaystyle\int_0^z t^{\alpha_1-1}(t - 1)^{\alpha_2-1}\, dt}{\displaystyle\int_1^\infty t^{\alpha_1-1}(t - 1)^{\alpha_2-1}\, dt}, \tag{15.25}$$

which obviously differs from (15.24). In particular, in the case of an equilateral triangle with side l, (15.25) reduces to

$$w = l\,\frac{\displaystyle\int_0^z \frac{dt}{\sqrt[3]{t^2(t - 1)^2}}}{\displaystyle\int_1^\infty \frac{dt}{\sqrt[3]{t^2(t - 1)^2}}}.$$

Example 3. In the general case, the upper half-plane is to be mapped onto an n-gon ($n \geqslant 4$) with interior angles $\alpha_1\pi, \ldots, \alpha_n\pi$ and side lengths $l_1, \ldots, l_n$, where as in Figure 15.3, we associate the angles $\alpha_{k-1}\pi$ and $\alpha_k\pi$ with the side l_k ($k = 2, \ldots, n$), and the angles $\alpha_n\pi$ and $\alpha_1\pi$ with the side l_1. Suppose we choose three arbitrary points a_1, a_2 and a_3 of the real axis, which are to go into the vertices w_1, w_2 and w_3, with angles $\alpha_1\pi$, $\alpha_2\pi$ and $\alpha_3\pi$. Then we must still determine $n - 3$ points $a_4, \ldots, a_n$ of the real axis, as well as the constant $C > 0$, which makes $n - 2$ unknowns in all. At first glance, it appears that there are too many equations available for determining these unknowns, i.e., the n equations expressing the side lengths in

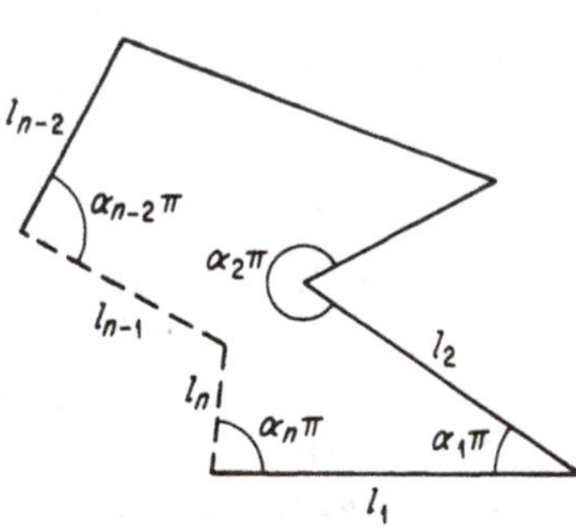

FIGURE 15.3

terms of appropriate integrals. However, it is immediately apparent from Figure 15.3 that once the angles of an n-gon are known, then specifying just $n - 2$ sides $l_1, \ldots, l_{n-2}$ uniquely determines the remaining two sides l_{n-1} and l_n.[10] In other words, there are actually only $n - 2$ independent equations of the form

$$C \int_{a_{k-1}}^{a_k} |(t - a_1)^{\alpha_1 - 1} \cdots (t - a_n)^{\alpha_n - 1}|\, dt = l_k \qquad (k = 1, \ldots, n - 2)$$

for determining the $n - 2$ unknowns $a_4, \ldots, a_n$, C, and these equations can actually be solved. This follows from the fact (not proved here) that apart from a trivial translation and rotation, (15.14) is the most general form of a function mapping the upper half-plane onto an n-gon (see M3, Sec. 47).

Problem 1. Justify the arbitrariness in the choice of a_1, a_2 and a_3 in Example 3. Can more points $a_4, \ldots, a_n$ be chosen arbitrarily?

Problem 2. Map the upper half-plane $\operatorname{Im} z > 0$ onto a rhombus with side length l and obtuse angle $\alpha\pi$.

Problem 3. Derive the formula for mapping the unit disk $|z| < 1$ onto the interior of a polygon.

Hint. Make a preliminary Möbius transformation mapping the upper half-plane onto the unit disk.

Problem 4. Show that the function

$$w = \int_0^z \frac{dt}{\sqrt{1 - t^4}}$$

maps the unit disk onto a square. What is the side length of this square?

Problem 5. Prove that the function

$$w = \int_0^z \frac{dt}{(1 - t^2)^{2/n}}$$

maps the unit disk onto a regular n-gon. What is the side length of this n-gon?

Problem 6. Onto what domain is the unit disk mapped by the function

$$w = \int_0^z \frac{(1 + t^n)^\lambda}{(1 - t^2)^{(2/n) + \lambda}}\, dt,$$

where

$$-1 < \lambda < 1 - \frac{2}{n}\ ?$$

Hint. The domain is starlike (see Sec. 37, Prob. 5).

---

[10] Obviously, specifying $n - 1$ angles of an n-gon uniquely determines the remaining angle.

CHAPTER 16

ANALYTIC CONTINUATION

74. ELEMENTS AND CHAINS

By the problem of analytic continuation of a single-valued function $f(z)$ defined on a set E we mean the problem of finding a domain D containing E and a function $\varphi(z)$ defined on D such that

1. $\varphi(z)$ is analytic on D;
2. $\varphi(z) = f(z)$ for all $z \in E$;
3. D is as large as possible.[1]

The function $\varphi(z)$ is then said to be an *analytic continuation* of $f(z)$ from E into D. If E has a limit point $z_0 \in D$, it follows from Theorem 10.8 that $\varphi(z)$ is unique.

Remark. Two other definitions of analytic continuation will be encountered below, but they can both be interpreted as variants of this basic definition.[2]

---

[1] Ultimately we will pursue this last condition even to the extent of allowing $\varphi(z)$ to be a multiple-valued function and D to be a "many-sheeted domain" (see Sec. 79). In this case, condition 2 is replaced by the requirement that some single-valued analytic branch of $\varphi(z)$ coincide with $f(z)$ on E.

[2] See Remark 2, p. 343 and the remark on p. 349.

Example 1. Let E be the real line $z = x$ $(y = 0)$, and let $f(x) = e^x$. Then

$$f(x) = \sum_{n=0}^{\infty} \frac{x^n}{n!} \qquad (|x| < \infty), \tag{16.1}$$

and the function

$$\varphi(z) = \sum_{n=0}^{\infty} \frac{z^n}{n!} \qquad (|z| < \infty) \tag{16.2}$$

obtained by replacing x by z in (16.1) is the unique analytic continuation of $f(x)$ into the whole finite plane. In fact, instead of the approach used in Sec. 29, we could have *defined* the complex exponential by the series (16.2). However, the treatment in Sec. 29 has the advantage of not requiring previous knowledge of complex series and the uniqueness theorem for analytic functions.

Example 2. Let E be the unit disk $|z| < 1$, and let

$$f(z) = \sum_{n=0}^{\infty} z^n \qquad (|z| < 1).$$

Then the function

$$\varphi(z) = \frac{1}{1 - z}$$

is the analytic continuation of $f(z)$ into the domain D consisting of the whole plane minus the point $z = 1$.

Example 3. Let E be the unit disk $|z| < 1$, and let

$$f(z) = \sum_{n=0}^{\infty} z^{2^n} \qquad (|z| < 1). \tag{16.3}$$

Then, as will be shown in Example 2, p. 355, there is no analytic continuation of $f(z)$ into a domain larger than E. This fact is summarized by saying that the circle $|z| = 1$ is the *natural boundary* of the series (16.3).

Next we consider the case where D intersects E but does not necessarily contain E. At the same time, we assume that E is itself a domain, thereby insuring that $E \cup D$ is a domain.

DEFINITION 1. *A set $\{G, f(z)\}$ consisting of a domain G and a single-valued analytic function defined on G is called an element, and G is called the domain of the element. Two elements $\{G, f(z)\}$ and $\{D, \varphi(z)\}$ are equal if and only if $G = D$, $f(z) \equiv \varphi(z)$.*

DEFINITION 2. *Each of two elements $\{G, f(z)\}$ and $\{D, \varphi(z)\}$ is said to be a direct analytic continuation of the other if $G \cap D \neq 0$ and if there exists a domain $g \subset G \cap D$ such that $f(z) = \varphi(z)$ for all $z \in g$.*

Remark 1. Let $\{G, f(z)\}$, $\{D, \varphi(z)\}$ and g be the same as in Definition 2, and let g' be the connected component of $G \cap D$ containing g (see p. 40). Then, by Theorem 10.8, $f(z)$ and $\varphi(z)$ coincide on g', and hence on $G \cap D$ if $G \cap D$ is connected. However, $f(z)$ and $\varphi(z)$ may not coincide on the other components of $G \cap D$ if $G \cap D$ is not connected.

Remark 2. Let $\{G_2, f_2(z)\}$ be a direct analytic continuation of $\{G_1, f_1(z)\}$ in the sense of Definition 2, and let

$$\varphi(z) = \begin{cases} f_1(z) & \text{if} \quad z \in G_1, \\ f_1(z) = f_2(z) & \text{if} \quad z \in g \subset G_1 \cap G_2, \\ f_2(z) & \text{if} \quad z \in G_2. \end{cases}$$

Then $\varphi(z)$ is an analytic continuation of $f(z)$ from G_1 (or G_2) into $G_1 \cup G_2$ in the sense of the definition on p. 341 if $G_1 \cap G_2$ is connected (see Prob. 4).

Example 1. Let G be the domain bounded by the ray $-\infty < x \leqslant 0$, $y = 0$ and D the domain bounded by the ray $0 \leqslant x < +\infty$, $y = 0$. Let

$$f(z) = \ln|z| + i\theta, \qquad \varphi(z) = \ln|z| + i\alpha, \tag{16.4}$$

where θ is the value of Arg z satisfying the condition $-\pi < \theta < \pi$ and α is the value satisfying the condition $0 < \alpha < 2\pi$. Each of the functions (16.4) is single-valued and analytic on its domain of definition. The set $G \cap D$ is nonempty and consists of two components, the upper half-plane Π_U and the lower half-plane Π_L. Moreover $\alpha = \theta$ in Π_U while $\alpha = \theta + 2\pi$ in Π_L, and hence $f(z) = \varphi(z)$ if $z \in \Pi_U$ but $f(z) \neq \varphi(z)$ if $z \in \Pi_L$. Therefore $\{G, f(z)\}$ and $\{D, \varphi(z)\}$ are direct analytic continuations of each other (with Π_U as the domain g figuring in Definition 2).

Example 2. Let G, D, $f(z)$ and $\varphi(z)$ be the same as in Example 1, and consider another function

$$\psi(z) = \ln|z| + i\beta$$

defined on D, where β is the value of Arg z satisfying the condition $-2\pi < \beta < 0$. Then $\alpha = \beta + 2\pi$ at every point of D, and hence $\varphi(z) \neq \psi(z)$ if $z \in D$. Therefore $\{D, \varphi(z)\}$ and $\{D, \psi(z)\}$ are not direct analytic continuations of each other. On the other hand, $\beta = \theta$ in the lower half-plane Π_L, and hence $f(z) = \psi(z)$ if $z \in \Pi_L$, so that $\{G, f(z)\}$ and $\{D, \psi(z)\}$ are direct analytic continuations of each other.

DEFINITION 3. *Let*

$$\{G_0, f_0(z)\}, \{G_1, f_1(z)\}, \ldots, \{G_n, f_n(z)\} \tag{16.5}$$

be a set of elements such that $\{G_k, f_k(z)\}$ *is a direct analytic continuation of* $\{G_{k-1}, f_{k-1}(z)\}$ *for every* $k = 1, \ldots, n$. *Then* (16.5) *is called a chain of* *elements joining $\{G_0, f_0(z)\}$ and $\{G_n, f_n(z)\}$. Each of the elements $\{G_0, f_0(z)\}$ and $\{G_n, f_n(z)\}$ is called an analytic continuation of the other.*[3]

Definition 3 immediately implies

THEOREM 16.1. *The relation leading from an element $\{G, f(z)\}$ to any of its analytic continuations is reflexive, symmetric and transitive.*[4] *More exactly,*

1. *$\{G, f\}$ is an analytic continuation of itself;*[5]
2. *If $\{D, \varphi\}$ is an analytic continuation of $\{G, f\}$, then $\{G, f\}$ is an analytic continuation of $\{D, \varphi\}$;*
3. *If $\{D, \varphi\}$ is an analytic continuation of $\{G, f\}$ and if $\{E, \psi\}$ is an analytic continuation of $\{D, \varphi\}$, then $\{E, \psi\}$ is an analytic continuation of $\{G, f\}$.*

Example. Let G_k be the domain

$$k\pi < \operatorname{Arg} z < (k+2)\pi \qquad (k = 0, \pm 1, \pm 2, \ldots), \tag{16.6}$$

and let $f_k(z)$ be the function

$$f_k(z) = \ln |z| + i\theta_k,$$

where $\theta_k = \theta_k(z)$ is the value of $\operatorname{Arg} z$ satisfying (16.6). Then any two elements $\{G_m, f_m\}$ and $\{G_n, f_n\}$ are analytic continuations of each other. In fact, suppose $n = m + p$, where $p > 0$ (say). Then $\{G_m, f_m\}, \ldots, \{G_n, f_n\}$ is a chain of $p + 1$ elements joining $\{G_m, f_m\}$ and $\{G_n, f_n\}$.

Problem 1. Suppose $f(z)$ can be continued analytically from a set $E \subset D$ with a limit point $z_0 \in D$ into the domain D itself. Let

$$f(z) = a_0 + a_1(z - z_0) + \cdots + a_n(z - z_0)^n + \cdots \qquad (|z - z_0| < \Delta_0)$$

be the power series expansion of $f(z)$ at z_0, where Δ_0 is the distance from z_0 to the boundary of D, and let $\{z_k\}$ be a sequence of points of E converging to z_0 (recall Sec. 9, Prob. 8). How can the coefficients $a_0, a_1, \ldots, a_n, \ldots$ be deduced from a knowledge of $f(z)$ on the set E?

Ans. Use the inductive formula

$$\begin{aligned} a_0 &= \lim_{k\to\infty} f(z_k), \\ a_n &= \lim_{k\to\infty} \frac{f(z_k) - a_0 - a_1(z_k - z_0) - \cdots - a_{n-1}(z_k - z_0)^{n-1}}{(z_k - z_0)^n} \\ (n &= 1, 2, \ldots). \end{aligned} \tag{16.7}$$

---

[3] The absence of the adjective "direct" is crucial here.

[4] Such a relation is called an *equivalence relation* (see e.g., G. Birkhoff and S. MacLane, *op. cit.*, p. 30).

[5] $\{G, f\}$ is shorthand for $\{G, f(z)\}$, and similarly for $\{D, \varphi\}$ and $\{E, \psi\}$.

Problem 2. Let $f(z)$, z_0, D and Δ_0 be the same as in the preceding problem, and let z' be a point in the disk $|z - z_0| < \Delta_0$. Then

$$f(z) = b_0 + b_1(z - z') + \cdots + b_n(z - z')^n + \cdots \qquad (|z - z'| < \Delta'),$$

where Δ' is the distance from z' to the boundary of D. Relate the coefficients $b_0, b_1, \ldots, b_n, \ldots$ to the coefficients $a_0, a_1, \ldots, a_n, \ldots$ figuring in (16.7).

Ans.

$$b_n = a_n + \frac{n+1}{1!} a_{n+1}(z' - z_0) + \frac{(n+2)(n+1)}{2!} a_{n+2}(z' - z_0)^2 + \cdots$$

$(n = 0, 1, 2, \ldots)$.

Problem 3. Interpret the proof of the uniqueness theorem for analytic functions (Theorem 10.8) from the standpoint of analytic continuation.

Problem 4. Generalize Remark 2, p. 343 to the case where $G_1 \cap G_2$ is not connected.

Hint. Recall footnote 1, p. 341.

Problem 5. Verify that the relation leading from a domain G to any of its conformal images (see p. 323) is an equivalence relation.

75. GENERAL AND COMPLETE ANALYTIC FUNCTIONS

Let F be a nonempty (finite or infinite) set of elements $\{G, f\}$. Then F is said to be *connected* if every pair of elements in F can be joined by a chain of elements all belonging to F. In particular, if F is connected, every element of F is an analytic continuation of every other element of F.

THEOREM 16.2. *Let F be a connected set of elements, and let D be the union of the domains of all the elements in F. Then D is itself a domain.*

Proof. Since F is nonempty, so is D. Moreover, if $z_0 \in D$, then z_0 belongs to at least one domain G such that $\{G, f\} \in F$. Therefore z_0 has a neighborhood contained in G and hence in D, i.e., D is open. To prove that D is connected, let z_0 and z^* be two distinct points of D. Then there are two elements $\{G_0, f_0\}$, $\{G^*, f^*\} \in F$ such that $z_0 \in G_0$, $z^* \in G^*$. Furthermore, F contains a chain

$$\{G_0, f_0\}, \{G_1, f_1\}, \ldots, \{G_n, f_n\} = \{G^*, f^*\}$$

joining $\{G_0, f_0\}$ and $\{G^*, f^*\}$. Since $\{G_{k-1}, f_{k-1}\}$ and $\{G_k, f_k\}$ are direct analytic continuations of each other, the set $G_{k-1} \cap G_k$ is nonempty $(k = 1, \ldots, n)$. Let z_k be any point of $G_{k-1} \cap G_k$, and let γ_k be a continuous curve contained in G_k joining z_k to z_{k+1} $(k = 1, \ldots, n-1)$. Then joining z_0 to z_1 by any curve $\gamma_0 \subset G_0$, z_0 to z_1 by any curve $\gamma_1 \subset G_1$, z_{n-1} to z_n by any curve $\gamma_{n-1} \subset G_{n-1}$, and finally z_n to z^* by any curve $\gamma_n \subset G_n$, we obtain a continuous curve contained in D joining z_0 to z^* (draw a picture!). Therefore D is connected as well as nonempty and open, and the proof is complete.

The following definitions form the basis for a rigorous theory of multiple-valued analytic functions:

DEFINITION 1. *A nonempty connected set of elements is called a general analytic function, said to be generated by any of its elements. The union of the domains of all the elements of a general analytic function F is called the domain of F.*

DEFINITION 2. *Given a general analytic function F, let z_0 be any point in the domain of F. Then by a value of F at z_0, written $F(z_0)$, is meant any value $f(z_0)$ where $\{G, f(z)\} \in F$, $z_0 \in G$.*[6]

DEFINITION 3. *A general analytic function which contains all analytic continuations of any of its elements is called a complete analytic function. The domain of a complete analytic function F is often called the domain of existence of F, and the elements of F are often called single-valued (analytic) branches of F.*

Remark. Roughly speaking, a complete analytic function is the result of unrestricted analytic continuation of some originally given sample of the function, and the domain of existence of a complete analytic function is the largest domain into which the given sample can ever be continued.

Example 1. Let G be a domain, and let $f(z)$ be single-valued and analytic on G. Then the set of all elements $\{K, f(z)\}$, where K is a disk contained in G, is a general analytic function F, with domain G. Moreover, $F(z)$ is single-valued and $F(z) = f(z)$.

Example 2. Let G be any simply connected domain not containing the origin. Then the complete analytic function generated by the element $\{G, \ln z\}$, where $\ln z$ is the principal value of the logarithm, is the multiple-valued function

$$F(z) = \operatorname{Ln} z,$$

whose domain of existence is the whole plane minus the origin.[7]

Example 3. Let F be a general analytic function whose domain is *simply connected*. Then, according to the *monodromy theorem*,[8] F must be single-valued.

Problem 1. Prove that every element belongs to a unique complete analytic function.

---

[6] Obviously, F is in general multiple-valued.

[7] Of course, single-valued branches of $\operatorname{Ln} z$ can also be defined on appropriate multiply connected domains.

[8] For the proof, see M3, Sec. 40.

Problem 2. Find the complete analytic function generated by the element $\{G, f(z)\}$, where G is the unit disk $|z| < 1$ and $f(z)$ is the function (16.3). What is the domain of existence of F?

Problem 3 (M3, Sec. 39). Given a meromorphic function $f(z)$, let G be any domain containing no poles of $f(z)$. What is the complete analytic function $F(z)$ generated by $\{G, f(z)\}$?

Ans. $F(z) = f(z)$.

76. ANALYTIC CONTINUATION ACROSS AN ARC

Given a domain G with boundary Γ, suppose every point $z_0 \in G$ can be joined to a point $\zeta \in \Gamma$ by a curve $\widehat{z_0\zeta} \subset G \cup \{\zeta\}$, i.e., by a curve entirely contained in G except for the end point ζ. Then ζ is said to be an *accessible point* of Γ. A set $E \subset \Gamma$ is said to be accessible if every point of E is accessible.

Example 1. Let G be the interior of a closed Jordan curve. Then it can be shown (see M3, Sec. 8) that every boundary point of G is accessible.

Example 2. Suppose we draw infinitely many line segments

$$x = \frac{1}{2n}, \quad 0 \le y \le \frac{3}{4} \qquad (n = 1, 2, \ldots)$$

and

$$x = \frac{1}{2n+1}, \quad \frac{1}{4} \le y \le 1 \qquad (n = 1, 2, \ldots)$$

in the closed square $0 \le x \le 1$, $0 \le y \le 1$, and then take G to be the domain whose boundary Γ consists of these segments together with the four sides of the square, as shown in Figure 16.1. Then no point of the segment OA is accessible, but every other point of Γ is accessible.

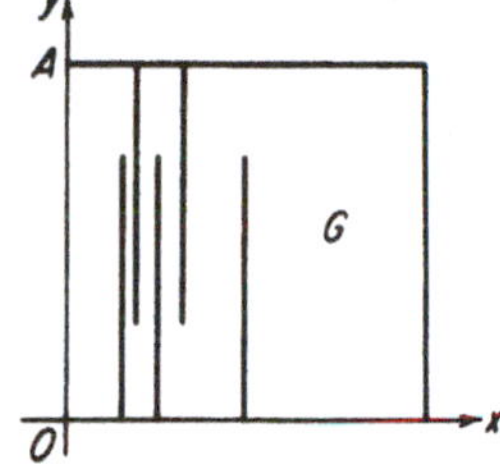

FIGURE 16.1

We now consider the problem of analytic continuation based on the use of "adjacent elements" sharing an appropriate boundary arc:

THEOREM 16.3. *Let $\{G_1, f_1\}$ and $\{G_2, f_2\}$ be two elements whose domains are disjoint but share an accessible Jordan boundary arc δ, where δ is open and rectifiable.*[9] *Suppose $f_1(z)$ is analytic on G_1 and continuous on $G_1 \cup \delta$, while $f_2(z)$ is analytic on G_2 and continuous on $G_2 \cup \delta$. Moreover, suppose*

[9] We call δ an arc to emphasize that its end points are distinct. By an *open arc* is meant an arc minus its end points (rectifiability of an open arc is defined in the obvious way, i.e., by including the end points).

$f_1(z)$ and $f_2(z)$ coincide on δ, so that

$$f_1(z) = f_2(z) \qquad \textit{if} \quad z \in \delta. \tag{16.8}$$

Then there exists a function $\varphi(z)$ analytic on $D = G_1 \cup \delta \cup G_2$ such that

$$\varphi(z) = \begin{cases} f_1(z) & \textit{if} \quad z \in G_1, \\ f_1(z) = f_2(z) & \textit{if} \quad z \in \delta, \\ f_2(z) & \textit{if} \quad z \in G_2. \end{cases} \tag{16.9}$$

Proof. The proof involves Morera's theorem (Theorem 8.6) and the generalized Cauchy integral theorem,[10] which states that if G is the interior of a closed rectifiable Jordan curve L and if $\varphi(z)$ is analytic on G and continuous on $\bar{G}$, then

$$\int_L \varphi(z)\, dz = 0. \tag{16.10}$$

First we note that D is a domain (here the accessibility of δ is vital!). Let L be an arbitrary closed rectifiable Jordan curve contained in D and traversed in the positive direction. If $L \cap \delta = 0$, then $L \subset G_1$ or $L \subset G_2$, and hence, by the ordinary form of Cauchy's integral theorem, (16.10) holds with $\varphi(z)$ given by (16.9). If $L \cap \delta \neq 0$, we divide L into two arcs L_1 and L_2 with end points $\zeta_1, \zeta_2 \in \delta$, as in Figure 16.2 (verify that this figure is sufficiently general for our purposes). It follows from the generalized Cauchy integral theorem that

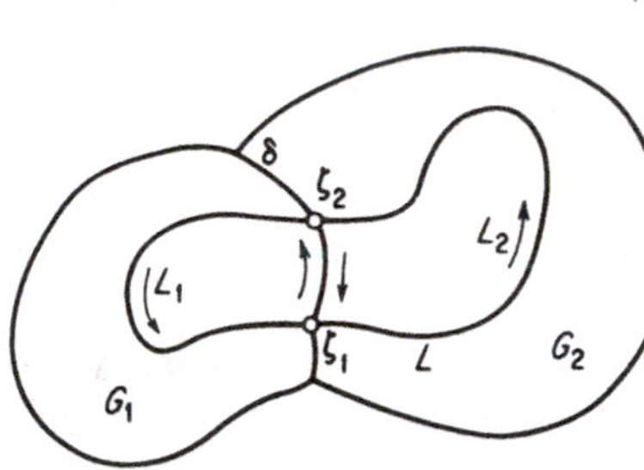

FIGURE 16.2

$$\int_{L_1 + \widehat{\zeta_1\zeta_2}} f_1(z)\, dz = \int_{L_2 + \widehat{\zeta_2\zeta_1}} f_2(z)\, dz = 0$$

($\widehat{\zeta_1\zeta_2} \subset \delta$), and hence

$$\int_{L_1 + \widehat{\zeta_1\zeta_2}} f_1(z)\, dz + \int_{L_2 + \widehat{\zeta_2\zeta_1}} f_2(z)\, dz = \int_{L_1} f_1(z)\, dz + \int_{L_2} f_2(z)\, dz = 0,$$

since (16.8) implies

$$\int_{\widehat{\zeta_1\zeta_2}} f_1(z)\, dz = -\int_{\widehat{\zeta_2\zeta_1}} f_2(z)\, dz.$$

[10] This result has already been mentioned in Sec. 37, Prob. 6. For the proof, see M3, Sec. 12.

But then, according to (16.9),

$$\int_{L_1+L_2} \varphi(z)\,dz = \int_L \varphi(z)\,dz = 0,$$

and hence, by Morera's theorem, $\varphi(z)$ is analytic on D, as asserted.

Definition. *Either of the elements* $\{G_1, f_1\}$ *and* $\{G_2, f_2\}$ *figuring in Theorem* 16.3 *is called a direct analytic continuation of the other* (*across the arc* δ).

Remark. Let $\{G_2, f_2\}$ be a direct analytic continuation of $\{G_1, f_1\}$ in the sense of the above definition, and let $\varphi(z)$ be given by (16.9). Then $\varphi(z)$ is an analytic continuation of $f(z)$ from G_1 (or G_2) into $G_1 \cup \delta \cup G_2$, in the sense of the definition on p. 341.

Problem 1. Give an example where G_1 and G_2 share an arc such that (16.8) fails to hold as well as an arc δ satisfying (16.8).

Problem 2. Generalize Theorem 16.3 to the case where δ is unbounded.

Hint. Assume that every finite subarc of δ is rectifiable.

Problem 3. Generalize Theorem 16.3 to the case where G_1 and G_2 overlap.

Hint. The function $\varphi(z)$ can now be double-valued on $G_1 \cap G_2$.

77. THE SYMMETRY PRINCIPLE

The following technique of analytic continuation is of great practical importance:

Theorem 16.4 (*Symmetry principle*).[11] *Let* G *be a domain whose boundary contains an accessible open line segment* δ, *and let* $w = f(z)$ *be analytic on* G *and continuous on* $G \cup \delta$. *Let* G^* *be the domain symmetric to* G *with respect to* δ,[12] *and suppose* G *and* G^* *are disjoint. Moreover, suppose* $\Delta = f(\delta)$ *is itself a line segment, and let* $f^*(z)$ *be the function defined on* $G^* \cup \delta$ *by the conditions*

1. $f^*(z) = f(z)$ *for all* $z \in \delta$;
2. $f^*(z^*)$ *and* $f(z)$ *are symmetric with respect to* Δ *for all* $z^* \in G^*$, *where* z^* *is the point symmetric to* $z \in G$ *with respect to* δ.

Then the elements $\{G, f\}$ *and* $\{G^*, f^*\}$ *are direct analytic continuations of each other across* δ.

---

[11] Often called *Schwarz's reflection principle*.

[12] More exactly, symmetric with respect to the straight line containing δ.

Proof. There is no loss of generality in assuming that δ and Δ are both segments of the real axis, as shown in Figure 16.3. In fact, this configuration can always be achieved by making suitable preliminary entire linear transformations in both the z and w-planes, and such transformations (and their inverses) preserve continuity, analyticity, accessibility and symmetry (recall Sec. 27). Thus we can write

$$z = x, \qquad y = \operatorname{Im} z = 0$$

on δ and

$$w = f(z) = u,$$
$$v = \operatorname{Im} f(z) = 0$$

on Δ. Then the points z^* and z symmetric with respect to δ are just complex conjugates of each other, and the same is true of the points w^* and w symmetric with respect to Δ:

$$z^* = \bar{z}, \qquad w^* = \bar{w}.$$

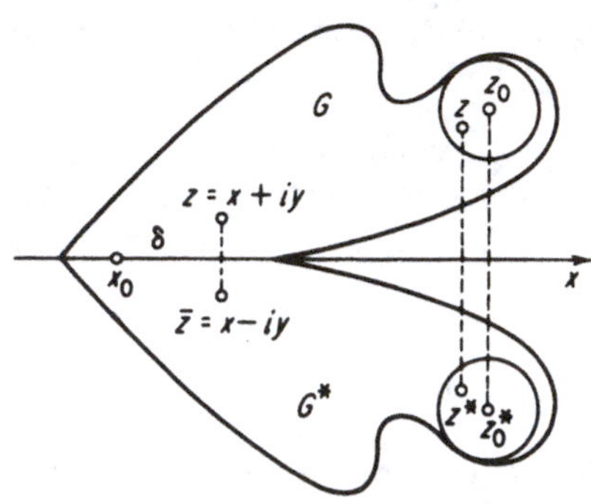

FIGURE 16.3

Given any point $z_0 \in G$, suppose G contains the neighborhood $|z - z_0| < \rho$ (see Figure 16.3). Being analytic on G, $f(z)$ has an expansion at z_0 of the form

$$f(z) = \sum_{n=0}^{\infty} a_n(z - z_0)^n \qquad (|z - z_0| < \rho).$$

But then

$$f^*(z^*) = \overline{f(z)} = \sum_{n=0}^{\infty} \bar{a}_n(\bar{z} - \bar{z}_0)^n \qquad (|z - z_0| < \rho),$$

i.e.,

$$f^*(z^*) = \sum_{n=0}^{\infty} \bar{a}_n(z^* - z_0^*)^n \qquad (|z^* - z_0^*| < \rho).$$

Therefore f^* is analytic on G^*,[13] since the point $z_0 \in G$ (and hence the point $z_0^* \in G^*$) is arbitrary.

Next we show that f^* is continuous on $G^* \cup \delta$ (it is obviously continuous on G^*). Let $z = x_0$ be any point on δ. Then

$$\lim_{\substack{z \to x_0 \\ z \in G}} f(z) = f(x_0) = u(x_0, 0),$$

since $f(z) = u(x, y) + iv(x, y)$ is real on δ and continuous on $G \cup \delta$,

[13] This is a situation where confusion can be avoided by omitting the argument of the function f^* (cf. p. 14).

and hence

$$\lim_{\substack{(x,y)\to(x_0,0)\\(x,y)\in G}} v(x, y) = 0.$$

It follows that

$$\lim_{\substack{z^*\to x_0\\z^*\in G^*}} f^*(z^*) = \lim_{\substack{z\to x_0\\z\in G}} \overline{f(z)} = \lim_{\substack{(x,y)\to(x_0,0)\\(x,y)\in G}} [u(x, y) - iv(x, y)] = u(x_0, 0)$$
$$= f(x_0) = f^*(x_0),$$

i.e., f^* is continuous on $G^* \cup \delta$, as asserted [how about the case where z^* approaches x_0 along δ itself?]. The rest of the proof is an immediate consequence of Theorem 16.3.

Example. Let G be the interior of the triangular contour ABC shown in Figure 16.4(a), and let $f(z)$ be analytic on G and continuous on $\bar{G}$. Suppose

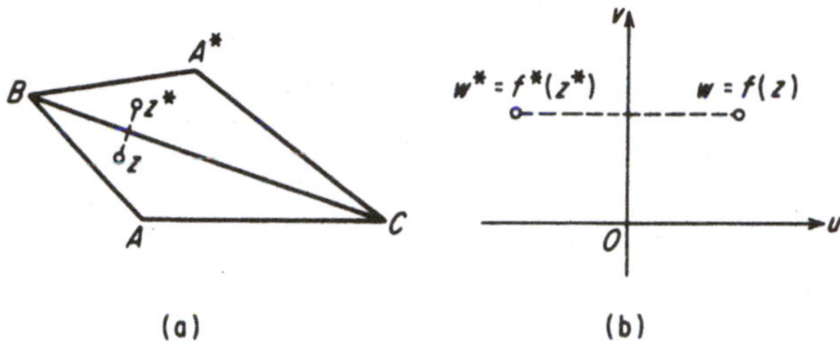

FIGURE 16.4

$f(z)$ takes purely imaginary values on the segment BC. Then $f(z)$ can be continued analytically across BC into the quadrilateral ABA^*C which is "twice as big" as the original domain G. Moreover, if z^* and z are symmetric with respect to BC, i.e., if z^* is the reflection of z in BC, then the image points $w = f(z)$ and $w^* = f^*(z^*)$ are symmetric with respect to the imaginary axis, as shown in Figure 16.4(b).

Remark. There is a natural generalization of the symmetry principle to the case where δ is a circular arc (see M3, Sec. 46).

Problem 1. Generalize Theorem 16.4 to the case where δ is an infinite line segment.

Problem 2. Generalize Theorem 16.4 to the case where the domains G and G^* overlap.

Problem 3. Prove that the function (15.5) maps the lower half-plane onto the rectangle with vertices $-K$, K, $K - iK'$, $-K - iK'$, where K and K' have the same meaning as in Sec. 72.

78. MORE ON SINGULAR POINTS

Given an element $\{G, f(z)\}$, let Γ be the boundary of the domain G. Then every point of Γ falls into one of the following two classes (*a priori*, either class can be empty):

1. Points $\zeta \in \Gamma$ for which there can be found a neighborhood $\mathcal{N}(\zeta)$ and a function $\varphi_\zeta(z)$ analytic on $\mathcal{N}(\zeta)$, i.e., an element $\{\mathcal{N}(\zeta), \varphi_\zeta(z)\}$ such that $\varphi_\zeta(z) = f(z)$ for all z in $\mathcal{N}(\zeta) \cap G$; such points are called *regular points* [of $\{G, f(z)\}$, or simply of $f(z)$].

2. Points $\zeta \in \Gamma$ for which no such neighborhood and analytic function can be found; such points are called *singular points* [of $\{G, f(z)\}$, or simply of $f(z)$].

It follows from the definition of a regular point that if $\zeta \in \Gamma$ is a regular point, then so is every point $\zeta' \in \mathcal{N}(\zeta) \cap \Gamma$. In fact, for $\mathcal{N}(\zeta')$ we need only choose any disk with center ζ' which is contained in $\mathcal{N}(\zeta)$, while for $\varphi_{\zeta'}(z)$ we need only choose the function $\varphi_\zeta(z)$ which is analytic on $\mathcal{N}(\zeta)$ and coincides with $f(z)$ on $\mathcal{N}(\zeta) \cap G$.[14] In particular, this means that if $f(z)$ has one regular point, then $f(z)$ has infinitely many regular points. On the other hand, $f(z)$ can have just one singular point.

It should be noted that every point $\zeta \in G$ has the characteristic property of a regular point, i.e., there exists a neighborhood $\mathcal{N}(\zeta)$ [in fact, any neighborhood of ζ contained in G] and a function $\varphi_\zeta(z)$ analytic on $\mathcal{N}(\zeta)$ [in fact, $f(z)$ itself] which coincides with $f(z)$ on $\mathcal{N}(\zeta) \cap G$. Therefore, extending our definition of a regular point to include interior points of G, we shall henceforth regard the points of the domain G itself as regular points of $f(z)$.

Example 1. Let G be the deleted neighborhood $0 < |z - \zeta| < R$, and let $f(z)$ be a function analytic on G. Then $f(z)$ has a regular point or a singular point at ζ in the sense of the above definition if and only if $f(z)$ has a regular point or a singular point in the sense of the definition on p. 234. The new definition is more general and will allow us to consider the case of nonisolated singular points (see Example 2, p. 355).

Example 2. Let G be the unit disk $|z| < 1$, and let

$$f(z) = \frac{1}{1 - z}.$$

---

[14] In a context like this, $\mathcal{N}(\zeta)$ will always denote one of the neighborhoods figuring in the definition of the regular point ζ.

For every point $\zeta \neq 1$ lying on the boundary of G (i.e., on the unit circle $|z| = 1$), we can find a neighborhood $\mathcal{N}(\zeta): |z - \zeta| < |1 - \zeta|$ and a function

$$\varphi_\zeta(z) = \frac{1}{1 - z}$$

analytic on $\mathcal{N}(\zeta)$ such that $\varphi_\zeta(z) = f(z)$ on $\mathcal{N}(\zeta) \cap K$. Therefore, every point $\zeta \neq 1$ is a regular point of $f(z)$. On the other hand, the point $\zeta = 1$ is a singular point of $f(z)$, since otherwise there would be a neighborhood $\mathcal{N}(1)$ and an analytic function $\varphi(z)$ coinciding with $f(z)$ on $E = \mathcal{N}(1) \cap K$. But then the limit

$$\varphi(1) = \lim_{\substack{z \to 1 \\ z \in E}} \varphi(z) = \lim_{\substack{z \to 1 \\ z \in E}} \frac{1}{1 - z}$$

would have to be finite, which is impossible.

Example 3. Given any element $\{G, f(z)\}$, let Γ be the boundary of G and suppose $f(z) \to \infty$ as $z \in G$ approaches an accessible point $\zeta \in \Gamma$. Then ζ is a singular point of $f(z)$ by exactly the same argument as in Example 2.

THEOREM 16.5. *Given a disk K: $|z - z_0| < R$ and an element $\{K, f(z)\}$, suppose every point of the circle $\Gamma: |z - z_0| = R$ is a regular point. Then there is a disk $K_1: |z - z_0| < R_1$ and another element $\{K_1, \varphi(z)\}$ such that $R_1 > R$ and $\varphi(z) = f(z)$ on K, i.e., $f(z)$ can be continued analytically into a larger concentric disk.*[15]

Proof. Let D be the union of K and all the neighborhoods $\mathcal{N}(\zeta)$, $\zeta \in \Gamma$, where $\mathcal{N}(\zeta)$ is any neighborhood figuring in the definition of the regular point ζ. Then D is obviously nonempty. Since K and all the $\mathcal{N}(\zeta)$ are open, so is D (recall Sec. 9, Prob. 5a). Moreover, D is connected, since any two points of D can be joined by a curve contained in D (why?). Therefore D is a domain.

Next let $\varphi_\zeta(z)$ have the same meaning as in the definition of the regular point ζ, and define the following function on D:

$$\varphi(z) = \begin{cases} f(z) & \text{if } z \in K, \\ \varphi_\zeta(z) & \text{if } z \in \mathcal{N}(\zeta). \end{cases} \tag{16.11}$$

As we now show, $\varphi(z)$ is single-valued and analytic on D. Since D contains points which simultaneously belong to a neighborhood $\mathcal{N}(\zeta)$ and to K, and also points which simultaneously belong to two neighborhoods $\mathcal{N}(\zeta')$ and $\mathcal{N}(\zeta'')$, we must verify that the definition (16.11) is consistent, in the sense that a point $z \in D$ will always be assigned the same value $\varphi(z)$ whether z is regarded as belonging to K, $\mathcal{N}(\zeta)$, $\mathcal{N}(\zeta')$ or $\mathcal{N}(\zeta'')$. There

---

[15] Elements like $\{K, f\}$ and $\{K_1, \varphi\}$, whose domains are disks, are often called *circular elements*.

are basically two cases to be considered:

1. Let $G = \mathcal{N}(\zeta) \cap K$. Then $\varphi_\zeta(z) = f(z)$ on G, by the very definition of the regular point ζ.
2. Let $G = \mathcal{N}(\zeta') \cap \mathcal{N}(\zeta'') \neq 0$ and let $E = G \cap K$ [see Figure 16.5, which illustrates the situation where neither neighborhood $\mathcal{N}(\zeta')$, $\mathcal{N}(\zeta'')$ is a subset of the other]. Then $\varphi_{\zeta'}(z) = f(z) = \varphi_{\zeta''}(z)$ on E, and hence by the uniqueness theorem for analytic functions (Theorem 10.8), $\varphi_{\zeta'}(z) = \varphi_{\zeta''}(z)$ on the whole set G, since $E \subset G$ obviously contains a limit point of G.

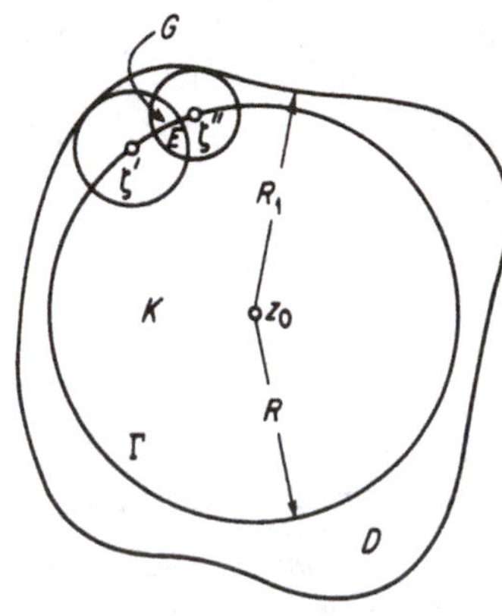

FIGURE 16.5

Thus the function $\varphi(z)$ defined by (16.11) is single-valued on D. Moreover $\varphi(z)$ is analytic on D, since, by construction, $\varphi(z)$ coincides with an analytic function [either $f(z)$ or some $\varphi_\zeta(z)$] in a neighborhood of every point of D [note that if $z \in D$, then there is a neighborhood $\mathcal{N}(z)$ such that $\mathcal{N}(z) \subset K$ or $\mathcal{N}(z) \subset \mathcal{N}(\zeta)$ for some $\zeta \in \Gamma$].

Finally let R_1 be the distance between z_0 and the boundary of D. Since every point is an interior point of D, R_1 must exceed the radius R of the original disk K. Therefore D contains the disk $K_1: |z - z_0| < R_1$, and the theorem is proved.

Given a power series

$$f(z) = a_0 + a_1(z - z_0) + \cdots + a_n(z - z_0)^n + \cdots \tag{16.12}$$

with radius of convergence R, the disk $K: |z - z_0| < R$ and the sum function $f(z)$ constitute a circular element $\{K, f(z)\}$. For simplicity, a regular or singular point of $\{K, f(z)\}$ will be called a regular or singular point of the series (16.12) or of its sum $f(z)$.

THEOREM 16.6. *The power series* (16.12) *has at least one singular point on its circle of convergence* $\Gamma: |z - z_0| = R$.

Proof. If the element $\{K, f(z)\}$, where K is the interior of Γ, has no singular points on Γ, it follows from Theorem 16.6 that $f(z)$ can be continued analytically into a larger disk $K_1: |z - z_0| < R_1$. But then the series (16.12) has a radius of convergence larger than R (why?), contrary to hypothesis.

Example 1. The geometric series

$$1 + z + z^2 + \cdots + z^n + \cdots$$

has circle of convergence $\Gamma:|z| = 1$ and sum $1/(1 - z)$. As required by Theorem 16.6, there is a singular point on Γ, namely, the point $z = 1$ (see Example 2, p. 352).

Example 2. We now show that *every* point on the circle of convergence $\Gamma:|z| = 1$ of the power series

$$f(z) = 1 + z^2 + z^4 + \cdots + z^{2^n} + \cdots$$

is a singular point. Suppose first that $z \to 1$ along the segment $\overline{01}$, i.e., along the radius of Γ contained in the nonnegative real axis. Then, since the nth partial sum

$$1 + x^2 + x^4 + \cdots + x^{2^n}$$

approaches $n + 1$ as $x \to 1-$, there is a $\delta(n) > 0$ such that

$$1 + x^2 + x^4 + \cdots + x^{2^n} > n$$

if $x > 1 - \delta(n)$. Therefore

$$f(x) = \sum_{k=0}^{\infty} x^{2^k} > \sum_{k=0}^{n} x^{2^k} > n$$

if $x > 1 - \delta(n)$, which implies

$$\lim_{x \to 1-} f(x) = \infty.$$

But then $z = 1$ is a singular point of $f(z)$ [recall Example 3, p. 353].

Next we note the identity

$$f(z) = z^2 + z^4 + \cdots + z^{2^n} + [1 + (z^{2^n})^2 + (z^{2^n})^4 + \cdots].$$

Since the term in brackets differs from the original series only by having z^{2^n} instead of z, we have

$$f(z) = z^2 + z^4 + \cdots + z^{2^n} + f(z^{2^n})$$

for any positive integer n. Consider all the 2^nth roots of unity, i.e., all the numbers

$$\sqrt[2^n]{1}, \tag{16.13}$$

which correspond to the vertices of a regular 2^n-gon inscribed in the unit circle. If ζ is one of these roots and if $z \to \zeta$ along the radius $\overline{0\zeta}$, then z^{2^n} obviously lies on the radius $\overline{01}$ and $z^{2^n} \to 1-$ as $z \to \zeta$. It follows that

$$\lim_{\substack{z \to \zeta \\ z \in \overline{0\zeta}}} f(z^{2^n}) = \infty,$$

and therefore

$$\lim_{\substack{z \to \zeta \\ z \in \overline{0\zeta}}} f(z) = \lim_{\substack{z \to \zeta \\ z \in \overline{0\zeta}}} [z^2 + z^4 + \cdots + z^{2^n} + f(z^{2^n})] = \infty.$$

Thus every root (16.13) is a singular point of $f(z)$, for all $n = 1, 2, \ldots$. Consequently, the set of singular points of $f(z)$ is *everywhere dense* in the unit circle Γ (i.e., every arc of Γ, however small, contains a singular point). But this means that *every* point of Γ is a singular point, since if $\zeta' \in \Gamma$ were a regular point, then, as on p. 352, some arc $\mathcal{N}(\zeta') \cap \Gamma$ would consist only of regular points. In particular, $f(z)$ cannot be continued analytically into any domain larger than the unit disk, as already noted in Example 3, p. 342.

Finally we generalize the concept of a singular point even further:

DEFINITION. *Let $F(z)$ be a complete analytic function.*[16] *Then by a singular point of $F(z)$ is meant a singular point of any element of $F(z)$.*

Example 1. If

$$F(z) = \frac{1}{z^{1/2} + 1},$$

then $z = 1$ is a singular point of $F(z)$, since it is a singular point of the single-valued branch of $F(z)$ for which $z^{1/2} = -1$. On the other hand, $z = 1$ belongs to the domain of existence of $F(z)$ [see Definition 3, p. 346], since $z = 1$ is a regular point of the single-valued branch of $F(z)$ for which $1^{1/2} = +1$.

Example 2. Suppose ζ is an isolated singular point of a complete analytic function $F(z)$, so that $F(z)$ is defined on some annulus $G: 0 < |z - \zeta| < R$. If $F(z)$ is single-valued on G, then ζ is an isolated singular point *of single-valued character*, as already studied in Sec. 56 (also recall Example 1, p. 352). More generally, $F(z)$ is multiple-valued on G, and then ζ is called an isolated singular point *of multiple-valued character*. Let D be the domain obtained from the disk $|z - \zeta| < R$ by deleting the line segment joining the origin to the point $z = R$ (thus D is a "slit disk"). Then it can be shown (see M3, Sec. 42) that there are just two possibilities:

1. $F(z)$ has precisely n distinct single-valued branches

$$\{D, f_0(z)\}, \{D, f_1(z)\}, \ldots, \{D, f_{n-1}(z)\}$$

on D such that

$$\begin{aligned} \lim_{\theta \to 2\pi-} f_{k-1}(re^{i\theta}) &= \lim_{\theta \to 0+} f_k(re^{i\theta}) \qquad (k = 1, \ldots, n-1), \\ \lim_{\theta \to 2\pi-} f_n(re^{i\theta}) &= \lim_{\theta \to 0+} f_0(re^{i\theta}) \end{aligned} \tag{16.14}$$

for every positive $r < R$. In this case, ζ is called a *branch point of order* $n - 1$ (cf. p. 76).

---

[16] We revert to the usual notation which does not distinguish between a function and its values at a point (cf. p. 14).

2. $F(z)$ has a countably infinite set of distinct single-valued branches

$$\{D, f_k(z)\} \qquad (k = 0, \pm 1, \pm 2, \ldots)$$

on D (and no others) such that

$$\lim_{\theta \to 2\pi -} f_{k-1}(re^{i\theta}) = \lim_{\theta \to 0+} f_k(re^{i\theta}) \qquad (k = 0, \pm 1, \pm 2, \ldots)$$

for every positive $r < R$. In this case, ζ is called a *logarithmic branch point* (cf. p. 129).

Example 3. Let $F(z)$ be a complete analytic function with a branch point ζ of order $n - 1$. Then ζ is called an *algebraic branch point* if $F(z)$ approaches a limit (finite or infinite) as $z \to \zeta$ (cf. p. 76). Otherwise ζ is called a *transcendental branch point*. For example, each of the functions $z^{1/n}$ and $\exp(z^{1/n})$ has an algebraic branch point of order $n - 1$ at the origin, while the function $\exp(z^{-1/n})$ has a transcendental branch point of order $n - 1$ at the origin.

Example 4. The considerations given in Example 2 and 3 are easily generalized to the case of branch points at ∞ by taking G to be the domain $|z| > R$ and D the domain obtained from G by deleting the infinite line segment joining the point $z = R$ to the point at infinity. For example, the function $\exp(z^{-1/n})$ has an algebraic branch point of order $n - 1$ at ∞.

Problem 1. Let G be the interior of a closed Jordan curve Γ. Prove that $\zeta \in \Gamma$ is a regular point of $\{G, f(z)\}$ if and only if $f(z)$ can be continued analytically across an arc $\delta \subset \Gamma$ containing ζ.

Problem 2 (M1, p. 387). Given a power series

$$f(z) = a_0 + a_1(z - z_0) + \cdots + a_n(z - z_0)^n + \cdots$$

with radius of convergence R, let ζ be a point of the circle Γ: $|z - z_0| = R$ and let z_1 be a point of the radius $\overline{z_0\zeta}$ distinct from z_0 and ζ. Prove that ζ is a singular point of $f(z)$ if

$$\Delta = R - |z_1 - z_0| = \frac{1}{\overline{\lim\limits_{n\to\infty}} \sqrt[n]{\dfrac{|f^{(n)}(z_1)|}{n!}}}$$

and a regular point if

$$\Delta < \frac{1}{\overline{\lim\limits_{n\to\infty}} \sqrt[n]{\dfrac{|f^{(n)}(z_1)|}{n!}}}.$$

Problem 3 (M1, p. 389). Prove the following result known as *Pringsheim's theorem*: Given a power series

$$f(z) = a_0 + a_1 z + \cdots + a_n z^n + \cdots$$

with radius of convergence 1, suppose the coefficients a_n are all nonnegative real numbers. Then $z = 1$ is a singular point of $f(z)$.

Problem 4. Prove that the circle $|z| = 1$ is the natural boundary (see p. 342) of the series

$$\sum_{n=0}^{\infty} z^{n!}.$$

Problem 5 (M1, p. 393). Prove that the point $z = 1$ is a singular point belonging to the domain of existence of the complete analytic function

$$F(z) = \frac{1}{\operatorname{Ln} z}.$$

Problem 6. Discuss the branch points considered in Secs. 20 and 32 from the standpoint of the general definition of branch points in Examples 2–4.

Problem 7. Find the singular points of the following functions:

a) $(\sqrt[6]{z})^4$; b) $(\sqrt[m]{z})^n$ (m, n positive integers); c) $\exp\left(\dfrac{1}{\sqrt{z} - 1}\right)$;

d) $\dfrac{\sin \sqrt{z}}{\sqrt{z}}$; e) $\tan \dfrac{1 - \sqrt{z}}{1 + \sqrt{z}}$; f) $e^{(\operatorname{Ln} z)^n}$ (n a positive integer);

g) $z^i = e^{i \operatorname{Ln} z}$; h) $\sin \operatorname{Ln} z$; i) $\dfrac{1}{z} \operatorname{Ln} \dfrac{1}{1 - z}$; j) $\dfrac{1}{z} \operatorname{Arc} \sin z$;

k) $\dfrac{1}{z^2} + \operatorname{Arc} \tan z$.

Ans.

a) Algebraic branch points of order 2 at $z = 0, \infty$;
c) Algebraic branch points of order 1 at $z = 0, \infty$; one of the branches has an essential singular point at $z = 1$;
e) Algebraic branch points of order 1 at $z = 0, \infty$; one of the branches has a limit point of poles at $z = 1$;
g) Logarithmic branch points at $z = 0, \infty$;
i) Logarithmic branch points at $z = 1, \infty$; every branch except one has a pole of order 1 at $z = 0$;
k) Logarithmic branch points at $z = \pm i$; every branch has a pole of order 2 at $z = 0$.

79. RIEMANN SURFACES

Finally we consider the problem of finding a "generalized domain" on which a given complete analytic function F is single-valued. This is done by the following simple construction: Let $\{G, f\}$ and $\{D, \varphi\}$ be any two elements of F such that $G \cap D \neq 0$ and think of G and D as being cut out of separate pieces of paper. Then "paste together" the components of $G \cap D$ on which $f(z)$ and $\varphi(z)$ coincide, while "leaving unpasted" the components on which $f(z)$ and $\varphi(z)$ differ (see Prob. 1). Doing this for every pair of elements of F, we obtain a "layered surface" covering every point z_0 in the domain of existence of F a number of times (possibly infinite) equal to the number of values F takes at z_0. This "many-sheeted structure" is called the *Riemann surface* of F.

Clearly, the effect of the construction just described is to make F single-valued on its Riemann surface. More precisely, a "point" on the Riemann surface can be interpreted as an ordered pair $(z, F(z))$, where z lies in the ordinary complex plane and $F(z)$ is a value of F at z. But equality of two such points $(z_1, F(z_1))$ and $(z_2, F(z_2))$ means not only that $z_1 = z_2$ but also that $F(z_1) = F(z_2)$!

More generally, we can define a complete analytic function F by using the definition of a direct analytic continuation given on p. 349 instead of that given on p. 342, or by using both definitions simultaneously (supply the details). Then in general two elements $\{G, f\}, \{D, \varphi\} \in F$ share suitable boundary arcs and overlap as well. Correspondingly, the prescription for constructing the Riemann surface of F involves pasting the boundaries of G and D together along every common arc δ on which $f(z)$ and $\varphi(z)$ coincide,[17] as well as pasting together those portions of the domains themselves on which $f(z) \equiv \varphi(z)$.

Remark. These heuristic considerations can be supported by perfectly rigorous mathematics, and in fact much of the modern theory of analytic functions is concerned with Riemann surfaces and related topics. The mathematical tools needed to pursue the subject in proper depth are rather sophisticated. From the standpoint of the advanced theory, the entities constructed here are "topological models of Riemann covering surfaces" (see M3, Sec. 35).

Example 1. Suppose F has a branch point of order 2 at a point ζ. To construct the Riemann surface of F in the vicinity of ζ, we start from three "replicas" D_0, D_1, D_2 of the slit disk figuring in Example 2, p. 356, all cut along the segment joining the points $z = 0$ and $z = R$. These domains have the appearance shown in Figure 16.6(a), where the upper and lower edges of the cut are suitably labelled in each case. Next we paste δ_0^- to δ_1^+, then δ_1^- to δ_2^+ and finally δ_2^- to δ_0^+ (this has the effect of joining D_0 to D_1, then D_1 to D_2 and finally D_2 back to D_0).[18] As a result, we obtain the "three-sheeted domain" shown in Figure 16.6(b). The intersections which come about as a result of "pasting D_2 back to D_0" have no mathematical meaning and should be disregarded.

---

[17] Here, as in Theorem 16.3, it is assumed that $f(z)$ is analytic on G and continuous on $G \cup \delta$, while $\varphi(z)$ is analytic on D and continuous on $D \cup \delta$. It is also assumed that δ has the properties needed to invoke Theorem 16.3 or one of its generalizations (see Sec. 76, Probs. 3, 4).

[18] The justification for this construction stems, of course, from the formulas (16.14) and Theorem 16.3.

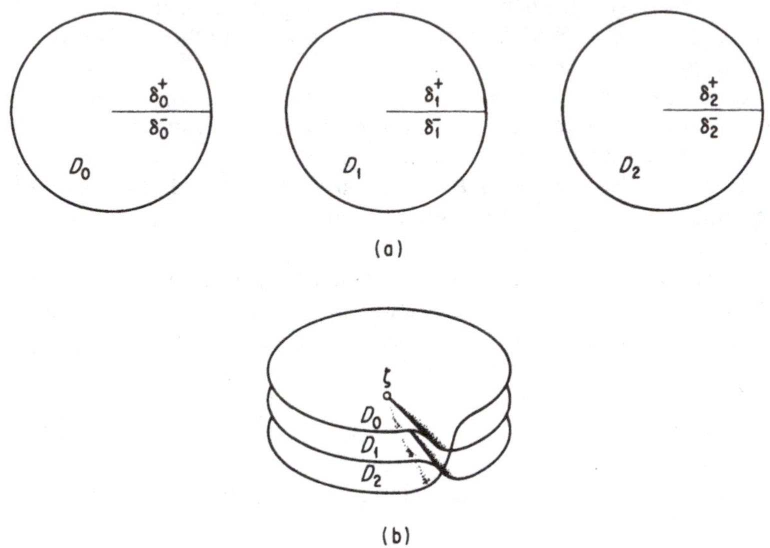

FIGURE 16.6

Example 2. Next we construct the Riemann surface of the function

$$w = F(z) = \sqrt[n]{z},$$

starting from n replicas $D_0, D_1, \ldots, D_{n-1}$ of the finite plane cut along the nonnegative real axis. Let δ_k^+ and δ_k^- denote the upper and lower edges of the nonnegative real axis regarded as the boundary of D_k, and let

$$f_k(z) = \sqrt[n]{|z|}\left(\cos\frac{\arg z + 2k\pi}{n} + i\sin\frac{\arg z + 2k\pi}{n}\right)$$

$$(k = 0, 1, \ldots, n-1),$$

so that every $\{D_k, f_k(z)\}$ is an element of $F(z)$.[19] We now paste δ_0^- to δ_1^+, δ_1^- to $\delta_2^+, \ldots, \delta_{n-2}^-$ to δ_{n-1}^+ and finally δ_{n-1}^- to δ_0^+, corresponding to the fact that $\{D_k, f_k(z)\}$ is a direct analytic continuation of $\{D_{k-1}, f_{k-1}(z)\}$ for every $k = 1, \ldots, n$ [with $\{D_n, f_n(z)\} = \{D_0, f_0(z)\}$]. The result is an n-sheeted Riemann surface, shown schematically in Figure 16.7 for the case $n = 4$. All n sheets come together at the branch points $z = 0$ and $z = \infty$.[20]

---

[19] In writing $\{D_k, f_k(z)\}$, there is really no need to distinguish between domains D_k with different subscripts.

[20] It is customary to include these points on the Riemann surface, although they do not belong to any of the domains D_k.

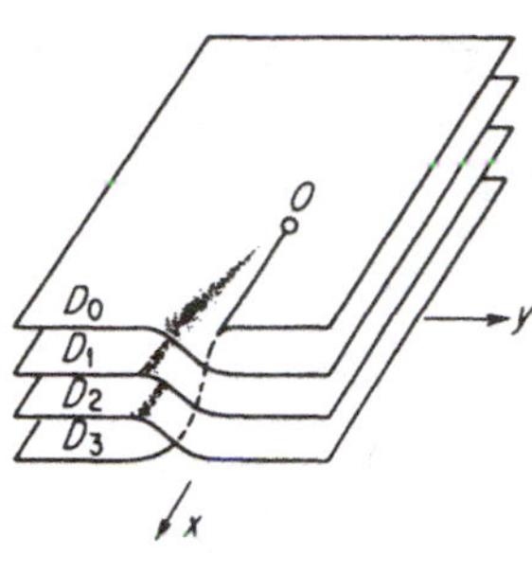

FIGURE 16.7

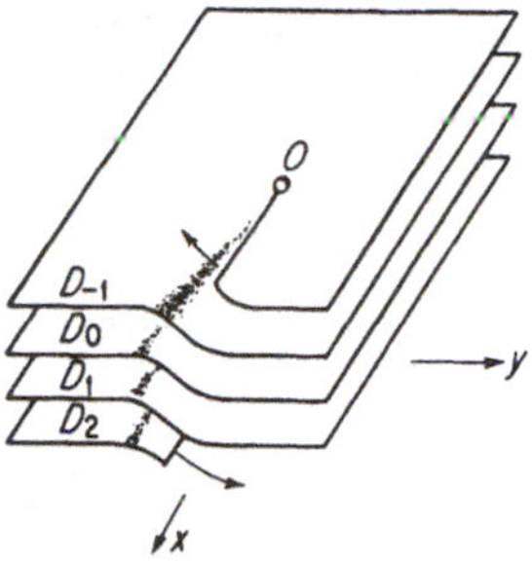

FIGURE 16.8

Example 3. Finally we construct the Riemann surface of the function

$$w = F(z) = \text{Ln } z,$$

starting from infinitely many replicas . . . , D_{-1}, D_0, D_1, . . . of the finite plane cut along the nonnegative real axis. The construction is the same as that given in the preceding example, except that now there is no "first sheet" and no "last sheet" which are pasted together at the end of the construction, and instead we obtain the infinite-sheeted Riemann surface shown schematically in Figure 16.8.

Problem 1. Can a complete analytic function $F(z)$ take the same value at several points of a Riemann surface all lying over the same point ζ in the z-plane? If so, how many such points are allowed?

Hint. Consider the function

$$F(z) = (z - \zeta)\sqrt{z}.$$

Comment. In constructing a Riemann surface we do not paste together separate points of the domains G and D of two elements $\{G, f\}$ and $\{D, \varphi\}$, but only common boundary arcs of G and D or whole domains (components of $G \cap D$).

Problem 2. Construct the Riemann surface of the function

$$w = F(z) = z + \sqrt{z^2 - 1},$$

which is the inverse of the Joukowski function of Sec. 22.

Problem 3. Construct the Riemann surface of the function

$$w = F(z) = \text{Arc sin } z.$$

Problem 4. Discuss the notion of a domain of univalence (see p. 73) from the standpoint of Riemann surfaces.

BIBLIOGRAPHY

Ahlfors, L. V., *Complex Analysis*, second edition, McGraw-Hill Book Co., New York (1966).

Churchill, R. V., *Complex Variables and Applications*, second edition, McGraw-Hill Book Co., New York (1960).

Copson, E. T., *An Introduction to the Theory of Functions of a Complex Variable*, Oxford University Press, London (1935).

Franklin, P., *Functions of a Complex Variable*, Prentice-Hall, Inc., Englewood Cliffs, N.J. (1958).

Hille, E., *Analytic Function Theory*, in two volumes, Ginn and Co., Boston (1959, 1962).

Knopp, K., *Theory of Functions* (translated by F. Bagemihl), in two volumes, Dover Publications, Inc., New York (1945, 1947).

Markushevich, A. I., *Theory of Functions of a Complex Variable* (translated by R. A. Silverman), in three volumes, Prentice-Hall, Inc., Englewood Cliffs, N.J. (1965, 1967).

Nehari, Z., *Conformal Mapping*, McGraw-Hill Book Co., New York (1952).

Nehari, Z., *Introduction to Complex Analysis*, Allyn and Bacon, Inc., Boston (1962).

Pennisi, L. L., *Elements of Complex Variables* (with the collaboration of L. I. Gordon and S. Lasher), Holt, Rinehart and Winston, Inc., New York (1963).

Phillips, E. G., *Functions of a Complex Variable with Applications*, Oliver and Boyd, London (1961).

Thron, W. J., *Introduction to the Theory of Functions of a Complex Variable*, John Wiley and Sons, Inc., New York (1953).

Titchmarsh, E. C., *The Theory of Functions*, second edition, Oxford University Press, London (1939).

INDEX

A

B

C

G

H

I

S

T

A CATALOG OF SELECTED

DOVER BOOKS

IN SCIENCE AND MATHEMATICS

Astronomy

BURNHAM'S CELESTIAL HANDBOOK, Robert Burnham, Jr. Thorough guide to the stars beyond our solar system. Exhaustive treatment. Alphabetical by constellation: Andromeda to Cetus in Vol. 1; Chamaeleon to Orion in Vol. 2; and Pavo to Vulpecula in Vol. 3. Hundreds of illustrations. Index in Vol. 3. 2,000pp. 6¹/₈ x 9¹/₄.
Vol. I: 0-486-23567-X
Vol. II: 0-486-23568-8
Vol. III: 0-486-23673-0

EXPLORING THE MOON THROUGH BINOCULARS AND SMALL TELESCOPES, Ernest H. Cherrington, Jr. Informative, profusely illustrated guide to locating and identifying craters, rills, seas, mountains, other lunar features. Newly revised and updated with special section of new photos. Over 100 photos and diagrams. 240pp. 8¹/₄ x 11. 0-486-24491-1

THE EXTRATERRESTRIAL LIFE DEBATE, 1750–1900, Michael J. Crowe. First detailed, scholarly study in English of the many ideas that developed from 1750 to 1900 regarding the existence of intelligent extraterrestrial life. Examines ideas of Kant, Herschel, Voltaire, Percival Lowell, many other scientists and thinkers. 16 illustrations. 704pp. 5³/₈ x 8¹/₂. 0-486-40675-X

THEORIES OF THE WORLD FROM ANTIQUITY TO THE COPERNICAN REVOLUTION, Michael J. Crowe. Newly revised edition of an accessible, enlightening book re-creates the change from an earth-centered to a sun-centered conception of the solar system. 242pp. 5³/₈ x 8¹/₂. 0-486-41444-2

ARISTARCHUS OF SAMOS: The Ancient Copernicus, Sir Thomas Heath. Heath's history of astronomy ranges from Homer and Hesiod to Aristarchus and includes quotes from numerous thinkers, compilers, and scholasticists from Thales and Anaximander through Pythagoras, Plato, Aristotle, and Heraclides. 34 figures. 448pp. 5³/₈ x 8¹/₂.
0-486-43886-4

A COMPLETE MANUAL OF AMATEUR ASTRONOMY: TOOLS AND TECHNIQUES FOR ASTRONOMICAL OBSERVATIONS, P. Clay Sherrod with Thomas L. Koed. Concise, highly readable book discusses: selecting, setting up and maintaining a telescope; amateur studies of the sun; lunar topography and occultations; observations of Mars, Jupiter, Saturn, the minor planets and the stars; an introduction to photoelectric photometry; more. 1981 ed. 124 figures. 25 halftones. 37 tables. 335pp. 6¹/₂ x 9¹/₄. 0-486-42820-8

AMATEUR ASTRONOMER'S HANDBOOK, J. B. Sidgwick. Timeless, comprehensive coverage of telescopes, mirrors, lenses, mountings, telescope drives, micrometers, spectroscopes, more. 189 illustrations. 576pp. 5⁵/₈ x 8¹/₄. (Available in U.S. only.)
0-486-24034-7

STAR LORE: Myths, Legends, and Facts, William Tyler Olcott. Captivating retellings of the origins and histories of ancient star groups include Pegasus, Ursa Major, Pleiades, signs of the zodiac, and other constellations. "Classic."—Sky & Telescope. 58 illustrations. 544pp. 5³/₈ x 8¹/₂. 0-486-43581-4

Chemistry

THE SCEPTICAL CHYMIST: THE CLASSIC 1661 TEXT, Robert Boyle. Boyle defines the term "element," asserting that all natural phenomena can be explained by the motion and organization of primary particles. 1911 ed. viii+232pp. $5^3/_8$ x $8^1/_2$.
0-486-42825-7

RADIOACTIVE SUBSTANCES, Marie Curie. Here is the celebrated scientist's doctoral thesis, the prelude to her receipt of the 1903 Nobel Prize. Curie discusses establishing atomic character of radioactivity found in compounds of uranium and thorium; extraction from pitchblende of polonium and radium; isolation of pure radium chloride; determination of atomic weight of radium; plus electric, photographic, luminous, heat, color effects of radioactivity. ii+94pp. $5^3/_8$ x $8^1/_2$. 0-486-42550-9

CHEMICAL MAGIC, Leonard A. Ford. Second Edition, Revised by E. Winston Grundmeier. Over 100 unusual stunts demonstrating cold fire, dust explosions, much more. Text explains scientific principles and stresses safety precautions. 128pp. $5^3/_8$ x $8^1/_2$. 0-486-67628-5

MOLECULAR THEORY OF CAPILLARITY, J. S. Rowlinson and B. Widom. History of surface phenomena offers critical and detailed examination and assessment of modern theories, focusing on statistical mechanics and application of results in mean-field approximation to model systems. 1989 edition. 352pp. $5^3/_8$ x $8^1/_2$. 0-486-42544-4

CHEMICAL AND CATALYTIC REACTION ENGINEERING, James J. Carberry. Designed to offer background for managing chemical reactions, this text examines behavior of chemical reactions and reactors; fluid-fluid and fluid-solid reaction systems; heterogeneous catalysis and catalytic kinetics; more. 1976 edition. 672pp. $6^1/_8$ x $9^1/_4$. 0-486-41736-0 $31.95

ELEMENTS OF CHEMISTRY, Antoine Lavoisier. Monumental classic by founder of modern chemistry in remarkable reprint of rare 1790 Kerr translation. A must for every student of chemistry or the history of science. 539pp. $5^3/_8$ x $8^1/_2$. 0-486-64624-6

MOLECULES AND RADIATION: An Introduction to Modern Molecular Spectroscopy. Second Edition, Jeffrey I. Steinfeld. This unified treatment introduces upper-level undergraduates and graduate students to the concepts and the methods of molecular spectroscopy and applications to quantum electronics, lasers, and related optical phenomena. 1985 edition. 512pp. $5^3/_8$ x $8^1/_2$. 0-486-44152-0

A SHORT HISTORY OF CHEMISTRY, J. R. Partington. Classic exposition explores origins of chemistry, alchemy, early medical chemistry, nature of atmosphere, theory of valency, laws and structure of atomic theory, much more. 428pp. $5^3/_8$ x $8^1/_2$. (Available in U.S. only.) 0-486-65977-1

GENERAL CHEMISTRY, Linus Pauling. Revised 3rd edition of classic first-year text by Nobel laureate. Atomic and molecular structure, quantum mechanics, statistical mechanics, thermodynamics correlated with descriptive chemistry. Problems. 992pp. $5^3/_8$ x $8^1/_2$.
0-486-65622-5

ELECTRON CORRELATION IN MOLECULES, S. Wilson. This text addresses one of theoretical chemistry's central problems. Topics include molecular electronic structure, independent electron models, electron correlation, the linked diagram theorem, and related topics. 1984 edition. 304pp. $5^3/_8$ x $8^1/_2$. 0-486-45879-2

Engineering

DE RE METALLICA, Georgius Agricola. The famous Hoover translation of greatest treatise on technological chemistry, engineering, geology, mining of early modern times (1556). All 289 original woodcuts. 638pp. $6^3/_4$ x 11. 0-486-60006-8

FUNDAMENTALS OF ASTRODYNAMICS, Roger Bate et al. Modern approach developed by U.S. Air Force Academy. Designed as a first course. Problems, exercises. Numerous illustrations. 455pp. $5^3/_8$ x $8^1/_2$. 0-486-60061-0

DYNAMICS OF FLUIDS IN POROUS MEDIA, Jacob Bear. For advanced students of ground water hydrology, soil mechanics and physics, drainage and irrigation engineering and more. 335 illustrations. Exercises, with answers. 784pp. $6^1/_8$ x $9^1/_4$. 0-486-65675-6

THEORY OF VISCOELASTICITY (SECOND EDITION), Richard M. Christensen. Complete consistent description of the linear theory of the viscoelastic behavior of materials. Problem-solving techniques discussed. 1982 edition. 29 figures. xiv+364pp. $6^1/_8$ x $9^1/_4$. 0-486-42880-X

MECHANICS, J. P. Den Hartog. A classic introductory text or refresher. Hundreds of applications and design problems illuminate fundamentals of trusses, loaded beams and cables, etc. 334 answered problems. 462pp. $5^3/_8$ x $8^1/_2$. 0-486-60754-2

MECHANICAL VIBRATIONS, J. P. Den Hartog. Classic textbook offers lucid explanations and illustrative models, applying theories of vibrations to a variety of practical industrial engineering problems. Numerous figures. 233 problems, solutions. Appendix. Index. Preface. 436pp. $5^3/_8$ x $8^1/_2$. 0-486-64785-4

STRENGTH OF MATERIALS, J. P. Den Hartog. Full, clear treatment of basic material (tension, torsion, bending, etc.) plus advanced material on engineering methods, applications. 350 answered problems. 323pp. $5^3/_8$ x $8^1/_2$. 0-486-60755-0

A HISTORY OF MECHANICS, René Dugas. Monumental study of mechanical principles from antiquity to quantum mechanics. Contributions of ancient Greeks, Galileo, Leonardo, Kepler, Lagrange, many others. 671pp. $5^3/_8$ x $8^1/_2$. 0-486-65632-2

STABILITY THEORY AND ITS APPLICATIONS TO STRUCTURAL MECHANICS, Clive L. Dym. Self-contained text focuses on Koiter postbuckling analyses, with mathematical notions of stability of motion. Basing minimum energy principles for static stability upon dynamic concepts of stability of motion, it develops asymptotic buckling and postbuckling analyses from potential energy considerations, with applications to columns, plates, and arches. 1974 ed. 208pp. $5^3/_8$ x $8^1/_2$. 0-486-42541-X

BASIC ELECTRICITY, U.S. Bureau of Naval Personnel. Originally a training course; best nontechnical coverage. Topics include batteries, circuits, conductors, AC and DC, inductance and capacitance, generators, motors, transformers, amplifiers, etc. Many questions with answers. 349 illustrations. 1969 edition. 448pp. $6^1/_2$ x $9^1/_4$. 0-486-20973-3

Mathematics

FUNCTIONAL ANALYSIS (Second Corrected Edition), George Bachman and Lawrence Narici. Excellent treatment of subject geared toward students with background in linear algebra, advanced calculus, physics and engineering. Text covers introduction to inner-product spaces, normed, metric spaces, and topological spaces; complete orthonormal sets, the Hahn-Banach Theorem and its consequences, and many other related subjects. 1966 ed. 544pp. $6^1/_8$ x $9^1/_4$. 0-486-40251-7

DIFFERENTIAL MANIFOLDS, Antoni A. Kosinski. Introductory text for advanced undergraduates and graduate students presents systematic study of the topological structure of smooth manifolds, starting with elements of theory and concluding with method of surgery. 1993 edition. 288pp. $5^3/_8$ x $8^1/_2$. 0-486-46244-7

VECTOR AND TENSOR ANALYSIS WITH APPLICATIONS, A. I. Borisenko and I. E. Tarapov. Concise introduction. Worked-out problems, solutions, exercises. 257pp. $5^5/_8$ x $8^1/_4$. 0-486-63833-2

AN INTRODUCTION TO ORDINARY DIFFERENTIAL EQUATIONS, Earl A. Coddington. A thorough and systematic first course in elementary differential equations for undergraduates in mathematics and science, with many exercises and problems (with answers). Index. 304pp. $5^3/_8$ x $8^1/_2$. 0-486-65942-9

FOURIER SERIES AND ORTHOGONAL FUNCTIONS, Harry F. Davis. An incisive text combining theory and practical example to introduce Fourier series, orthogonal functions and applications of the Fourier method to boundary-value problems. 570 exercises. Answers and notes. 416pp. $5^3/_8$ x $8^1/_2$. 0-486-65973-9

COMPUTABILITY AND UNSOLVABILITY, Martin Davis. Classic graduate-level introduction to theory of computability, usually referred to as theory of recurrent functions. New preface and appendix. 288pp. $5^3/_8$ x $8^1/_2$. 0-486-61471-9

AN INTRODUCTION TO MATHEMATICAL ANALYSIS, Robert A. Rankin. Dealing chiefly with functions of a single real variable, this text by a distinguished educator introduces limits, continuity, differentiability, integration, convergence of infinite series, double series, and infinite products. 1963 edition. 624pp. $5^3/_8$ x $8^1/_2$. 0-486-46251-X

METHODS OF NUMERICAL INTEGRATION (SECOND EDITION), Philip J. Davis and Philip Rabinowitz. Requiring only a background in calculus, this text covers approximate integration over finite and infinite intervals, error analysis, approximate integration in two or more dimensions, and automatic integration. 1984 edition. 624pp. $5^3/_8$ x $8^1/_2$. 0-486-45339-1

INTRODUCTION TO LINEAR ALGEBRA AND DIFFERENTIAL EQUATIONS, John W. Dettman. Excellent text covers complex numbers, determinants, orthonormal bases, Laplace transforms, much more. Exercises with solutions. Undergraduate level. 416pp. $5^3/_8$ x $8^1/_2$. 0-486-65191-6

RIEMANN'S ZETA FUNCTION, H. M. Edwards. Superb, high-level study of landmark 1859 publication entitled "On the Number of Primes Less Than a Given Magnitude" traces developments in mathematical theory that it inspired. xiv+315pp. $5^3/_8$ x $8^1/_2$. 0-486-41740-9

CALCULUS OF VARIATIONS WITH APPLICATIONS, George M. Ewing. Applications-oriented introduction to variational theory develops insight and promotes understanding of specialized books, research papers. Suitable for advanced undergraduate/graduate students as primary, supplementary text. 352pp. 5³/₈ x 8¹/₂. 0-486-64856-7

MATHEMATICIAN'S DELIGHT, W. W. Sawyer. "Recommended with confidence" by *The Times Literary Supplement,* this lively survey was written by a renowned teacher. It starts with arithmetic and algebra, gradually proceeding to trigonometry and calculus. 1943 edition. 240pp. 5³/₈ x 8¹/₂. 0-486-46240-4

ADVANCED EUCLIDEAN GEOMETRY, Roger A. Johnson. This classic text explores the geometry of the triangle and the circle, concentrating on extensions of Euclidean theory, and examining in detail many relatively recent theorems. 1929 edition. 336pp. 5³/₈ x 8¹/₂. 0-486-46237-4

COUNTEREXAMPLES IN ANALYSIS, Bernard R. Gelbaum and John M. H. Olmsted. These counterexamples deal mostly with the part of analysis known as "real variables." The first half covers the real number system, and the second half encompasses higher dimensions. 1962 edition. xxiv+198pp. 5³/₈ x 8¹/₂. 0-486-42875-3

CATASTROPHE THEORY FOR SCIENTISTS AND ENGINEERS, Robert Gilmore. Advanced-level treatment describes mathematics of theory grounded in the work of Poincaré, R. Thom, other mathematicians. Also important applications to problems in mathematics, physics, chemistry and engineering. 1981 edition. References. 28 tables. 397 black-and-white illustrations. xvii + 666pp. 6¹/₈ x 9¹/₄. 0-486-67539-4

COMPLEX VARIABLES: Second Edition, Robert B. Ash and W. P. Novinger. Suitable for advanced undergraduates and graduate students, this newly revised treatment covers Cauchy theorem and its applications, analytic functions, and the prime number theorem. Numerous problems and solutions. 2004 edition. 224pp. 6¹/₂ x 9¹/₄. 0-486-46250-1

NUMERICAL METHODS FOR SCIENTISTS AND ENGINEERS, Richard Hamming. Classic text stresses frequency approach in coverage of algorithms, polynomial approximation, Fourier approximation, exponential approximation, other topics. Revised and enlarged 2nd edition. 721pp. 5³/₈ x 8¹/₂. 0-486-65241-6

INTRODUCTION TO NUMERICAL ANALYSIS (2nd Edition), F. B. Hildebrand. Classic, fundamental treatment covers computation, approximation, interpolation, numerical differentiation and integration, other topics. 150 new problems. 669pp. 5³/₈ x 8¹/₂. 0-486-65363-3

MARKOV PROCESSES AND POTENTIAL THEORY, Robert M. Blumental and Ronald K. Getoor. This graduate-level text explores the relationship between Markov processes and potential theory in terms of excessive functions, multiplicative functionals and subprocesses, additive functionals and their potentials, and dual processes. 1968 edition. 320pp. 5³/₈ x 8¹/₂. 0-486-46263-3

ABSTRACT SETS AND FINITE ORDINALS: An Introduction to the Study of Set Theory, G. B. Keene. This text unites logical and philosophical aspects of set theory in a manner intelligible to mathematicians without training in formal logic and to logicians without a mathematical background. 1961 edition. 112pp. 5³/₈ x 8¹/₂. 0-486-46249-8

INTRODUCTORY REAL ANALYSIS, A.N. Kolmogorov, S. V. Fomin. Translated by Richard A. Silverman. Self-contained, evenly paced introduction to real and functional analysis. Some 350 problems. 403pp. $5^3/_8$ x $8^1/_2$. 0-486-61226-0

APPLIED ANALYSIS, Cornelius Lanczos. Classic work on analysis and design of finite processes for approximating solution of analytical problems. Algebraic equations, matrices, harmonic analysis, quadrature methods, much more. 559pp. $5^3/_8$ x $8^1/_2$. 0-486-65656-X

AN INTRODUCTION TO ALGEBRAIC STRUCTURES, Joseph Landin. Superb self-contained text covers "abstract algebra": sets and numbers, theory of groups, theory of rings, much more. Numerous well-chosen examples, exercises. 247pp. $5^3/_8$ x $8^1/_2$. 0-486-65940-2

QUALITATIVE THEORY OF DIFFERENTIAL EQUATIONS, V. V. Nemytskii and V.V. Stepanov. Classic graduate-level text by two prominent Soviet mathematicians covers classical differential equations as well as topological dynamics and ergodic theory. Bibliographies. 523pp. $5^3/_8$ x $8^1/_2$. 0-486-65954-2

THEORY OF MATRICES, Sam Perlis. Outstanding text covering rank, nonsingularity and inverses in connection with the development of canonical matrices under the relation of equivalence, and without the intervention of determinants. Includes exercises. 237pp. $5^3/_8$ x $8^1/_2$. 0-486-66810-X

INTRODUCTION TO ANALYSIS, Maxwell Rosenlicht. Unusually clear, accessible coverage of set theory, real number system, metric spaces, continuous functions, Riemann integration, multiple integrals, more. Wide range of problems. Undergraduate level. Bibliography. 254pp. $5^3/_8$ x $8^1/_2$. 0-486-65038-3

MODERN NONLINEAR EQUATIONS, Thomas L. Saaty. Emphasizes practical solution of problems; covers seven types of equations. ". . . a welcome contribution to the existing literature. . . ."—*Math Reviews*. 490pp. $5^3/_8$ x $8^1/_2$. 0-486-64232-1

MATRICES AND LINEAR ALGEBRA, Hans Schneider and George Phillip Barker. Basic textbook covers theory of matrices and its applications to systems of linear equations and related topics such as determinants, eigenvalues and differential equations. Numerous exercises. 432pp. $5^3/_8$ x $8^1/_2$. 0-486-66014-1

LINEAR ALGEBRA, Georgi E. Shilov. Determinants, linear spaces, matrix algebras, similar topics. For advanced undergraduates, graduates. Silverman translation. 387pp. $5^3/_8$ x $8^1/_2$. 0-486-63518-X

MATHEMATICAL METHODS OF GAME AND ECONOMIC THEORY: Revised Edition, Jean-Pierre Aubin. This text begins with optimization theory and convex analysis, followed by topics in game theory and mathematical economics, and concluding with an introduction to nonlinear analysis and control theory. 1982 edition. 656pp. $6^1/_8$ x $9^1/_4$. 0-486-46265-X

SET THEORY AND LOGIC, Robert R. Stoll. Lucid introduction to unified theory of mathematical concepts. Set theory and logic seen as tools for conceptual understanding of real number system. 496pp. $5^3/_8$ x $8^1/_4$. 0-486-63829-4

TENSOR CALCULUS, J.L. Synge and A. Schild. Widely used introductory text covers spaces and tensors, basic operations in Riemannian space, non-Riemannian spaces, etc. 324pp. $5^3/_8$ x $8^1/_4$. 0-486-63612-7

ORDINARY DIFFERENTIAL EQUATIONS, Morris Tenenbaum and Harry Pollard. Exhaustive survey of ordinary differential equations for undergraduates in mathematics, engineering, science. Thorough analysis of theorems. Diagrams. Bibliography. Index. 818pp. $5^3/_8$ x $8^1/_2$. 0-486-64940-7

INTEGRAL EQUATIONS, F. G. Tricomi. Authoritative, well-written treatment of extremely useful mathematical tool with wide applications. Volterra Equations, Fredholm Equations, much more. Advanced undergraduate to graduate level. Exercises. Bibliography. 238pp. $5^3/_8$ x $8^1/_2$. 0-486-64828-1

FOURIER SERIES, Georgi P. Tolstov. Translated by Richard A. Silverman. A valuable addition to the literature on the subject, moving clearly from subject to subject and theorem to theorem. 107 problems, answers. 336pp. $5^3/_8$ x $8^1/_2$. 0-486-63317-9

INTRODUCTION TO MATHEMATICAL THINKING, Friedrich Waismann. Examinations of arithmetic, geometry, and theory of integers; rational and natural numbers; complete induction; limit and point of accumulation; remarkable curves; complex and hypercomplex numbers, more. 1959 ed. 27 figures. xii+260pp. $5^3/_8$ x $8^1/_2$. 0-486-42804-8

THE RADON TRANSFORM AND SOME OF ITS APPLICATIONS, Stanley R. Deans. Of value to mathematicians, physicists, and engineers, this excellent introduction covers both theory and applications, including a rich array of examples and literature. Revised and updated by the author. 1993 edition. 304pp. $6^1/_8$ x $9^1/_4$. 0-486-46241-2

CALCULUS OF VARIATIONS, Robert Weinstock. Basic introduction covering isoperimetric problems, theory of elasticity, quantum mechanics, electrostatics, etc. Exercises throughout. 326pp. $5^3/_8$ x $8^1/_2$. 0-486-63069-2

THE CONTINUUM: A CRITICAL EXAMINATION OF THE FOUNDATION OF ANALYSIS, Hermann Weyl. Classic of 20th-century foundational research deals with the conceptual problem posed by the continuum. 156pp. $5^3/_8$ x $8^1/_2$. 0-486-67982-9

CHALLENGING MATHEMATICAL PROBLEMS WITH ELEMENTARY SOLUTIONS, A. M. Yaglom and I. M. Yaglom. Over 170 challenging problems on probability theory, combinatorial analysis, points and lines, topology, convex polygons, many other topics. Solutions. Total of 445pp. $5^3/_8$ x $8^1/_2$. Two-vol. set.
Vol. I: 0-486-65536-9 Vol. II: 0-486-65537-7

INTRODUCTION TO PARTIAL DIFFERENTIAL EQUATIONS WITH APPLICATIONS, E. C. Zachmanoglou and Dale W. Thoe. Essentials of partial differential equations applied to common problems in engineering and the physical sciences. Problems and answers. 416pp. $5^3/_8$ x $8^1/_2$. 0-486-65251-3

STOCHASTIC PROCESSES AND FILTERING THEORY, Andrew H. Jazwinski. This unified treatment presents material previously available only in journals, and in terms accessible to engineering students. Although theory is emphasized, it discusses numerous practical applications as well. 1970 edition. 400pp. $5^3/_8$ x $8^1/_2$. 0-486-46274-9

Math—Decision Theory, Statistics, Probability

INTRODUCTION TO PROBABILITY, John E. Freund. Featured topics include permutations and factorials, probabilities and odds, frequency interpretation, mathematical expectation, decision-making, postulates of probability, rule of elimination, much more. Exercises with some solutions. Summary. 1973 edition. 247pp. $5^3/_8 \times 8^1/_2$. 0-486-67549-1

STATISTICAL AND INDUCTIVE PROBABILITIES, Hugues Leblanc. This treatment addresses a decades-old dispute among probability theorists, asserting that both statistical and inductive probabilities may be treated as sentence-theoretic measurements, and that the latter qualify as estimates of the former. 1962 edition. 160pp. $5^3/_8 \times 8^1/_2$. 0-486-44980-7

APPLIED MULTIVARIATE ANALYSIS: Using Bayesian and Frequentist Methods of Inference, Second Edition, S. James Press. This two-part treatment deals with foundations as well as models and applications. Topics include continuous multivariate distributions; regression and analysis of variance; factor analysis and latent structure analysis; and structuring multivariate populations. 1982 edition. 692pp. $5^3/_8 \times 8^1/_2$. 0-486-44236-5

LINEAR PROGRAMMING AND ECONOMIC ANALYSIS, Robert Dorfman, Paul A. Samuelson and Robert M. Solow. First comprehensive treatment of linear programming in standard economic analysis. Game theory, modern welfare economics, Leontief input-output, more. 525pp. $5^3/_8 \times 8^1/_2$. 0-486-65491-5

PROBABILITY: AN INTRODUCTION, Samuel Goldberg. Excellent basic text covers set theory, probability theory for finite sample spaces, binomial theorem, much more. 360 problems. Bibliographies. 322pp. $5^3/_8 \times 8^1/_2$. 0-486-65252-1

GAMES AND DECISIONS: INTRODUCTION AND CRITICAL SURVEY, R. Duncan Luce and Howard Raiffa. Superb nontechnical introduction to game theory, primarily applied to social sciences. Utility theory, zero-sum games, n-person games, decision-making, much more. Bibliography. 509pp. $5^3/_8 \times 8^1/_2$. 0-486-65943-7

INTRODUCTION TO THE THEORY OF GAMES, J. C. C. McKinsey. This comprehensive overview of the mathematical theory of games illustrates applications to situations involving conflicts of interest, including economic, social, political, and military contexts. Appropriate for advanced undergraduate and graduate courses; advanced calculus a prerequisite. 1952 ed. x+372pp. $5^3/_8 \times 8^1/_2$. 0-486-42811-7

FIFTY CHALLENGING PROBLEMS IN PROBABILITY WITH SOLUTIONS, Frederick Mosteller. Remarkable puzzlers, graded in difficulty, illustrate elementary and advanced aspects of probability. Detailed solutions. 88pp. $5^3/_8 \times 8^1/_2$. 0-486-65355-2

PROBABILITY THEORY: A CONCISE COURSE, Y. A. Rozanov. Highly readable, self-contained introduction covers combination of events, dependent events, Bernoulli trials, etc. 148pp. $5^3/_8 \times 8^1/_4$. 0-486-63544-9

THE STATISTICAL ANALYSIS OF EXPERIMENTAL DATA, John Mandel. First half of book presents fundamental mathematical definitions, concepts and facts while remaining half deals with statistics primarily as an interpretive tool. Well-written text, numerous worked examples with step-by-step presentation. Includes 116 tables. 448pp. $5^3/_8 \times 8^1/_2$. 0-486-64666-1

Math—Geometry and Topology

ELEMENTARY CONCEPTS OF TOPOLOGY, Paul Alexandroff. Elegant, intuitive approach to topology from set-theoretic topology to Betti groups; how concepts of topology are useful in math and physics. 25 figures. 57pp. 5³/₈ x 8¹/₂. 0-486-60747-X

A LONG WAY FROM EUCLID, Constance Reid. Lively guide by a prominent historian focuses on the role of Euclid's Elements in subsequent mathematical developments. Elementary algebra and plane geometry are sole prerequisites. 80 drawings. 1963 edition. 304pp. 5³/₈ x 8¹/₂. 0-486-43613-6

EXPERIMENTS IN TOPOLOGY, Stephen Barr. Classic, lively explanation of one of the byways of mathematics. Klein bottles, Moebius strips, projective planes, map coloring, problem of the Koenigsberg bridges, much more, described with clarity and wit. 43 figures. 210pp. 5³/₈ x 8¹/₂. 0-486-25933-1

THE GEOMETRY OF RENÉ DESCARTES, René Descartes. The great work founded analytical geometry. Original French text, Descartes's own diagrams, together with definitive Smith-Latham translation. 244pp. 5³/₈ x 8¹/₂. 0-486-60068-8

EUCLIDEAN GEOMETRY AND TRANSFORMATIONS, Clayton W. Dodge. This introduction to Euclidean geometry emphasizes transformations, particularly isometries and similarities. Suitable for undergraduate courses, it includes numerous examples, many with detailed answers. 1972 ed. viii+296pp. 6¹/₈ x 9¹/₄. 0-486-43476-1

EXCURSIONS IN GEOMETRY, C. Stanley Ogilvy. A straightedge, compass, and a little thought are all that's needed to discover the intellectual excitement of geometry. Harmonic division and Apollonian circles, inversive geometry, hexlet, Golden Section, more. 132 illustrations. 192pp. 5³/₈ x 8¹/₂. 0-486-26530-7

THE THIRTEEN BOOKS OF EUCLID'S ELEMENTS, translated with introduction and commentary by Sir Thomas L. Heath. Definitive edition. Textual and linguistic notes, mathematical analysis. 2,500 years of critical commentary. Unabridged. 1,414pp. 5³/₈ x 8¹/₂. Three-vol. set.
Vol. I: 0-486-60088-2 Vol. II: 0-486-60089-0 Vol. III: 0-486-60090-4

SPACE AND GEOMETRY: IN THE LIGHT OF PHYSIOLOGICAL, PSYCHOLOGICAL AND PHYSICAL INQUIRY, Ernst Mach. Three essays by an eminent philosopher and scientist explore the nature, origin, and development of our concepts of space, with a distinctness and precision suitable for undergraduate students and other readers. 1906 ed. vi+148pp. 5³/₈ x 8¹/₂. 0-486-43909-7

GEOMETRY OF COMPLEX NUMBERS, Hans Schwerdtfeger. Illuminating, widely praised book on analytic geometry of circles, the Moebius transformation, and two-dimensional non-Euclidean geometries. 200pp. 5³/₈ x 8¹/₄. 0-486-63830-8

DIFFERENTIAL GEOMETRY, Heinrich W. Guggenheimer. Local differential geometry as an application of advanced calculus and linear algebra. Curvature, transformation groups, surfaces, more. Exercises. 62 figures. 378pp. 5³/₈ x 8¹/₂. 0-486-63433-7

History of Math

THE WORKS OF ARCHIMEDES, Archimedes (T. L. Heath, ed.). Topics include the famous problems of the ratio of the areas of a cylinder and an inscribed sphere; the measurement of a circle; the properties of conoids, spheroids, and spirals; and the quadrature of the parabola. Informative introduction. clxxxvi+326pp. $5^3/_8$ x $8^1/_2$. 0-486-42084-1

A SHORT ACCOUNT OF THE HISTORY OF MATHEMATICS, W. W. Rouse Ball. One of clearest, most authoritative surveys from the Egyptians and Phoenicians through 19th-century figures such as Grassman, Galois, Riemann. Fourth edition. 522pp. $5^3/_8$ x $8^1/_2$. 0-486-20630-0

THE HISTORY OF THE CALCULUS AND ITS CONCEPTUAL DEVELOPMENT, Carl B. Boyer. Origins in antiquity, medieval contributions, work of Newton, Leibniz, rigorous formulation. Treatment is verbal. 346pp. $5^3/_8$ x $8^1/_2$. 0-486-60509-4

THE HISTORICAL ROOTS OF ELEMENTARY MATHEMATICS, Lucas N. H. Bunt, Phillip S. Jones, and Jack D. Bedient. Fundamental underpinnings of modern arithmetic, algebra, geometry and number systems derived from ancient civilizations. 320pp. $5^3/_8$ x $8^1/_2$. 0-486-25563-8

THE HISTORY OF THE CALCULUS AND ITS CONCEPTUAL DEVELOPMENT, Carl B. Boyer. Fluent description of the development of both the integral and differential calculus—its early beginnings in antiquity, medieval contributions, and a consideration of Newton and Leibniz. 368pp. $5^3/_8$ x $8^1/_2$. 0-486-60509-4

GAMES, GODS & GAMBLING: A HISTORY OF PROBABILITY AND STATISTICAL IDEAS, F. N. David. Episodes from the lives of Galileo, Fermat, Pascal, and others illustrate this fascinating account of the roots of mathematics. Features thought-provoking references to classics, archaeology, biography, poetry. 1962 edition. 304pp. $5^3/_8$ x $8^1/_2$. (Available in U.S. only.) 0-486-40023-9

OF MEN AND NUMBERS: THE STORY OF THE GREAT MATHEMATICIANS, Jane Muir. Fascinating accounts of the lives and accomplishments of history's greatest mathematical minds—Pythagoras, Descartes, Euler, Pascal, Cantor, many more. Anecdotal, illuminating. 30 diagrams. Bibliography. 256pp. $5^3/_8$ x $8^1/_2$. 0-486-28973-7

HISTORY OF MATHEMATICS, David E. Smith. Nontechnical survey from ancient Greece and Orient to late 19th century; evolution of arithmetic, geometry, trigonometry, calculating devices, algebra, the calculus. 362 illustrations. 1,355pp. $5^3/_8$ x $8^1/_2$. Two-vol. set. Vol. I: 0-486-20429-4 Vol. II: 0-486-20430-8

A CONCISE HISTORY OF MATHEMATICS, Dirk J. Struik. The best brief history of mathematics. Stresses origins and covers every major figure from ancient Near East to 19th century. 41 illustrations. 195pp. $5^3/_8$ x $8^1/_2$. 0-486-60255-9

Physics

OPTICAL RESONANCE AND TWO-LEVEL ATOMS, L. Allen and J. H. Eberly. Clear, comprehensive introduction to basic principles behind all quantum optical resonance phenomena. 53 illustrations. Preface. Index. 256pp. 5³/₈ x 8¹/₂. 0-486-65533-4

QUANTUM THEORY, David Bohm. This advanced undergraduate-level text presents the quantum theory in terms of qualitative and imaginative concepts, followed by specific applications worked out in mathematical detail. Preface. Index. 655pp. 5³/₈ x 8¹/₂. 0-486-65969-0

ATOMIC PHYSICS (8th EDITION), Max Born. Nobel laureate's lucid treatment of kinetic theory of gases, elementary particles, nuclear atom, wave-corpuscles, atomic structure and spectral lines, much more. Over 40 appendices, bibliography. 495pp. 5³/₈ x 8¹/₂. 0-486-65984-4

A SOPHISTICATE'S PRIMER OF RELATIVITY, P. W. Bridgman. Geared toward readers already acquainted with special relativity, this book transcends the view of theory as a working tool to answer natural questions: What is a frame of reference? What is a "law of nature"? What is the role of the "observer"? Extensive treatment, written in terms accessible to those without a scientific background. 1983 ed. xlviii+172pp. 5³/₈ x 8¹/₂. 0-486-42549-5

AN INTRODUCTION TO HAMILTONIAN OPTICS, H. A. Buchdahl. Detailed account of the Hamiltonian treatment of aberration theory in geometrical optics. Many classes of optical systems defined in terms of the symmetries they possess. Problems with detailed solutions. 1970 edition. xv + 360pp. 5³/₈ x 8¹/₂. 0-486-67597-1

PRIMER OF QUANTUM MECHANICS, Marvin Chester. Introductory text examines the classical quantum bead on a track: its state and representations; operator eigenvalues; harmonic oscillator and bound bead in a symmetric force field; and bead in a spherical shell. Other topics include spin, matrices, and the structure of quantum mechanics; the simplest atom; indistinguishable particles; and stationary-state perturbation theory. 1992 ed. xiv+314pp. 6¹/₈ x 9¹/₄. 0-486-42878-8

LECTURES ON QUANTUM MECHANICS, Paul A. M. Dirac. Four concise, brilliant lectures on mathematical methods in quantum mechanics from Nobel Prize-winning quantum pioneer build on idea of visualizing quantum theory through the use of classical mechanics. 96pp. 5³/₈ x 8¹/₂. 0-486-41713-1

THIRTY YEARS THAT SHOOK PHYSICS: THE STORY OF QUANTUM THEORY, George Gamow. Lucid, accessible introduction to influential theory of energy and matter. Careful explanations of Dirac's anti-particles, Bohr's model of the atom, much more. 12 plates. Numerous drawings. 240pp. 5³/₈ x 8¹/₂. 0-486-24895-X

ELECTRONIC STRUCTURE AND THE PROPERTIES OF SOLIDS: THE PHYSICS OF THE CHEMICAL BOND, Walter A. Harrison. Innovative text offers basic understanding of the electronic structure of covalent and ionic solids, simple metals, transition metals and their compounds. Problems. 1980 edition. 582pp. 6¹/₈ x 9¹/₄. 0-486-66021-4

HYDRODYNAMIC AND HYDROMAGNETIC STABILITY, S. Chandrasekhar. Lucid examination of the Rayleigh-Benard problem; clear coverage of the theory of instabilities causing convection. 704pp. $5^3/_8$ x $8^1/_4$. 0-486-64071-X

INVESTIGATIONS ON THE THEORY OF THE BROWNIAN MOVEMENT, Albert Einstein. Five papers (1905–8) investigating dynamics of Brownian motion and evolving elementary theory. Notes by R. Fürth. 122pp. $5^3/_8$ x $8^1/_2$. 0-486-60304-0

THE PHYSICS OF WAVES, William C. Elmore and Mark A. Heald. Unique overview of classical wave theory. Acoustics, optics, electromagnetic radiation, more. Ideal as classroom text or for self-study. Problems. 477pp. $5^3/_8$ x $8^1/_2$. 0-486-64926-1

GRAVITY, George Gamow. Distinguished physicist and teacher takes reader-friendly look at three scientists whose work unlocked many of the mysteries behind the laws of physics: Galileo, Newton, and Einstein. Most of the book focuses on Newton's ideas, with a concluding chapter on post-Einsteinian speculations concerning the relationship between gravity and other physical phenomena. 160pp. $5^3/_8$ x $8^1/_2$. 0-486-42563-0

PHYSICAL PRINCIPLES OF THE QUANTUM THEORY, Werner Heisenberg. Nobel Laureate discusses quantum theory, uncertainty, wave mechanics, work of Dirac, Schroedinger, Compton, Wilson, Einstein, etc. 184pp. $5^3/_8$ x $8^1/_2$. 0-486-60113-7

ATOMIC SPECTRA AND ATOMIC STRUCTURE, Gerhard Herzberg. One of best introductions; especially for specialist in other fields. Treatment is physical rather than mathematical. 80 illustrations. 257pp. $5^3/_8$ x $8^1/_2$. 0-486-60115-3

AN INTRODUCTION TO STATISTICAL THERMODYNAMICS, Terrell L. Hill. Excellent basic text offers wide-ranging coverage of quantum statistical mechanics, systems of interacting molecules, quantum statistics, more. 523pp. $5^3/_8$ x $8^1/_2$. 0-486-65242-4

THEORETICAL PHYSICS, Georg Joos, with Ira M. Freeman. Classic overview covers essential math, mechanics, electromagnetic theory, thermodynamics, quantum mechanics, nuclear physics, other topics. First paperback edition. xxiii + 885pp. $5^3/_8$ x $8^1/_2$.
0-486-65227-0

PROBLEMS AND SOLUTIONS IN QUANTUM CHEMISTRY AND PHYSICS, Charles S. Johnson, Jr. and Lee G. Pedersen. Unusually varied problems, detailed solutions in coverage of quantum mechanics, wave mechanics, angular momentum, molecular spectroscopy, more. 280 problems plus 139 supplementary exercises. 430pp. $6^1/_2$ x $9^1/_4$.
0-486-65236-X

THEORETICAL SOLID STATE PHYSICS, Vol. 1: Perfect Lattices in Equilibrium; Vol. II: Non-Equilibrium and Disorder, William Jones and Norman H. March. Monumental reference work covers fundamental theory of equilibrium properties of perfect crystalline solids, non-equilibrium properties, defects and disordered systems. Appendices. Problems. Preface. Diagrams. Index. Bibliography. Total of 1,301pp. $5^3/_8$ x $8^1/_2$. Two volumes. Vol. I: 0-486-65015-4 Vol. II: 0-486-65016-2

WHAT IS RELATIVITY? L. D. Landau and G. B. Rumer. Written by a Nobel Prize physicist and his distinguished colleague, this compelling book explains the special theory of relativity to readers with no scientific background, using such familiar objects as trains, rulers, and clocks. 1960 ed. vi+72pp. $5^3/_8$ x $8^1/_2$. 0-486-42806-0